Perspectives in Neural Computing

Springer
*London*
*Berlin*
*Heidelberg*
*New York*
*Barcelona*
*Hong Kong*
*Milan*
*Paris*
*Singapore*
*Tokyo*

*Also in this series:*

Gustavo Deco and Dragan Obradovic
An Information-Theoretic Approach to Neural Computing
0-387-94666-7

Achilleas Zapranis and Apostolos-Paul Refenes
Principles of Neural Model Identification, Selection and Adequacy
1-85233-139-9

Walter J. Freeman
Neurodynamics: An Exploration in Mesoscopic Brain Dynamics
1-85233-616-1

H. Malmgren, M Borga and L. Niklasson (Eds)
Artificial Neural Networks in Medicine and Biology
1-85233-289-1

Mark Girolami
Advances in Independent Component Analysis
1-85233-263-8

Robert M. French and Jacques P. Sougné (Eds)
Connectionist Models of Learning, Development and Evolution
1-85233-354-5

Roberto Tagliaferri and Maria Marinaro (Eds)
Neural Nets WIRN Vietri-01
1-85233-505-X

Jimmy Shadbolt and John G. Taylor (Eds)
Neural Networks and the Financial Markets
1-85233-531-9

*Related Title*

Robert Hecht-Nielsen and Thomas McKenna (Eds)
Computational Models for Neuroscience: Human Cortical Information Processing
1-85233-593-9

Artur S. d'Avila Garcez, Krysia B. Broda and
Dov M. Gabbay

# Neural-Symbolic Learning Systems

## Foundations and Applications

Springer

Artur S. d'Avila Garcez, MEng, MSc, DIC, PhD
Department of Computing, City University, Northampton Square,
London EC1V 0HB

Krysia B. Broda, BSc, BA, MSc, PhD
Department of Computing, Imperial College, 180 Queen's Gate, London SW7 2BZ

Dov M. Gabbay, FRSC, FAvH, FRSA
Department of Computer Science, King's College, Strand, London WC2R 2LS

*Series Editor*
J.G. Taylor, BA, BSc, MA, PhD, FlnstP
Centre for Neural Networks, Department of Mathematics, King's College,
Strand, London WC2R 2LS, UK

British Library Cataloguing in Publication Data
d'Avila Garcez, Artur S.
Neural-symbolic learning systems : foundations and
applications. - (Perspectives in neural computing)
1.Neural networks (Computer Science)
I.Title II.Broda, Krysia III.Gabbay, Dov M., 1945-
006.3'2
ISBN 1852335122

Library of Congress Cataloging-in-Publication Data
d'Avila Garcez, Artur S., 1970-
Neural-symbolic learning systems : foundations and applications / Artur S. d'Avila
Garcez, Krysia B. Broda, and Dov. M. Gabbay.
p. cm. -- (Perspectives in neural computing)
Includes bibliographical references and index.
ISBN 1-85233-512-2 (alk. paper)
1. Neural networks (Computer science) 2. Artificial intelligence. I. Broda, Krysia,
1949- II. Gabbay, Dov M., 1945- III. Title. IV. Series.
QA76.87 .D38 2002
006.3'2--dc21 2002070771



Perspectives in Neural Computing ISSN 1431-6854

ISBN 1-85233-512-2 Springer-Verlag London Berlin Heidelberg
a member of BertelsmannSpringer Science+Business Media GmbH
http://www.springer.co.uk


Printed in Great Britain



Typesetting: Ian Kingston Editorial Services, Nottingham, UK
Printed and bound by the Athenæum Press Ltd., Gateshead, Tyne & Wear
34/3830-543210 Printed on acid-free paper SPIN 10838895

*To Simone Garcez – A.G.*

*To Parimal Pal – K.B.*

*To Lydia Rivlin – D.G.*

# Preface

Computing Science and Artificial Intelligence are concerned with producing devices that help or replace human beings in their daily activities. To be successful, adequate modelling of these activities needs to be carried out. This urgent need for models has given rise to new disciplines as well as accelerating the development of old disciplines. Among these are Logic and Computation, Neural Networks, Fuzzy Systems and Causal Probabilistic Networks.

Learning Systems play a central role in this new landscape, comprising collections of techniques developed within each of the above disciplines. Notwithstanding this, human beings do not use these techniques in isolation, but in an integrated way. Therefore the study of how these techniques could complement each other – and the understanding of the role of each of them in a particular application – is bound to be the way forward towards the development of more effective intelligent systems.

In this book, we present the key theory and algorithms that form the core of the integration of Artificial Neural Networks and Logic Programming. By combining Neural Networks and Logic Programming into *Neural-Symbolic Learning Systems*, our aim is to benefit from the advantages presented by the Connectionist and Symbolist paradigms of Artificial Intelligence. The book starts with an intuitive presentation of Neural-Symbolic Learning Systems, illustrated with a worked example. It then presents a concise introduction to Logic Programming and Artificial Neural Networks from a neural-symbolic integration perspective. Basic concepts from other disciplines are introduced as the need arises, focusing on their relevance to neural-symbolic integration. The book is then divided into three main parts, presenting recent advances on the theoretical foundations of (*i*) Knowledge Representation and Learning, (*ii*) Knowledge Extraction and (*iii*) Inconsistency Handling in neural-symbolic systems.

The book presents a balance between theory and practice, with each part containing the results of applications using real-world problems in diverse areas such as: DNA Sequence Analysis, Power Systems Fault Diagnosis and Software Requirements Specifications. It will be valuable to researchers and graduate students in the areas of Engineering, Computer Science, Artificial Intelligence, Machine Learning and Neurocomputing, who require a comprehensive introduction to Neural-Symbolic Learning Systems, as well as knowledge of the latest research issues relevant to the field. It will also be of use to intelligent systems practitioners and to those interested in the applications of hybrid systems of Artificial Intelligence.

## Acknowledgements

We would like to thank Professor John Taylor, editor of the series, the late Professor Imad Torsun, Dr Gerson Zaverucha, Dr Luis Alfredo Carvalho, Dr Ruy Milidiu, Professor Valmir Barbosa, Dr Alberto de Souza, Dr Odinaldo Rodrigues, Dr Sanjay Modgil and Dr Luis Lamb for their valuable comments and suggestions. We also would like to thank Rosie Kemp and the entire production team at Springer-Verlag for their effective assistance in all phases of this project. Artur Garcez was partially funded by the Brazilian Research Council CAPES and by the British Research Council EPSRC, while he was a PhD student and then a Research Associate at Imperial College. He is grateful to Professor Jeffrey Kramer, Professor Bashar Nuseibeh, Dr Alessandra Russo and the Requirements Engineering group at Imperial College, Department of Computing, for many useful discussions.

Artur would like to thank his wife, Simone, for her loving care and understanding; his parents, Lucia and Cicero; his sister, Cristiane; and his friends for their continuous support and encouragement. Krysia would like to thank Paul for his patience and encouragement. Dov would like to thank his wife, Lydia, for many years of support and understanding.

London, May 2002

*Artur Garcez, Krysia Broda, Dov Gabbay*

# About the Authors

**Artur Garcez** is a Lecturer in the Department of Computing at City University, London, and a Research Consultant to Performance Systems and Methods Ltd. He is known for his work on the theoretical foundations of Neural-Symbolic Integration: knowledge insertion, revision and extraction; and its applications to Machine Learning, Bioinformatics and Software Engineering. He holds an M.Eng. in Computing Engineering from the Pontifical Catholic University of Rio de Janeiro, Brazil, an M.Sc. in Computing and Systems Engineering from the Federal University of Rio de Janeiro, Brazil, and a Ph.D. in Computing from Imperial College, London, UK.

**Krysia Broda** is a Senior Lecturer in the Department of Computing at Imperial College, London. She holds a Ph.D. in Automated Reasoning from Imperial College. She has wide experience on Logic Programming and a number of recent publications on Labelled Deductive Systems (LDS) for non-standard logics such as Resource Logics and Substructural and Fuzzy Logics, and on transformation and abduction methods in LDS. Her current research interests include knowledge representation in neural networks and knowledge extraction from trained neural networks, and its applications to goal-directed reactive robots.

**Dov Gabbay** is Augustus de Morgan Professor of Logic at the Department of Computer Science and the Department of Philosophy at King's College, London. He received a B.Sc. in Mathematics and Physics, a M.Sc. in Logic, and a Ph.D. in Logic from the Hebrew University, Jerusalem. He has initiated several research areas in Logic and Computation, having made significant contributions to the areas of Temporal, Multimodal and Monmonotonic Logics and their applications; Logic and Language; and the theory of inference systems via Labelled Deductive Systems. He is Fellow of the Royal Society of Canada, and the editor of several international Logic journals and several multi-volume Logic handbooks.

# Table of Contents

## Part II. Knowledge Extraction from Neural Networks



## Part III. Knowledge Revision in Neural Networks

# 1. Introduction and Overview

This book is about the integration of neural networks and symbolic rules. While symbolic artificial intelligence assumes that the *mind* is the focus of intelligence, and thus intelligent behaviour emerges from complex symbol processing mechanisms, connectionist artificial intelligence admits that intelligence lies in the *brain*, and therefore tries to model it by simulating its electrochemical neuronal structures. Clearly, such structures are capable of learning and performing the higher level cognitive tasks that human beings are accustomed to, as well as lower level, everyday cognitive activities. In this framework, the role of symbolic computation is to provide the system with the background information needed for the learning process, as well as to provide us with the information needed for understanding the system, since high-level cognitive tasks are much more clearly digested by human beings as symbols and operations over symbols rather than in the form of interconnected neurons.

## 1.1 Why Integrate Neurons and Symbols?

Human cognition successfully integrates the connectionist and symbolic paradigms of Artificial Intelligence (AI). Yet the modelling of cognition develops these separately in neural computation and symbolic logic/AI areas. There is now a movement towards a fruitful mid-point between these extremes, in which the study of logic is combined with connectionism. It is essential that these be integrated, thereby enabling a technology for building better intelligent systems.

The aim of *Neural-Symbolic Integration* is to explore and exploit the advantages that each paradigm presents. Among the advantages of artificial neural networks are massive parallelism, inductive learning and generalisation capabilities. On the other hand, symbolic systems can explain their inference process, e.g. through automated theorem proving, and use powerful declarative languages for *Knowledge Representation*.

In this book, we explore the synergies of neural-symbolic integration from the following perspective. We use a *Neural Network* to simulate a given task.

The network is obtained by being programmed (set up) and/or by somehow adapting and generalising over well-known situations (learning). The network is the mechanism to execute the task, while symbolic logic enables the necessary interaction between the network and the outside world.

"It is generally accepted that one of the main problems in building Expert Systems (which are responsible for the industrial success of Artificial Intelligence) lies in the process of knowledge acquisition, known as the *Knowledge Acquisition Bottleneck*" [LD94]. An alternative is the automation of this process using *Machine Learning* techniques (see [Mit97, Rus96]). Symbolic machine learning methods are usually more effective if they can exploit background knowledge (incomplete domain theory). In contrast, neural networks have been successfully applied as a learning method from examples only (data learning) [TBB+91, SMT91]. As a result, the integration of theory and data learning in neural networks seems to be a natural step towards more powerful training mechanisms.[TS94a]

The *Inductive Learning* task employed in symbolic machine learning is to find hypotheses that are consistent with a background knowledge to explain a given set of examples. In general, these hypotheses are definitions of concepts described in some logical language. The examples are descriptions of instances and non-instances of the concept to be learned, and the background knowledge provides additional information about the examples and the concepts' domain knowledge [LD94].

In contrast with (symbolic) learning systems, the learning of (numeric) neural networks implicitly encodes patterns and their generalisations in the networks' weights, so reflecting the statistical properties of the trained data [BL96]. It has been indicated that neural networks can outperform symbolic learning systems, especially when data are noisy[1] [TBB+91]. This result, due also to the massively parallel architecture of neural networks, contributed decisively to the growing interest in combining, and possibly integrating, neural and symbolic learning systems (see [Kur97] for a clarifying treatment of the suitability of neural networks for the representation of symbolic knowledge).

> "We believe that neural networks are capable of more than pattern recognition; they can also perform higher cognitive tasks which are fundamentally rule-governed. Further, we believe that they can perform higher cognitive tasks better if they incorporate rules rather than eliminate them. A number of well known cognitive models, particularly of language, have been criticised for going too far in eliminating rules in fundamentally rule-governed domains. We argue that with a suitable choice of high-level, rule governed task, representation, processing architecture, and learning algorithm, neural networks

[1] We say that data are noisy when some of its variables are corrupted or missing altogether.

> can represent and learn rules involving higher-level categories while simultaneously learning those categories. The resulting network can exhibit better learning and task performance than neural networks that do not incorporate rules, and have capabilities that go beyond that of purely symbolic rule-learning algorithms." Paul Smolensky [MMS92]

According to Minsky, "Both kinds of intelligent computational systems, symbolic and connectionist, have virtues and deficiencies. It is very important to integrate them, through neural-symbolic systems, in order to explore the capabilities each one possesses"[Min91]. In this sense, the aim of such integration is twofold: while logic can benefit from neural networks' successful applications on various knowledge domains, neural network learning and generalisation processes can be rigorously studied and explained by logic.

> "There is still a feeling that something is wrong with agent systems and artificial intelligence systems even in cases where the system gives the right answer. There are cases where the 'human computer', slow as it is, gives the correct answer immediately while the agent system may take some time to find it. Something must be wrong. Why are we faster? Is it the way we perceive the rules as opposed to the way we represent them in the agent system? Do we know immediately which rule to use? We must look for the 'correct' representation in the sense that it mirrors the way we perceive and apply the rules. Humans use an overall impression to make decisions about how to go about finding answers to queries. Neural networks are in a better position to model this capacity. In fact, we believe that every agent system should have a neural net component." Dov Gabbay [Gab98]

It would certainly be rewarding if the gap between the study of (symbolic) artificial intelligence and the study of (numerical) artificial neural networks could be reduced. This might suggest massively parallel formal systems that prescribe how to reason, in a certain domain, in a way that is enlightening from the point of view of actual practice. On the other hand, logic may be a very useful tool in helping to explain neural networks' inference process, as well as in formalising their learning and generalisation mechanisms.

## 1.2 Strategies of Neural-Symbolic Integration

According to [Hil95], *Neural-Symbolic Systems* can be divided into: *Unification Systems* and *Hybrid Systems* (see Figure 1.1). The first category comprises connectionist systems that perform some kind of symbolic computation. The second category contains systems that present a logical as well as a connectionist component, which interact with each other.

The principle underlying *Unification Strategies* is that all the functionality of symbol processing arises from neuronal structures and processes. Unification strategies comprise *Neuronal Modelling* (or Neuroscience) and *Connectionist Logic Systems*. The first group investigates the relationship between specific cognitive tasks and biological reality. By developing computational models of the cognitive tasks of the brain, it attempts to understand how the brain works by building on its cellular units: the neurons [OM00]. The second group, Connectionist Logic Systems, is concerned about the development of models of artificial neural networks that can compute complex symbolic processes in parallel. In this group, the representation of symbolic knowledge in neural networks can be either *localised* or *distributed*. In a localised representation, each neuron is a concept, while in a distributed representation, the most elementary concepts arise from the interaction of many processing elements. In both cases, Connectionist Logic Systems might use either *energy minimisation* or *propagation of activation* as the mechanism of inference. Examples of Connectionist Logic Systems by energy minimisation are [Bal86, Pin95, NR92, Vin94]. Among the Connectionist Logic Systems by propagation of activation are [Sha88, HK92, Sun95, HK94].

Differently from Connectionist Logic Systems, *Hybrid Systems*, in general, combine a symbolic component with a connectionist one. Hybrid Systems can be classified in many different ways: by the application domain; the symbolic and connectionist models used; the functionality of the symbolic and neural components of the system; etc. Following Medsker [Med94], we use the degree of interaction between the symbolic and neural components of the system as a classification scheme for Hybrid Systems.

1. *Stand-Alone Models*: There is no interaction between the symbolic and neural components. Both can be used in the same application in order to compare efficiency.
2. *Loosely Coupled Models*: There is a weak interaction between the components. Neural and Symbolic modules perform specific tasks within the system and communicate via data files. We include here, for example, the systems in which a neural network pre-processes data to be used by an expert system.
3. *Tightly Coupled Models*: There is a strong interaction between the components. Communications between the neural and symbolic modules of the system occur via data structures stored in memory.
4. *Fully Integrated Models*: Data structures and processing are not shared between the components by function, but are part of a unique system with a dual (neural and symbolic) nature. Gallant's Connectionist Expert System [Gal88] is the seminal work towards fully integrated models. Other examples include [GO93, Fu91, MR91, KP92, TS94a].

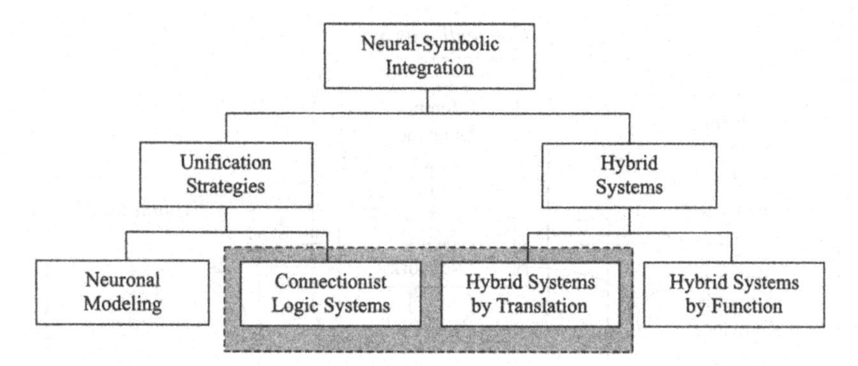


**Fig. 1.1.** A classification of neural-symbolic systems.

Stand-Alone, Loosely Coupled and Tightly Coupled models all belong to the group of *Hybrid Systems by Function*, according to Hilario's classification [Hil95] (see Figure 1.1). Fully Integrated models, on the other hand, fall into the class of *Hybrid Systems by Translation.* Similarly to Connectionist Logic Systems, Hybrid Systems by Translation use, in general, translation methods from symbolic knowledge (usually production rules) to connectionist models and vice versa. However, while Connectionist Logic Systems are mainly concerned about performing (high-level) symbolic computation in massively parallel models, Hybrid Systems by Translation are concerned about the use of symbolic knowledge to help the process of (low-level) inductive learning, by serving either as background knowledge or as an explanation for the learning method applied. In other words, Connectionist Logic Systems are mainly concerned about the benefits that Neural Networks can bring to Logic, such as massive parallelism, while Hybrid Systems by Translation are rather concerned about the benefits that Logic can bring to Neural Networks, such as learning with background knowledge or rule extraction.

## 1.3 Neural-Symbolic Learning Systems

In this book, we want to go one step further in the process of neural-symbolic integration, by exploring some of the features of Connectionist Logic Systems (CLSs) and Hybrid Systems by Translation (HSTs). While CLSs are, in general, provably equivalent to a logical formalism, HSTs lack such a fundamental property. On the other hand, CLSs present none or very limited learning capabilities, one of the most important features of HSTs and, indeed, of artificial neural networks. We argue that one can combine features of CLSs and HSTs in order to achieve both equivalence and learning capability in a fully integrated framework. We argue, thus, for the creation of a new category

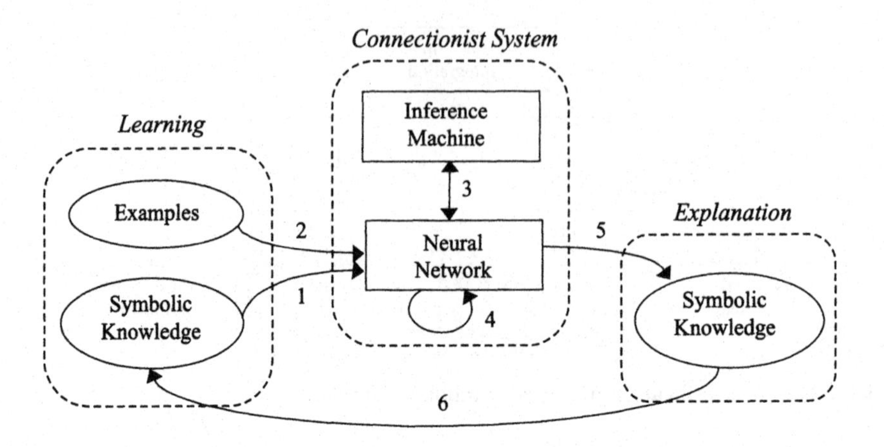


**Fig. 1.2.** Neural-symbolic learning systems.

in Hilario's classification scheme of Figure 1.1, and call it *Neural-Symbolic Learning Systems*.

*Neural-Symbolic Learning Systems* contain six main phases: (1) *symbolic knowledge insertion*; (2) *inductive learning with examples*; (3) *massively parallel deduction*; (4) *theory fine-tuning*; (5) *symbolic knowledge extraction*; and (6) *feedback* (see Figure 1.2). In phase (1), a (symbolic) background knowledge is translated into the initial architecture of a neural network by some *Translation Algorithm*. In phase (2), such a network should be able to be trained with examples efficiently, thus refining the initial (incomplete) theory given as *background knowledge*. For example, differently from most CLSs, the network could be trained using the Backpropagation learning algorithm. In phase (3), the network must be able to be used as a massively parallel computational model of the logical consequences of the theory encoded in it. This is so because, as opposed to most HSTs, the Translation Algorithm of a Neural-Symbolic Learning System must be provably correct. In phase (4), the information obtained with the computation carried out in phase 3 may help in fine-tuning the network to represent the knowledge domain better. This mechanism can be used, for example, to solve inconsistencies between the background knowledge and the training examples. In phase (5), the results of refining the network should be explained by the extraction of a revised (symbolic) knowledge from it. As with the insertion of rules, the *Extraction Algorithm* of a Neural-Symbolic Learning System must be provably correct, so that each rule extracted is guaranteed to be encoded in the network. Finally, in phase (6), the knowledge extracted may be analysed by an expert to decide whether it should feed the system once more, closing the learning cycle.

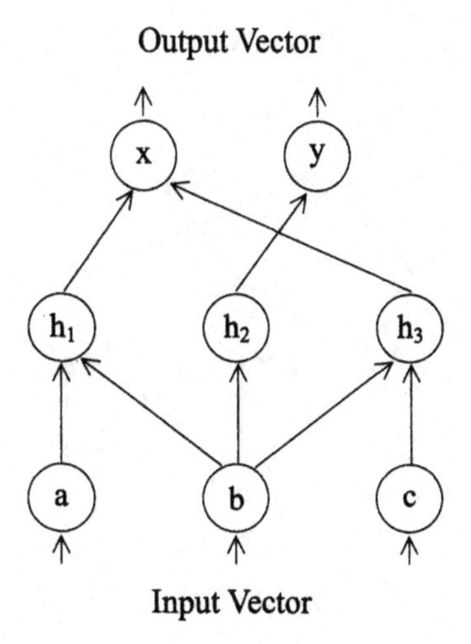


**Fig. 1.3.** Translating a symbolic knowledge into a neural network.

A typical application of Neural-Symbolic Learning Systems is in safety-critical domains, such as fault diagnosis systems, where the neural network can detect a fault quickly, triggering safety procedures, while the knowledge extracted from it can justify the fault later on. If mistaken, the information can be used to fine tune the learning system.

## 1.4 A Simple Example

In order to give the reader an overview of the sequence of processes 1 to 6 of Neural-Symbolic Learning Systems (see Figure 1.2), we give a simple illustrative example.

**Phase 1**: Let us assume that the following incomplete theory is given as background knowledge, in the form of a logic program, $\mathcal{P} = \{a \wedge b \rightarrow x; b \wedge c \rightarrow x; b \rightarrow y\}$. A way of translating $\mathcal{P}$ into a neural network $\mathcal{N}$ is to associate $a$, $b$ and $c$ with input neurons, and $x$ and $y$ with output neurons. A layer of hidden neurons is then necessary to allow $\mathcal{N}$ to capture the relationships between $\{a, b, c\}$ and $\{x, y\}$ of $\mathcal{P}$. In fact, a hidden neuron of $\mathcal{N}$ for each rule of $\mathcal{P}$ is sufficient. The network should look like that of Figure 1.3, where hidden neuron $h_1$ should represent rule $a \wedge b \rightarrow x$, hidden neuron $h_2$ should represent $b \rightarrow y$, and hidden neuron $h_3$ should represent $b \wedge c \rightarrow x$.

Output neuron $y$ should only be activated if input neuron $b$ is activated. Similarly, output neuron $x$ should only be activated if $a$ and $b$ are activated, or if $b$ and $c$ are activated. In other words, the weights of the connections of the network must be set up such that hidden neuron $h_1$ performs a logical

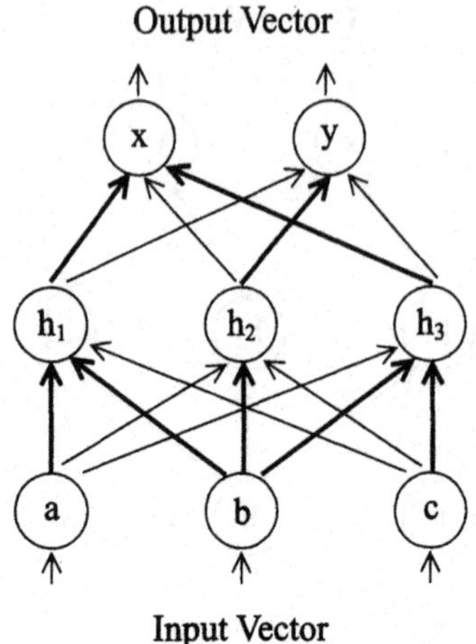


**Fig. 1.4.** Performing inductive learning with examples.

*and* between inputs $a$ and $b$, hidden neuron $h_2$ is activated if, and only if, $b$ is activated, and hidden neuron $h_3$ performs a logical *and* between inputs $b$ and $c$. Similarly, output neuron $x$ must perform a logical *or* between $h_1$ and $h_3$, while $y$ should be activated if, and only if, $h_2$ is activated. If this is the case, we should be able to show that $\mathcal{P}$ and $\mathcal{N}$ are, in fact, equivalent.

**Phase 2**: Our next step is to perform inductive learning with examples in $\mathcal{N}$. The idea is to change the values of the weights of the network according to a set of training examples (input and output vectors), using some neural learning algorithm. In order to do so, we fully connect $\mathcal{N}$ as in Figure 1.4, so that it can learn new relations between $\{a, b, c\}$ and $\{x, y\}$ in addition to the ones already inserted in it by $\mathcal{P}$. In principle, this process also allows the network to change its background knowledge.

**Phase 3**: Let us assume that the set of training examples is such that it does not change the background knowledge, but only expands it. Although we do not know yet which is the new knowledge encoded in the network, we should be able to compute its logical consequences in parallel, using the network. For example, the network might have learned the rule $a \wedge c \rightarrow x$. As a result, $x$ could be derived from $a$ and $c$, as well as from $a$ and $b$ or from $b$ and $c$. Also, $y$ would still be derivable from $b$.

**Phase 4**: Now, suppose that having $x$ and $y$ together is not desirable; that is, suppose that $x \wedge y \rightarrow \bot$ is an integrity constraint of the application domain. The choice between $x$ and $y$ may depend, though, on extra-logical considerations. Assume that, as a matter of fact, $x$ is preferred to $y$. Neglecting, for the time being, many important aspects of theory revision, the conflict may be adjudicated by evolving the neural network, as depicted in Figure 1.5. The idea is to add a hidden neuron ($h_4$), responsible for blocking

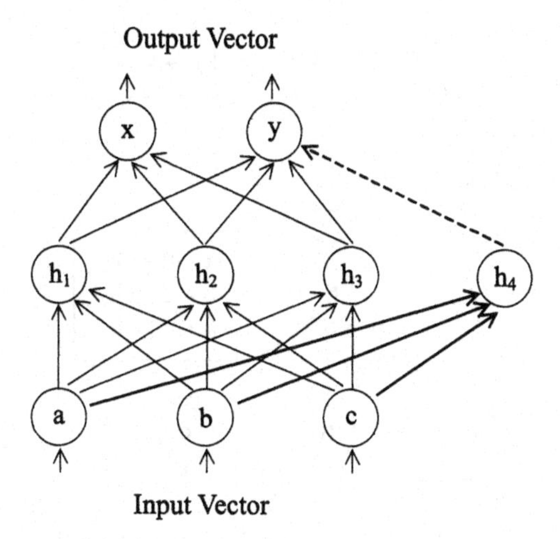


**Fig. 1.5.** Fine-tuning the network.

the activation of $y$ whenever $x$ is activated. Clearly, the same process could be used to solve inconsistencies of the form $z \wedge \neg z \rightarrow \bot$.

**Phase 5**: Our next task is to try to extract the (refined) knowledge encoded in the network so that we can explain its answers by inspecting the symbolic knowledge that it encodes. In the case of binary inputs, an option is to use brute force, and check all the possible combinations of input vectors. However, in real-world applications, hundreds of input neurons may be necessary, and the problem may be intractable. The challenge here, therefore, is to consider a small number of input vectors and still make sure that the extraction of rules is correct. In order to do so, we need, in general, to decompose the network into sub-networks, and to extract rules that map $\{a, b, c\}$ into $\{h_1, h_2, h_3, h_4\}$, and $\{h_1, h_2, h_3, h_4\}$ into $\{x, y\}$. However, since a network's behaviour is not equivalent to the behaviour of its parts grouped together, and since neurons $h_1, h_2, h_3$ and $h_4$ do not represent concepts, but rules, we need to be especially careful when deriving the final set of rules of a trained network in order to maintain correctness.

We also want to be able to extract rules that reflect the process of generalisation of the network, which occurs during learning, as opposed to rules that account for the network's training set and background knowledge only. For example, a possible generalisation is the rule $a \wedge c \rightarrow x$, which, together with rules $a \wedge b \rightarrow x$ and $b \wedge c \rightarrow x$ of the background knowledge, could be simplified to derive $2(abc) \rightarrow x$, indicating that any two of the concepts $a$, $b$ and $c$ would imply $x$. This so-called *M of N* rule could be encoded in neuron $h_2$, as exemplified in Figure 1.6.

**Phase 6**: When background knowledge is translated into a neural network, it is possible to create a neat network structure. However, when the

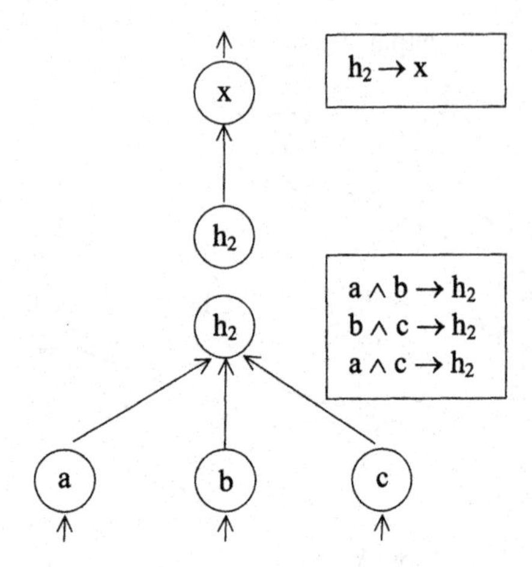


**Fig. 1.6.** Extracting rules from trained networks.

network is trained with examples, it will most probably lose its well-behaved structure. The task of rule extraction is supposed to discover the knowledge refined by the network. As a result, a new neat network structure could be created by the translation of the new knowledge into the network. In addition, new examples could be trained in such a network, and so on, thus closing the cycle of theory refinement.

In this example, for the sake of simplicity, we have neglected many important aspects of neural-symbolic integration. They will be discussed in detail in the rest of the book. The purpose of this example was to provide the reader with an intuitive presentation of Neural-Symbolic Learning Systems, as well as with an introductory description of the technical material that will follow.

## 1.5 How to Read this Book

Following a brief presentation of the relevant definitions and results on Inductive Learning, Artificial Neural Networks, Logic Programming and Nonmonotonic Reasoning, and Belief Revision, this book is divided into three parts:

- *Knowledge Refinement in Neural Networks;*
- *Knowledge Extraction from Neural Networks; and*
- *Knowledge Revision in Neural Networks.*

The first part is based on Holldobler and Kalinke's work on CLSs [HK94, HKS99] and on Towell and Shavlik's work on HSTs [Sha96, TS94a, TS94b]. We have carefully chosen the above approaches because Holldobler and

Kalinke, differently from most CLSs, use a neat and simple model of neural networks to compute one of the standard semantics of Logic Programming, thus facilitating the inclusion of learning capabilities into the model, while Towell and Shavlik's Knowledge-based Artificial Neural Networks (KBANN), and its subsequent developments (e.g. [OS93, Opi95]), have been empirically shown to be superior to some of the main neural, symbolic and hybrid systems, being, to the best of our knowledge, the most effective HST to date.

In *Part I*, we will present important theoretical results on Neural-Symbolic Learning Systems – such as the proof of soundness of the Translation Algorithm – and empirically investigate their efficiency by applying them to real world problems of Computational Biology and Fault Diagnosis. We will also compare results with some of the main neural and symbolic systems of Machine Learning.

The second part of the book deals with the problem of symbolic knowledge extraction from trained neural networks. The subject has grown beyond the study of neural-symbolic integration systems and is now a research area on its own (see, for example, [Cra96, Mai98, Mai97, Set97a, Thr94]). Although knowledge extraction is an integral part of Neural-Symbolic Learning Systems, we have tried as much as possible to present the subject independently from the previous chapters. Our approach builds upon Fu's extraction method [Fu94] and some features of KBANN's $M$ of $N$ method [TS93].

In *Part II*, we will present new theoretical results on knowledge extraction from trained neural networks, culminating with the proof of soundness of the Extraction Algorithm, which we believe should be the minimum requirement of any method of rule extraction. We will also present empirical evidence of the performance of the extraction method, by applying it to the problems of Computational Biology and Fault Diagnosis used in Part I to investigate the performance of the learning systems.

The third part of this book tackles the problem of theory revision in neural networks. Theory revision may be necessary as a result of the presence of inconsistencies between a symbolic theory and the result of learning from examples. To the best of our knowledge, this book contains the first account and treatment of the subject.

In *Part III*, we will present a one-shot learning algorithm, which will allow neural networks to evolve incrementally from an undesirable stable state into a new stable state. This method, here called Minimal Learning, can be used to fine-tune the network's answers after learning with examples takes place, but, most importantly, it can be applied to solve inconsistencies in the answers computed by the network. The study of how to deal with inconsistencies can be carried out independently of *Part II* of this book.

We conclude the book by presenting a number of challenges and open problems of neural-symbolic learning systems.

## 1.6 Summary

We have mentioned that the aim of Neural-Symbolic Integration is twofold: while logic may benefit from the successful application of neural networks on various knowledge domains, the learning and generalisation processes of neural networks may be rigorously studied and explained by logic. We should be able to explore both directions of this "equivalence" relation, taking into consideration aspects such as learning capability and massively parallel deduction and correctness. Notwithstanding this, finding the balance between what we would like to be able to represent in a neural network and what a neural network naturally represents and learns is a difficult task. It all depends on how we want to benefit from the integration of neurons and symbols.

In this monograph, we are committed to high learning performance, for we believe this is the most important asset of artificial neural networks. Thus we concentrate on single hidden layer networks and Backpropagation – the neural learning combination most commonly successfully applied in industry. We then follow the idea of finding the best symbolic representation for such a neural model. The closest match was the class of grounded extended logic programs of Gelfond and Lifschitz [GL91], augmented with the metalevel superiority relations of Nute's Defeasible Logic [Nut94]. The use of more expressive logics, such as first-order logic, would require a proportionally more complex neural model, which could result in a degradation of learning performance. In other words, the limits of synergetic neural-symbolic integration depend on the objectives of the application. Use it to simplify, not to complicate. Be guided by the application and its needs.

# 2. Background

In this chapter we introduce some basic concepts used throughout this book. We also present basic definitions and results on Inductive Learning and Artificial Neural Networks. In order that the book be reasonably self-contained, we include an introductory section on Logic Programming and Nonmonotonic Reasoning, together with some results in Belief Revision of logic program theories. References to introductory work in each of these areas are provided at the beginning of each section. Detailed background material and related work will be presented throughout the book when needed.

## 2.1 General Preliminaries

We need to assert some basic concepts that will be used throughout this book (see [DP90], [Ger93] and [PY73] for details). $\aleph$ and $\Re$ will denote the sets of natural and real numbers, respectively.

**Definition 2.1.1.** *A* partial order *is a reflexive, transitive and antisymmetric binary relation on a set.*

**Definition 2.1.2.** *A partial order $\preceq$ on a set $X$ is* total *if for every $x, y \in X$, either $x \preceq y$ or $y \preceq x$. Sometimes, $\preceq$ is also called a* linear order, *or simply a* chain.

As usual, $x \prec y$ abbreviates $x \preceq y$ and $x \neq y$.

**Definition 2.1.3.** *In a partially ordered set $[X, \preceq]$, $x$ is the* immediate predecessor *of $y$ if $x \prec y$ and there is no element $z$ in $X$ such that $x \prec z \prec y$. The inverse relation is called the* immediate successor.

**Definition 2.1.4.** *Let $X$ be a set and $\preceq$ a partial ordering on $X$. Let $x \in X$.*

- *$x$ is* minimal *if there is no element $y \in X$ such that $y \prec x$.*
- *$x$ is a* minimum *if for all elements $y \in X, x \preceq y$. If such an $x$ exists, then $x$ is unique and will be denoted by $inf(X)$.*
- *$x$ is* maximal *if there is no element $y \in X$ such that $x \prec y$.*

- *$x$ is a* maximum *if for all elements $y \in X, y \preceq x$. If such an $x$ exists, then $x$ is unique and will be denoted by $sup(X)$.*

A maximum (minimum) element is also maximal (minimal), but is, in addition, comparable to every other element. This property and antisymmetry leads directly to the demonstration of the uniqueness of $inf(X)$ and $sup(X)$.

**Definition 2.1.5.** *Let $[X, \preceq]$ be a partially ordered set and let $Y \subseteq X$.*

- *An element $x \in X$ is an* upper bound *of $Y$ if $y \preceq x$ for all $y \in Y$ (a* lower bound *is defined dually).*
- *$x$ is called the* least upper bound *(lub) of $Y$ if $x$ is an upper bound of $Y$ and $x \preceq z$ for all upper bounds $z$ of $Y$ (the* greatest lower bound *(glb) is defined dually).*
- *If any two elements $x$ and $y$ in $X$ have a least upper bound, denoted by $x \vee y$ and read as "$x$* join *$y$", and a greatest lower bound, denoted by $x \wedge y$ and read as "$x$* meet *$y$", then $X$ is called a* lattice.
- *If $lub(Y)$ and $glb(Y)$ exist for all $Y \subseteq X$ then $X$ is called a* complete lattice.
- *A lattice $L$ is* distributive *if $x \vee (y \wedge z) = (x \vee y) \wedge (x \vee z)$ and $x \wedge (y \vee z) = (x \wedge y) \vee (x \wedge z)$, for all $x, y, z \in L$.*

**Definition 2.1.6.** *Let $U$ be a set and $f : U \times U \rightarrow \Re$ be a function satisfying the following conditions:*

- $f(x, y) \geq 0$,
- $f(x, y) = 0 \leftrightarrow x = y$,
- $f(x, y) = f(y, x)$, *and*
- $f(x, y) \leq f(x, z) + f(z, y)$.

$f$ is called a *metric* on $U$[1]. A *metric space* is a tuple $\langle U, f \rangle$. A metric $f$ on $U$ is *bounded* iff for some constant $k \in \Re$, $f(x, y) \leq k$, for all $x, y \in U$.

## 2.2 Inductive Learning

*Learning*, one of the basic attributes of intelligent comportment, can be defined as the change of behaviour motivated by changes in the environment in order to perform better in different knowledge domains [Mic87]. Learning strategies can be classified as: learning from instruction, learning by deduction, learning by analogy, learning from examples and learning by observation and discovery; the latter two are forms of *Inductive Learning*. [Rus96] is a

[1] $f$ is sometimes called a *distance function*.

good introductory text on Machine Learning, while [Mit97] contains some new hybrid models of learning, in addition to the traditional, purely symbolic, paradigm.

The task of (symbolic) *Inductive Learning* is to find hypotheses that are consistent with background knowledge to explain a given set of examples. In general, those hypotheses are definitions of concepts described in some logical language, the examples are descriptions of instances and non-instances of the concept to be learned, and the background knowledge gives additional information about the examples and the concepts' domain knowledge (see [LD94]). A general formulation for inductive learning is as follows: Given a background knowledge or an initial set of beliefs ($\mathcal{B}$) and a set of positive ($e^+$) and negative ($e^-$) examples of some concept, find a hypothesis ($h$) in some knowledge representation language ($\mathcal{L}$) such that $\mathcal{B} \cup h \vdash_{\mathcal{L}} e^+$ and $\mathcal{B} \cup h \nvdash_{\mathcal{L}} e^-$, for each element of $e^-$.

The study of Inductive Learning dates back to the 1960s [HMS66]. One of its most successful areas is called *Case Based Reasoning (CBR)*, which is best instantiated in Quinlan's ID3 algorithm [Qui86] and Michalski's AQ15 [MMHL86]. More recently, the importance of adding background knowledge to help the learning process, depending on the application at hand, has been highlighted (see e.g. [HN94] and [PK92]), and the area of *Inductive Logic Programming (ILP)* has flourished (A comprehensive survey on ILP can be found in [MR94]). In [LD94], CBR and ILP are integrated by means of the LINUS system, which translates a fragment of first-order logic into attribute-value examples in order to apply, for instance, a decision tree generation learning algorithm like ID3. The result of learning can then be translated back into the original language.

If a learning algorithm fails to extract any pattern from a set of examples, one cannot expect it to be able to extrapolate (or generalise) to examples it has not seen. Finding a pattern means being able to describe a large number of cases in a concise way. A general principle of inductive learning, often called *Ockham's razor* is thus as follows: *the most likely hypothesis is the simplest one that is consistent with all observations.* Unfortunately, finding the smallest representation (e.g. the smallest decision tree) is an intractable problem, and usually the use of some heuristics is needed. In [GL97], some conditions for Ockham's Razor applicability are discussed. In a more general perspective, the work of Valiant discusses the general theory of learning [Val84].

## 2.3 Neural Networks

Differently from (symbolic) machine learning, (numeric) neural networks perform inductive learning in such a way that the statistical characteristics of

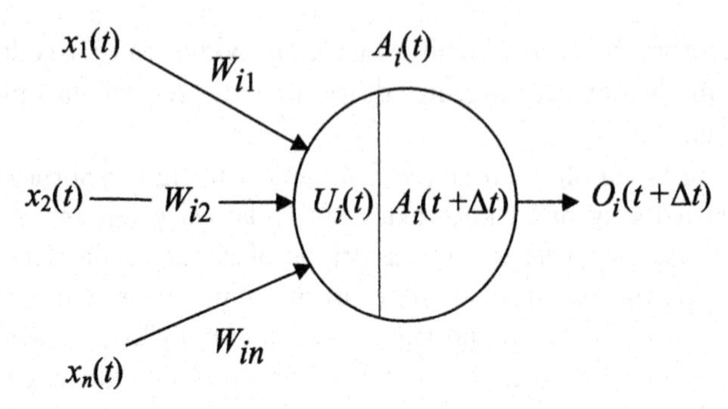


**Fig. 2.1.** The processing unit or neuron.

the data are encoded in their sets of weights. Connectionist *Inductive Learning* is also concerned with the problem of how to *change*. However, it assumes that "it is much easier to learn if the system responds to changes in a graded, proportional matter, instead of radically altering the way it behaves. These graded changes allow the system to try out various new ways of processing. By exploring lots of little changes, the system can evaluate and strengthen those that improve performance, while abandoning those that do not" [OM00].

Good introductory texts on neural networks can be found in [HKP91] and [Tay93]. More advanced texts are [Hay99a] and [BL96].

### 2.3.1 Architectures

An artificial neural network is a directed graph. A unit in this graph is characterised, at time $t$, by its input vector $I_i(t)$, its input potential $U_i(t)$, its activation state $A_i(t)$, and its output $O_i(t)$. The units (neurons) of the network are interconnected via a set of directed and weighted connections. If there is a connection from unit $i$ to unit $j$, then $W_{ji} \in \Re$ denotes the *weight* associated with such a connection.

We start by characterising the neuron's *functionality* (see Figure 2.1). The *activation state* of a neuron $i$ at time $t$ ($A_i(t)$) is a bounded real or integer number. The output of neuron $i$ at time $t$ ($O_i(t)$) is given by the *output rule* $f_i$, such that $O_i(t) = f_i(A_i(t))$. The input potential of neuron $i$ at time $t$ ($U_i(t)$) is obtained by applying the *propagation rule* of neuron $i$ ($g_i$) such that $U_i(t) = g_i(I_i(t), W_i)$, where $I_i(t)$ contains the input signals $(x_1(t), x_2(t), ..., x_n(t))$ to neuron $i$ at time $t$, and $W_i$ denotes the weight vector $(W_{i1}, W_{i2}, ..., W_{in})$ to neuron $i$. Finally, the neuron's new activation state $A_i(t+\Delta t)$ is given by its *activation rule* $h_i$, which is a function of the neuron's current activation state and input potential, i.e. $A_i(t+\Delta t) = h_i(A_i(t), U_i(t))$, and the neuron's new output value $O_i(t+\Delta t) = f_i(A_i(t+\Delta t))$.

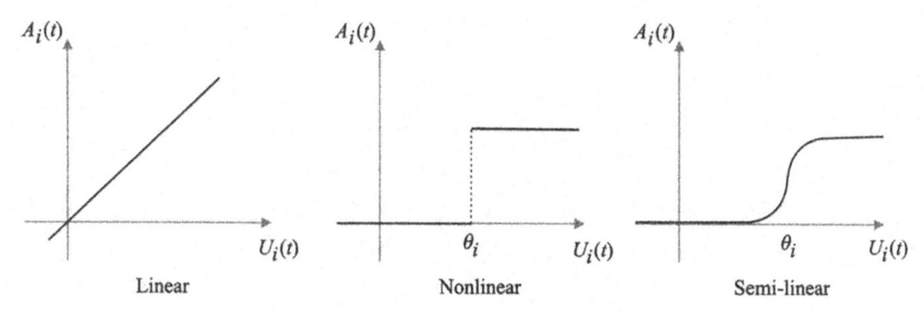


**Fig. 2.2.** Activation functions.

There are three basic kinds of *activation functions* ($h_i$): linear, non-linear and semi-linear. Neurons with linear, non-linear (binary) and semi-linear activation functions are called linear, non-linear (binary) and semi-linear neurons, respectively. In Figure 2.2, $\theta_i$ is known as the *threshold* of the neuron's activation function.

In general, $h_i$ does not depend on the previous activation state of the unit, that is, $A_i(t + \Delta t) = h_i(U_i(t))$, the propagation rule $g_i$ is a weighted sum, such that $U_i(t) = \sum_j W_{ij}x_j(t)$, and the output rule $f_i$ is given by the identity function, i.e. $O_i(t) = A_i(t)$. In addition, most neural models also have a *learning rule*, responsible for changing the weights of the network, and thus allowing it to perform inductive learning.

The units of a neural network can be organised in layers. A *n-layer feedforward network* $N$ is an acyclic graph. $N$ consists of a sequence of layers and connections between successive layers, containing one input layer, $n - 2$ hidden layers and one output layer, where $n \geq 2$. When $n = 3$, we say that $N$ is a *single hidden layer network.* When each unit occurring in the $i$th layer is connected to each unit occurring in the $i + 1$st layer, we say that $N$ is a *fully connected network* (see Figure 2.3).

The most interesting properties of a neural network do not arise from the functionality of each neuron, but from the collective effect resulting from the interconnection of units. Let $r$ and $s$ be the number of units occurring in the input and output layer, respectively. A multilayer feedforward network $N$ computes a function $f : \Re^r \rightarrow \Re^s$ as follows. The input vector is presented to the input layer at time $t_1$ and propagated through the hidden layers to the output layer. At each time point, all units update their input potential and activation state synchronously. At time $t_n$ the output vector is read off the output layer.

In this book, we concentrate on single hidden layer feedforward networks. We do so because of the following result.

**Theorem 2.3.1.** *[Cyb89] Let $h : \Re \rightarrow \Re$ be a continuous and semi-linear function. Let $\varepsilon \in \Re^+$ and $n \in \aleph$. Given a continuous and real-valued function*

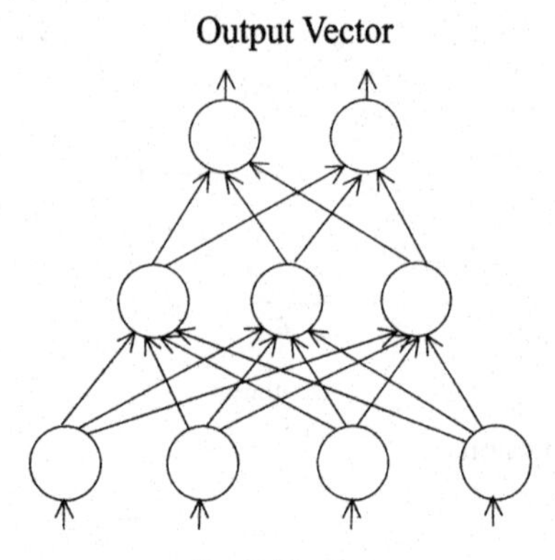


**Fig. 2.3.** A typical feedforward neural network.

*g on* $\mathbb{I}^n = [0,1]^n$*, there exist a finite K and parameters* $\alpha_j \in \Re, \theta_j \in \Re$ *and* $y_j \in \Re^n (1 \leq j \leq K)$*, such that if* $f(x) = \sum_{j=1}^{K} \alpha_j \cdot h(y_j x + \theta_j), x \in \mathbb{I}^n$*, then* $|f(x) - g(x)| < \varepsilon$ *for all* $x \in \mathbb{I}^n$*.*

In other words, by making $\alpha_j$ the weight from the $j$th hidden neuron to the output neuron, single hidden layer feedforward networks can approximate any (Borel measurable) function arbitrarily well, regardless of the dimension of the input space $n$. Independently, a similar result was proved in [HSW89]. In this sense, single hidden layer networks are *universal approximators* of virtually any function of interest (see [HSW89]).

We also use *bipolar* semi-linear activation functions $h(x)$ with inputs in $\{-1, 1\}$. The reasons for this will become clear in Section 3.1. The following systems of equations describe the dynamics of the networks that we use here.

$$\begin{aligned} n_1 &= h(W^1_{11} i_1 + W^1_{12} i_2 + \cdots + W^1_{1p} i_p - \theta_{n_1}) \\ n_2 &= h(W^1_{21} i_1 + W^1_{22} i_2 + \cdots + W^1_{2p} i_p - \theta_{n_2}) \\ &\vdots \\ n_r &= h(W^1_{r1} i_1 + W^1_{r2} i_2 + \cdots + W^1_{rp} i_p - \theta_{n_r}) \end{aligned} \tag{2.1}$$

$$\begin{aligned} o_1 &= h(W^2_{11} n_1 + W^2_{12} n_2 + \cdots + W^2_{1r} n_r - \theta_{o_1}) \\ o_2 &= h(W^2_{21} n_1 + W^2_{22} n_2 + \cdots + W^2_{2r} n_r - \theta_{o_2}) \\ &\vdots \\ o_q &= h(W^2_{q1} n_1 + W^2_{q2} n_2 + \cdots + W^2_{qr} n_r - \theta_{o_q}) \end{aligned} \tag{2.2}$$

where $\mathbf{i} = (i_1, i_2, ..., i_p)$ is the network's input vector $(i_{j(1 \leq j \leq p)} \in \{-1, 1\})$, $\mathbf{o} = (o_1, o_2, ..., o_q)$ is its output vector $(o_{j(1 \leq j \leq q)} \in [-1, 1])$, $\mathbf{n} = (n_1, n_2, ..., n_r)$ is the hidden layer vector $(n_{j(1 \leq j \leq r)} \in [-1, 1])$, $\theta_{n_j(1 \leq j \leq r)}$ is the $j$th hidden neuron threshold $(\theta_{n_j} \in \Re)$, $\theta_{o_j(1 \leq j \leq q)}$ is the $j$th output neuron threshold $(\theta_{o_j} \in \Re)$, $-\theta_{n_j}$ (resp. $-\theta_{o_j}$) is called the *bias* of the $j$th hidden neuron (resp. output neuron), $W^1_{ij(1 \leq i \leq r, 1 \leq j \leq p)}$ is the weight of the connection from the $j$th neuron in the input layer to the $i$th neuron in the hidden layer $(W^1_{ij} \in \Re)$, $W^2_{ij(1 \leq i \leq q, 1 \leq j \leq r)}$ is the weight of the connection from the $j$th neuron in the hidden layer to the $i$th neuron in the output layer $(W^2_{ij} \in \Re)$, and finally $h(x) = \frac{2}{1+e^{-\beta x}} - 1$ is the standard bipolar (semi-linear) activation function, where $\beta$ $(\beta \in \Re^+)$ defines the steepness of $h(x)$. Notice that for each output $o_j (1 \leq j \leq q)$ in $\mathbf{o}$ we have $o_j = h(\sum_{i=1}^{r}(W^2_{ji}.h(\sum_{k=1}^{p}(W^1_{ik}.i_k) - \theta_{n_i})) - \theta_{o_j})$.[2]

### 2.3.2 Learning Strategy

A neural network's *learning* process (or *training*) is done by successively changing its weights in order to approximate the function $f$ computed by it to a desired function $g$. In *supervised learning,* one attempts to estimate the unknown function $g$ from *examples* (input and output patterns) presented to the network. The idea is to *minimise the error* associated with the set of examples by performing small changes to the network's weights.

In the case of *Backpropagation* [RHW86], the learning process occurs as follows: given a set of input patterns $\mathbf{i}^i$ and corresponding target vectors $\mathbf{t}^i$, the network's outputs $\mathbf{o}^i = f(\mathbf{i}^i)$ may be compared with the targets $\mathbf{t}^i$, and an error such as

$$Err(\mathbf{W}) = \frac{1}{2}\sum_i(\mathbf{o}^i - \mathbf{t}^i)^2 \tag{2.3}$$

can be computed. This error depends on the set of example ((**i**, **t**) pairs) and may be minimised by *gradient descent*, i.e. by the iterative application of changes

$$\Delta\mathbf{W} = -\eta \cdot \nabla_{\mathbf{W}} \tag{2.4}$$

to the weight vector **W**, where

$$\nabla_{\mathbf{W}} = \left(\frac{\partial Err(\mathbf{W})}{\partial W_{11}}, \frac{\partial Err(\mathbf{W})}{\partial W_{12}}, ..., \frac{\partial Err(\mathbf{W})}{\partial W_{ij}}\right)$$

The computation of $\nabla_{\mathbf{W}}$ is not obvious for a network with hidden units. However, in the famous paper "*Learning Internal Representations by Error*

[2] Whenever it is unnecessary to differentiate between the hidden and output layers, we refer to the weights in the network as $W_{ij}$ only. Similarly, we refer to the network's thresholds in general as $\theta_i$ only.

*Propagation*" [RHW86], Rumelhart, Hinton and Williams presented a simple and efficient way of computing such derivatives.[3] They have shown that a backward pass of $\mathbf{o}^i - \mathbf{t}^i$ through the network, analogous to the forward propagation of $\mathbf{i}^i$, allows the recursive computation of $\nabla_{\mathbf{W}}$. The idea is that, in the forward pass through the network, one should also calculate the derivative of $h_k(x)$ for each neuron $k$, $dk = h'_k(U_k(t))$. For each output neuron $o$, one simply calculates $\partial o = (\mathbf{o}^i - \mathbf{t}^i)do$. One can then compute weight changes for all connections that feed into the output layer. For each connection $W_{oj}$, $\Delta W_{oj} = -\eta \cdot \partial o \cdot O_j(t)$. After this is done, $\partial j = (W_{oj} \cdot \partial o) \cdot dj$ can be calculated for each hidden neuron $j$. This propagates the errors back one layer, and for each connection $W_{ji}$, from the input neurons $i$ to $j$, $\Delta W_{ji} = -\eta \cdot \partial j \cdot O_i(t)$. Of course, the same process could be repeated for many hidden layers. This procedure is called the *generalised delta rule*, and it is very useful because the backward pass has the same computational complexity as the forward pass.

In the above procedure, $\eta$ is called the *learning rate*. True gradient descent requires infinitesimal steps to be taken ($\eta \approx 0$). However, the larger the constant $\eta$, the larger the change in the weights, and thus a good choice of $\eta$ is responsible for faster convergence. The design challenge here is how to choose the learning rate as large as possible without leading to oscillation. A variation of standard backpropagation allows the adaptation of this parameter during learning. In this case, $\eta$ is typically large at the beginning of learning, and decreases as the network approaches a minimum of the error surface.

Backpropagation training may lead to a local rather than a global error minimum. In an attempt to ameliorate this problem and also improve training time, the *term of momentum* can be added to the learning process. The term of momentum allows a network to respond not only to the local gradient, but also to recent trends in the error surface, acting as a low pass filter.

Momentum is added to backpropagation learning by making weight changes equal to the sum of a fraction of the last weight change and the new change suggested by the backpropagation rule. Equation 2.5 shows how backpropagation with momentum is expressed mathematically.

$$\Delta \mathbf{W}(i) = -\eta \cdot \nabla_{\mathbf{W}(i)} + \alpha \Delta \mathbf{W}(i-1) \tag{2.5}$$

where $\alpha \Delta W(i-1)$ is the term of momentum and $0 < \alpha < 1$ is the momentum constant. Typically, $\alpha = 0.9$.

Another learning difficulty is known as the *problem of symmetry*. "If all weights start out with equal values and if the solution requires that unequal weights be developed, the system can never learn" [RHW86]. This is so because the error backpropagated is proportional to the actual values of the

[3] The term "backpropagation" appears to have evolved after 1985. However, the basic idea of backpropagation was first described by Werbos in his PhD thesis [Wer74].

weights. Symmetry breaking is achieved by starting the system with small random weights.

If the application at hand contains too many degrees of freedom and too few training data, backpropagation can merely "memorise" the data. This behaviour is known as *overfitting*. The ultimate measure of success, therefore, should not be how closely the network approximates the training data, but how well it accounts for yet unseen cases, i.e. how well the network generalises to new data. In order to evaluate the network's *generalisation*, the set of examples is commonly partitioned into a *training set* and a *testing set*.

Finally, it should be noted that a network's learning capability and its activation functions are closely related. Linear neurons possess less *learning capability* than non-linear ones, because their hidden layers act only as a multiplier of the input vector. As a result, complex functions cannot be learned (see [DS96], [Gar96] and [SS95]). On the other hand, semi-linear activation functions are continuous and derivable, which is an important property for the use of backpropagation.

Summarising, there are three major issues one must address when applying neural networks in a given domain: the *representation* problem (model, architecture and size), the *learning* problem (time and training method) and the *generalisation* problem (performance).

All the networks in this book were trained using standard backpropagation with momentum, the most commonly successfully used learning algorithm. The reader is referred to [CR95, RHW86] for the backpropagation algorithm and its variations and improvements; to [LGT96] for a discussion on the problem of overfitting; and to [HK93, Wil97] for incremental learning algorithms, whose concept we apply in Chapter 7.

### 2.3.3 Recurrent Networks

In this section, we introduce the basic concepts of recurrent neural networks, also known as *feedback networks*. For a neural network to represent and generalise *temporal sequences*, it is necessary that it presents a recurrent architecture, i.e. an architecture with *cycles*, as the network of Figure 2.4 illustrates [HKP91].

*Backpropagation* can be extended to recurrent networks by performing the propagation of the error through the inverted connections. The *Recurrent Backpropagation* algorithm, as well as considerations on the problems of *caotic cycles* and *synchronisation*, can be found in [HKP91]. Alternatively, *Backpropagation through Time* [Wer90] extends standard *Backpropagation* to recurrent networks by *unfolding* the temporal sequences represented in the network into a multi-layered feedforward network. For each time step, a layer is added to the network's topology.

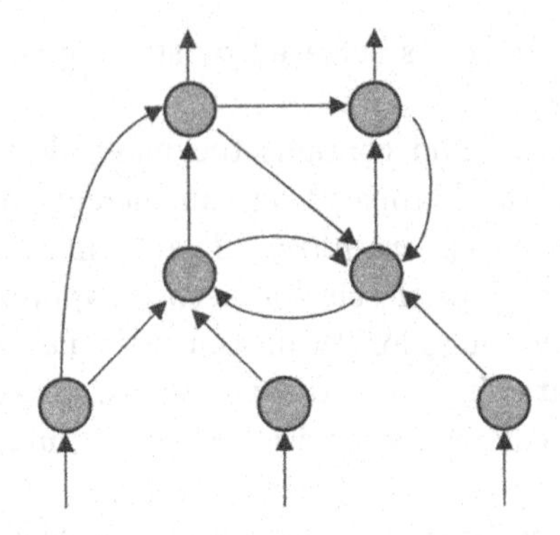

Fig. 2.4. A recurrent neural network.

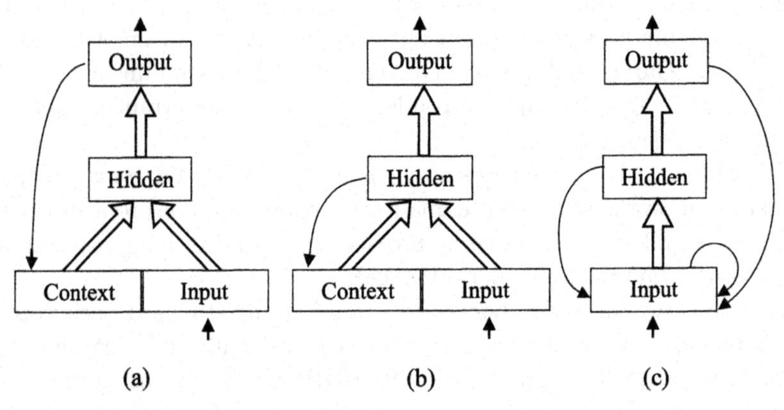


Fig. 2.5. Partially recurrent networks.

A simpler way of representing temporal sequences in neural networks is to use *partially recurrent networks*. In such networks, the majority of the connections are *feedforward*, and only certain connections are allowed to be *feedback*. The idea is to allow the network to recall its recent past, without complicating its learning process. Furthermore, if such feedback connections have fixed, predefined weights, then standard *Backpropagation* can be used [HKP91]. Figure 2.5 shows different partially recurrent architectures. A number of *context* neurons, composing either part of a layer or a whole layer, are responsible for synchronising the feedback signal. The remainder of the network is allowed to work asynchronously. Processing at time $t$ depends on the *input* of the network at $t$, and on the *context* of the network, obtained from processing at time $t-1$.

In the network of Figure 2.5(a), proposed in [Jor86], the context units at time $t$ simply contain a copy of the activation states of the output units at time $t-1$. Only feedforward connections are allowed to change during learning. As a result, standard *Backpropagation* can be applied for learning, during which the context units are treated as normal inputs. Figure 2.5(b) shows the architecture proposed in [Elm90]. It differs from the network of Figure 2.5(a), called *Jordan Network*, only in that now the hidden layer, and not the output layer, provides the network's feedback. Similar behaviours can be obtained by either of the networks. Finally, Figure 2.5(c) presents an architecture in which every input unit is also a context unit. Feedback is allowed from any neuron in the network to any neuron in its input layer. The signal propagated from the input to the hidden layer is a weighted sum of the network's feedback and its input signal, which is computed in its own input layer [HKP91].

In this book, we use feedforward networks and partially recurrent networks, sometimes without differentiating context units from input units. We train all the networks using standard *Backpropagation*, since all the recurrent connections contain fixed weights $W_r = 1$, with the sole purpose of ensuring that the output feed the input in the next learning or recall process, as in *Jordan Networks.* During learning, any input signal takes precedence over the feedback (i.e. the feedback is only used for learning if the input signal is not present). During recall, the feedback takes precedence over any input signal. This process will be described in more detail in Chapter 3. The reader is referred to [Jor86, Kol94, HKP91] for details on *recurrent networks.*

## 2.4 Logic Programming

In this section we introduce the basic concepts of logic programming, interspersed with illustrating examples.

### 2.4.1 What is Logic Programming?

The basis of logic programming comprises two main concepts:

- A logic program, which is a theory written in the language of first-order Horn clause logic, and
- A specialised execution procedure that allows deductions to be made from the theory in order to derive answers to queries.

*Example 2.4.1.* Consider the following clauses (1) to (4):

```
(1) arc(a,b).
(2) arc(b,c).
(3) path(X,Y) <- arc(X,Y).
(4) path(X,Y) <- arc(X,Z), path(Z,Y).
```

Clauses (1)–(4) form a logic program in which $X$, $Y$ and $Z$ are implicitly universally quantified variables. The reader can think of this theory as a description of a digraph, where $arc(u, v)$ holds if there is a direct link between nodes $u$ and $v$ and $path(u, v)$ holds if there is a path of one or more arcs between nodes $u$ and $v$. Clause (3) is read as "there is a path from $X$ to $Y$ if there is an arc from $X$ to $Y$", and Clause (4) as "there is a path from $X$ to $Y$ if there is an arc from $X$ to $Z$ and a path from $Z$ to $Y$".

Each clause is an implication which has an atomic conclusion and an antecedent consisting of a conjunction of none or more conditions, also called *literals*, each of which is either an *atom* (as in the example) or a negated atom (e.g. $\neg arc(a, b)$). Clause (1) is an implication with an empty conjunction for its body which, by convention, evaluates to true and is usually omitted. Such clauses are also called *facts* and may occasionally be equivalently written as in $arc(a, b) \leftarrow true$. Typical queries to this program, which can be answered by computation, include $path(a, b)$ (is there a path between nodes $a$ and $b$), $path(a, X)$ (find $X$ such that there is a path between $a$ and $X$) and $path(X, Y)$ (find $X$ and $Y$ such that there is a path between them). The answers to these queries using the above program are, respectively, {yes}, $\{X = b\}$ or $\{X = c\}$, and $\{X = a, Y = b\}$, $\{X = b, Y = c\}$ or $\{X = a, Y = c\}$. The actual process of computation used to obtain these answers is briefly described at the end of this section.

Each of the clauses (3) and (4) in the program given in Example 2.4.1 represents a set of ground clauses formed by substituting for each variable a ground term from some fixed language $\mathcal{L}$, a process called *instantiation*. Clauses (1) and (2) are already ground. It is assumed that the language $\mathcal{L}$ includes at least the constant functors occurring in the program, which here are $a$, $b$ and $c$.

*Example 2.4.2.* In Example 2.4.1 let the fixed language $\mathcal{L}$ be exactly the set of terms $\{a, b, c\}$. There are then 9 ground instances of clause (3), namely:

```
path(a,a)<-arc(a,a),      path(a,b)<-arc(a,b),
path(a,c)<-arc(a,c),      path(b,a)<-arc(b,a),
path(b,b)<-arc(b,b),      path(b,c)<-arc(b,c),
path(c,a)<-arc(c,a),      path(c,b)<-arc(c,b),
path(c,c)<-arc(c,c)
```

and 27 of clause (4). Clauses (1) and (2) each have 1 ground instance (itself).

Let $\mathcal{L}$ be a fixed language of all ground terms built from a finite set of functors; then the *ground instances* of a clause $C$ is the set of clauses obtained from $C$ by simultaneously substituting for each variable (if any) a term from $\mathcal{L}$.[4] The set of ground instances of a program $\mathcal{P}$ is the union of the ground

[4] For each variable the same substitution is made for all occurrences of the variable.

instances of each clause in $\mathcal{P}$. If the language contains only constant functors then the ground instances of $\mathcal{P}$ will be a finite set. This will be the case in this book. Furthermore, in this book, unless stated otherwise, we will make the assumption that a program $\mathcal{P}$ has been already instantiated (or grounded) and thus all of its clauses are ground (or propositional).

In this section and the next we use standard notation and terminology of [Llo87] with the exception that general logic programs are called normal logic programs in [Llo87].

**Definition 2.4.1.** *A* Definite Logic Program $\mathcal{P}$ *is a finite set of clauses of the form* $A_0 \leftarrow A_1, ..., A_n$*, where each* $A_i$ *is an atom.* $A_0$ *is called the* head *(or the* consequent*) of the clause, and* $A_1, ..., A_n$ *(the comma represents conjunction) is called the* body *(or the* antecedent*) of the clause.*

Let $\mathcal{P}$ be a definite program. $B_{\mathcal{P}}$ will denote the set of atoms occurring in $\mathcal{P}$ and is called the *Herbrand base* of $\mathcal{P}$. An *(Herbrand) interpretation* (or *valuation*) is a mapping from propositional variables in $B_{\mathcal{P}}$ to $\{true, false\}$. An interpretation $\mathcal{I}$ is extended to map clauses to $\{true, false\}$ as follows: a clause $A_0 \leftarrow A_1, ..., A_n$ is mapped to *true* either if $A_0$ is mapped to *true* or if one of $A_i$ is mapped to *false*. A *model* for $\mathcal{P}$ is a Herbrand interpretation which maps all clauses in $\mathcal{P}$ to *true*. An interpretation $\mathcal{I}$ is often represented by the subset of atoms in $B_{\mathcal{P}}$ that it maps to *true*. $M_{\mathcal{P}}$ will denote the least *Herbrand model* of $\mathcal{P}$; that is, the smallest subset of $B_{\mathcal{P}}$ that represents an interpretation that is a model of $\mathcal{P}$.

*Example 2.4.3.* The set of atoms, or $B_{\mathcal{P}}$, for the program $\mathcal{P}$ of Example 2.4.2 is

$$\left\{\begin{array}{l} path(a,a),\ path(a,b),\ path(a,c),\ path(b,a),\ path(b,b),\\ path(b,c),\ path(c,a),\ path(c,b),\ path(c,c),\ arc(a,a),\\ arc(a,b),\ arc(a,c),\ arc(b,a),\ arc(b,b),\ arc(b,c),\\ arc(c,a),\ arc(c,b),\ arc(c,c) \end{array}\right\}$$

The interpretation $\mathcal{I}_1$ represented by

$$\left\{\begin{array}{l} path(a,a),\ path(a,b),\ path(b,c),\ path(a,c),\\ arc(a,a),\ arc(a,b),\ arc(b,c) \end{array}\right\}$$

is a model of $\mathcal{P}$, whereas $\mathcal{I}_2 = \{arc(a,a)\}$ is not; for instance, the clause $path(a,a) \leftarrow arc(a,a)$ is mapped to *false* by $\mathcal{I}_2$. The minimum model of $\mathcal{P}$ is

$$\{arc(a,b),\ arc(b,c),\ path(a,b),\ path(b,c),\ path(a,c)\}.$$

### 2.4.2 Fixpoints and Definite Programs

The following result dates back to 1976 [vEK76]. The interest in this result arises from the fact that, for a definite program $\mathcal{P}$, the collection of all Herbrand interpretations forms a complete lattice and there is a monotonic mapping associated with $\mathcal{P}$ defined on this lattice.

**Proposition 2.4.1.** (Model Intersection Property) *Let $\mathcal{P}$ be a definite program and $\{M_i\}$ be a non-empty set of Herbrand models for $\mathcal{P}$. Then $\cap_i M_i$ is also a Herbrand model for $\mathcal{P}$.*

In the following, we recall some concepts and results concerning monotonic mappings and their fixpoints.

**Definition 2.4.2.** *Let $[L, \preceq]$ be a complete lattice and $T : L \rightarrow L$ be a mapping.*

- *$T$ is* monotonic *if $x \preceq y \rightarrow T(x) \preceq T(y)$.*
- *Let $Y \subseteq L$. $Y$ is* directed *if every finite subset of $Y$ has an upper bound in $Y$.*
- *$T$ is* continuous *if for every directed subset $Y$ of $L$, $T(lub(Y)) = lub(T(Y))$.*

**Definition 2.4.3.** *Let $[L, \preceq]$ be a complete lattice and $T : L \rightarrow L$ be a mapping. $a \in L$ is the* least fixpoint *of $T$ if $a$ is a fixpoint of $T$ (i.e. $T(a) = a$) and for all fixpoints $b$ of $T$, $a \preceq b$. Similarly, $a \in L$ is the* greatest fixpoint *of $T$ if $a$ is a fixpoint of $T$ and for all fixpoints $b$ of $T$, $b \preceq a$.*

**Proposition 2.4.2.** *Let $L$ be a complete lattice and $T : L \rightarrow L$ be monotonic. $T$ has a least fixpoint ($lfp(T)$) and a greatest fixpoint ($gfp(T)$).*

Let us now recall some elementary properties of ordinal numbers in order to define the concept of *ordinal powers* of $T$. The first ordinal (0) is defined to be $\emptyset$. Then we define $1 = \{\emptyset\} = \{0\}$, $2 = \{\emptyset, \{\emptyset\}\} = \{0, 1\}$, $3 = \{\emptyset, \{\emptyset\}, \{\emptyset, \{\emptyset\}\}\} = \{0, 1, 2\}$, and so on. The first infinite ordinal is $\omega = \{0, 1, 2, ...\}$, the set of all non-negative integers. We can specify an ordering $\prec$ on the collections of all ordinals by defining $\alpha \prec \beta$ iff $\alpha \in \beta$. If $\alpha$ is an ordinal, the successor of $\alpha$ is the ordinal $\alpha + 1 = \alpha \cup \{\alpha\}$, which is the least ordinal greater than $\alpha$. An ordinal $\alpha$ is said to be a limit ordinal if it is not the successor of any ordinal. The smallest limit ordinal (apart from 0) is $\omega$. After $\omega$, comes $\omega + 1 = \omega \cup \{\omega\}$, and so on. The next limit ordinal is $\omega 2$, which is the set containing all $n \in \omega$ and all $\omega + n$. After $\omega 2$ comes $\omega 2 + 1, \omega 2 + 2, ..., \omega 3, \omega 3 + 1, ....$

**Definition 2.4.4.** *Let $L$ be a complete lattice and $T : L \rightarrow L$ be monotonic. Then we define:*

$T \uparrow 0 = inf(L)$;

$T \uparrow \alpha = T(T \uparrow (\alpha - 1))$, *if* $\alpha$ *is a successor ordinal;*
$T \uparrow \alpha = lub\{T \uparrow \beta \mid \beta \prec \alpha)$, *if* $\alpha$ *is a limit ordinal;*
$T \downarrow 0 = sup(L)$;
$T \downarrow \alpha = T(T \downarrow (\alpha - 1))$, *if* $\alpha$ *is a successor ordinal;*
$T \downarrow \alpha = glb\{T \downarrow \beta \mid \beta \prec a)$, *if* $\alpha$ *is a limit ordinal.*

**Proposition 2.4.3.** *Let* $L$ *be a complete lattice and* $T : L \rightarrow L$ *be continuous. Then* $lfp(T) = T \uparrow \omega$.

Let $\mathcal{P}$ be a definite program. Then $2^{B_\mathcal{P}}$, which is the set of all Herbrand interpretations of $\mathcal{P}$, is a complete lattice under the partial order of set inclusion $\subseteq$. The top element of this lattice is $B_\mathcal{P}$ and the bottom element is $\emptyset$.

**Definition 2.4.5.** (Immediate Consequence Operator) *Let* $\mathcal{P}$ *be a definite program. The mapping* $T_\mathcal{P} : 2^{B_\mathcal{P}} \rightarrow 2^{B_\mathcal{P}}$ *is defined as follows. Let* $I$ *be a Herbrand interpretation; then* $T_\mathcal{P}(I) = \{A \in B_\mathcal{P} \mid A \leftarrow A_1, ..., A_n$ *is a clause in* $\mathcal{P}$ *and* $\{A_1, ..., A_n\} \subseteq I\}$.

$T_\mathcal{P}$ provides the link between the declarative and the procedural semantics of $\mathcal{P}$. Clearly, $T_\mathcal{P}$ is monotonic. Therefore, Herbrand interpretations that are models can be characterised in terms of $T_\mathcal{P}$.

**Proposition 2.4.4.** *Let* $\mathcal{P}$ *be a definite program and* $I$ *a Herbrand interpretation of* $\mathcal{P}$. *Then the mapping* $T_\mathcal{P}$ *is continuous and* $I$ *is a model of* $\mathcal{P}$ *iff* $T_\mathcal{P}(I) \subseteq I$.

**Proposition 2.4.5.** (Fixpoint Characterisation of Least Herbrand Model) *Let* $\mathcal{P}$ *be a definite program.* $M_\mathcal{P} = lfp(T_\mathcal{P}) = T_\mathcal{P} \uparrow \omega$.

Note that, as $B_\mathcal{P}$ is finite, the lattice is also finite, and there is some $n \in \omega$ such that $T_\mathcal{P} \uparrow n = T_\mathcal{P} \uparrow n + 1$, and hence $T_\mathcal{P} \uparrow \omega$ will be equal to $T_\mathcal{P} \uparrow n$, for some successor ordinal $n \in \omega$.

*Example 2.4.4.* For the program of Example 2.4.2

$T_\mathcal{P} \uparrow 0 = \emptyset$
$T_\mathcal{P} \uparrow 1 = \{arc(a,b),\ arc(b,c)\}$
$T_\mathcal{P} \uparrow 2 = \{arc(a,b),\ arc(b,c),\ path(a,b),\ path(b,c)\}$
$T_\mathcal{P} \uparrow 3 = \{arc(a,b),\ arc(b,c),\ path(a,b),\ path(b,c), path(a,c)\}$
$T_\mathcal{P} \uparrow 4 = T_\mathcal{P} \uparrow 3 = T_\mathcal{P} \uparrow \omega$

$$T_\mathcal{P} \downarrow 0 = \left\{ \begin{array}{l} arc(a,a),\ arc(a,b),\ arc(a,c),\ arc(b,a),\ arc(b,b), \\ arc(b,c),\ arc(c,a),\ arc(c,b),\ arc(c,c),\ path(a,a), \\ path(a,b),\ path(a,c),\ path(b,a),\ path(b,b), \\ path(b,c),\ path(c,a),\ path(c,b),\ path(c,c) \end{array} \right\}$$

$$T_{\mathcal{P}} \downarrow 1 = \left\{ \begin{array}{l} arc(a,b),\ arc(b,c),\ path(a,a),\ path(a,b), \\ path(a,c),\ path(b,a),\ path(b,b),\ path(b,c), \\ path(c,a),\ path(c,b),\ path(c,c) \end{array} \right\}$$

$$T_{\mathcal{P}} \downarrow 2 = \left\{ \begin{array}{l} arc(a,b),\ arc(b,c),\ path(a,a),\ path(a,b), \\ path(a,c),\ path(b,a),\ path(b,b),\ path(b,c) \end{array} \right\}$$

$$T_{\mathcal{P}} \downarrow 3 = \left\{ \begin{array}{l} arc(a,b),\ arc(b,c),\ path(a,a),\ path(a,b), \\ path(a,c),\ path(b,c) \end{array} \right\}$$

$$T_{\mathcal{P}} \downarrow 4 = \{arc(a,b),\ arc(b,c),\ path(a,b),\ path(b,c),\ path(a,c)\}$$

$$T_{\mathcal{P}} \downarrow 5 = T_{\mathcal{P}} \downarrow 4 = T_{\mathcal{P}} \downarrow \omega$$

Notice that $T_{\mathcal{P}} \uparrow \omega = T_{\mathcal{P}} \downarrow \omega$ and is also the minimal Herbrand model of $\mathcal{P}$.

The declarative semantics of logic programs is given by the notions of least Herbrand models and describes what is logically implied by a program. On the other hand, the procedural semantics describes what can be computed by the program using simple deduction from the given facts. Although, according to Proposition 2.4.5, it is possible to use the result of evaluating $T_{\mathcal{P}} \uparrow \omega$ to decide whether a particular atom belongs to the minimal model of $\mathcal{P}$ and hence whether it is logically implied by $\mathcal{P}$, there is an alternative computation process to answer queries. This process is exemplified next for definite programs (here restricted to ground clauses). It operates backwards from the query as follows: for the query "is there a path from $a$ to $c$", $path(a,c)$, the computation will answer "yes" if there is a clause whose head is the goal $path(a,c)$ and for whose body the computation process can also answer "yes". That is, the computation process must answer "yes" for each atom in the body. For the query $path(a,c)$ there are four clauses that might allow the answer "yes", namely *instances* of clauses (3) and (4) whose head is $path(a,c)$. The instance derived from (3) would require the answer "yes" to $arc(a,c)$, which is not so. The instances derived from (4) require "yes" to be the answer for at least one of the queries $arc(a,a), path(a,c)$, or $arc(a,b), path(b,c)$ or $arc(a,c), path(c,c)$. Only the second of these has a chance of success and it eventually requires $arc(b,c)$ to be answered "yes", which will be the case if there is a clause whose head is $arc(b,c)$. In this program there is, namely clause (2). The computation process described above is called *query evaluation* for definite logic programs. The following proposition states the relationship between the computation process and the declarative semantics of logic programs.

**Proposition 2.4.6.** (Correctness of the Computation Process) *Let $\mathcal{P}$ be a definite program and $\mathcal{Q}$ be a query of one or more atoms, all in the language $\mathcal{L}$; then the answer to the query $\mathcal{Q}$ is "yes" if, and only if, each atom of $\mathcal{Q}$ belongs to $M_{\mathcal{P}}$.*

## 2.5 Nonmonotonic Reasoning

The field of *Nonmonotonic Reasoning* grew out of attempts to capture the essential aspects of commonsense (practical) reasoning. It has resulted in a number of important formalisms, the most well known of them being the Circumscription method of McCarthy [McC80], the Default Theory of Reiter [Rei80], and Autoepistemic Logic of Moore [Moo85] (see [Ant97] for an introduction to the subject, and [MT93a] for a thorough study).

Nonmonotonicity is used for reasoning with incomplete information. If, later, more information becomes available, it may turn out that some conclusions are no longer justified, and must be withdrawn. The standard, and now ubiquitous, example for this case is that, if we learn that Tweety is a bird, we conclude that it can fly, but if we subsequently find out that Tweety is a penguin, we withdraw that conclusion. This use of logic, called *Belief Revision*, is clearly nonmonotonic.

One way of classifying different semantics is by studying which properties they satisfy. A very strong property of inference relations is monotonicity. An inference relation is monotonic if it satisfies $\Gamma \vdash \Psi$ implies $\Gamma \cup \delta \vdash \Psi$. The *Negation as Finite Failure* rule [Cla78] and the *Closed World Assumption* (CWA) [Rei80] introduce nonmonotonicity into Logic Programming when deriving negative literals. When incorporated into query evaluation the first of these rules allows one to conclude the answer to query $\neg A$ to be "yes", (we say that the query *succeeds*) if, and only if, the answer to the query $A$ is "no" (we say the query "fails"). For example, if $\mathcal{P}$ is the program $\{p \leftarrow \neg q\}$, then $p$ will succeed as it reduces to the query $\neg q$, which succeeds because the query $q$ fails, whereas if $\mathcal{P}$ is augmented with the fact $q$, then it does not entail $p$ any more since the query $q$ will succeed and $\neg q$ will fail. In the following we will use the symbol $\sim$ to represent negation in logic programs, also called *Default Negation.* The symbol $\neg$ will be reserved for classical negation.

The CWA is the declarative counterpart of negation as failure and an example of its use is the process of booking a flight. Assume that you want to know whether there is a flight from London to Rio de Janeiro on 10 December. Assume that the database of your travel agent does not contain such a flight. You will be informed that there is no flight from London to Rio on that date. In order to be able to jump to this conclusion, the travel agent had to assume that all flights from London to Rio were listed on the database (i.e. that the database is complete). If, later, a new flight is entered in the database then the earlier conclusion is now incorrect (see [Luk90] and [Bre90]).

### 2.5.1 Stable Models and Acceptable Programs

One of the striking features of *Logic Programming* (see [Llo87]) is that it can naturally support nonmonotonic reasoning – by means of negative literals, as we saw in the example above. Many concepts introduced in the area of non-monotonic reasoning have a natural counterpart within logic programming in spite of its limited syntax.

**Definition 2.5.1.** *A* general clause *is a rule of the form* $A_0 \leftarrow L_1, ..., L_n$, *where* $A_0$ *is an atom and* $L_i$ $(1 \leq i \leq n)$ *is a literal. A* general logic program *is a finite set of general clauses.*

So far, we have seen that if $\mathcal{P}$ is a definite logic program then the least Herbrand model of $\mathcal{P}$ exists, and its classical (two-valued) semantics can be defined as the least fixpoint of a meaning operator $T_{\mathcal{P}}$. The semantics is obtained by lattice-theoretic considerations which require $T_{\mathcal{P}}$ to be monotonic. However, if $\mathcal{P}$ is a general logic program then $T_{\mathcal{P}}$ may not be monotonic and, consequently, the existence of a least fixpoint of $T_{\mathcal{P}}$ cannot be guaranteed.

The immediate consequence operator $T_{\mathcal{P}}$ was defined for definite programs in Definition 2.4.5. The same definition is used for general programs, where $\sim A$ is mapped to $false$ ($true$) by $\mathcal{I}$ if $A$ is mapped to $true$ ($false$).

For example, the program $\mathcal{P} = \{p \leftarrow \sim q\}$ has two minimal Herbrand models, $\{p\}$ and $\{q\}$, and neither of them is a least model of $\mathcal{P}$. The first is a fixpoint of $T_{\mathcal{P}}$, but not the second and, for example, $T_{\mathcal{P}}(\{p\}) = \{p\}$, whereas $T_{\mathcal{P}}(\{p, q\}) = \emptyset$. For general programs $\mathcal{P}$, $T_{\mathcal{P}}$ may even have no fixpoint at all, for example when $\mathcal{P} = \{p \leftarrow \sim p\}$.

There is no general agreement upon answering the question of what the standard model of a general program is. Several possibilities have been suggested, none of which is completely satisfactory. We give a brief description of some of the plausible answers to this question that have been suggested in the literature [AB94].

**Definition 2.5.2.** (Supported Interpretations) *An interpretation $I$ is called supported if $A_0 \in I$ implies that for some (general) clause $A_0 \leftarrow L_1, ..., L_n \in \mathcal{P}$ we have that $I \models L_1 \wedge ... \wedge L_n$. Intuitively, $L_1, ..., L_n$ is an explanation for $A_0$.*

**Proposition 2.5.1.** *$I$ is a supported model of $\mathcal{P}$ iff $T_{\mathcal{P}}(I) = I$.*

Thus, in view of the observation on the behaviour of the $T_{\mathcal{P}}$ operator, we see that for some programs no supported models exist, e.g. when $\mathcal{P} = \{p \leftarrow \sim p\}$. On the other hand, the interpretation $\mathcal{I} = \{p, r\}$ is a supported model of the program $\mathcal{P} = \{p \leftarrow \sim q; q \leftarrow \sim r; r\}$. For the program $\mathcal{P} = p \leftarrow \sim q$ the model $\{p\}$ is supported but $\{q\}$ is not. One possible approach is to accept that some programs have no natural supported model and to identify classes of programs for which a natural supported model exists. This is considered next. The following definition will also be important for the sequel.

**Definition 2.5.3.** (Dependency) *Consider a program $\mathcal{P}$. The dependency graph $D_{\mathcal{P}}$ for $\mathcal{P}$ is a directed graph with signed edges. Its nodes are the literals occurring in $\mathcal{P}$. For every clause in $\mathcal{P}$ with $p$ in its head and $q$ as a positive (resp. negative) literal in its body, there is a positive (resp. negative) edge $(p, q)$ in $D_{\mathcal{P}}$.*

- *We say that $p$* uses *(or refers to) $q$* positively (resp. negatively).
- *We say that $p$* depends positively (resp. negatively) *on $q$ if there is a path in $D_{\mathcal{P}}$ from $p$ to $q$ with only positive edges (resp. at least one negative edge).*
- *We say that $p$* depends evenly (resp. oddly) *on $q$ if there is a path in $D_{\mathcal{P}}$ from $p$ to $q$ with an even (resp. odd) number of negative edges.*

**Definition 2.5.4.** (Locally Stratified Programs) *A program $\mathcal{P}$ is called* Locally stratified *if no cycle with a negative edge exists in its dependency graph.*

In other words, a program is locally stratified if no recursion through a negated literal is used in it (see [Prz88]). The standard model $M_{\mathcal{P}}$ is not uniquely characterised, though, because for some locally stratified programs more than one supported model exists. For example, consider the program $\{p \leftarrow q; q \leftarrow p\}$, which is obviously locally stratified but has two supported models, $\{p, q\}$ and $\emptyset$.

In [GL88], Gelfond and Lifschitz introduced the important notion of *stable models*, by using the intuition of rational beliefs from autoepistemic logic.

**Definition 2.5.5.** (Gelfond-Lifschitz Transformation) *Let $\mathcal{P}$ be a grounded logic program. Given a set $I$ of atoms from $\mathcal{P}$, let $\mathcal{P}_I$ be the program obtained from $\mathcal{P}$ by deleting:* (1) *each rule that has a negative literal $\sim A$ in its body with $A \in I$, and* (2) *all the negative literals in the bodies of the remaining rules.*

Clearly, $\mathcal{P}_I$ is a positive program, so that $\mathcal{P}_I$ has a unique minimal Herbrand model. If this model coincides with $I$ then we say that $I$ is a stable model of $\mathcal{P}$.

**Definition 2.5.6.** (Stable Models) *A Herbrand interpretation $I$ of a program $\mathcal{P}$ is called stable iff $T_{\mathcal{P}_I}(I) = I$.*

The intuition behind the definition of a stable model is given in [GL88] as follows: consider a rational agent with a set of beliefs $I$ and a set of premises $\mathcal{P}$. Then, any clause that has a literal $\sim A$ in its body when $A \in I$ is useless, and may be removed from $\mathcal{P}$. Moreover, any literal $\sim A$ with $A \notin I$ is trivial, and may be deleted from the clauses in which it appears in $\mathcal{P}$. This yields the simplified (definite) program $\mathcal{P}_I$, and if $I$ happens to be precisely the set of atoms that follows logically from the simplified set of premises, then the set of beliefs $I$ is stable. Hence, stable models are possible sets of belief a rational agent might hold.

The following result clarifies the relation between stable models, minimal models and locally stratified programs.

**Theorem 2.5.1.** *Consider a general program $\mathcal{P}$. Any stable model of $\mathcal{P}$ is a minimal model of $\mathcal{P}$. If $\mathcal{P}$ is locally stratified, then it has a unique stable model.*

*Example 2.5.1.* For the program $\mathcal{P} = \{p \leftarrow\sim q; q \leftarrow\sim r; r\}$ and interpretations $I$, the following programs $\mathcal{P}_I$ are obtained. The last of these is a stable model of $\mathcal{P}$.

$$I = \emptyset, \mathcal{P}_I = \{p, q, r\} \qquad I = \{p, q\}, \mathcal{P}_I = \{q, r\}$$
$$I = \{p, q, r\}, \mathcal{P}_I = \{r\} \quad I = \{p, r\}, \mathcal{P}_I = \{p, r\}$$

The program $\{p \leftarrow\sim q; q \leftarrow\sim p\}$ has two stable models, namely $\{p\}$ and $\{q\}$, whereas the program $\{p \leftarrow\sim p\}$ has none.

Example 2.5.1 showed that stable model semantics can also allow more than one stable model, or none at all. This reflects some uncertainty about the conclusions that should be drawn from a program. In some cases, a local uncertainty can destroy too much information. For example, if $\mathcal{P}$ is a stratified program in which the variable $p$ does not occur, then $\mathcal{P} \cup \{p \leftarrow\sim p\}$ has no stable model. Thus, the information contained in $\mathcal{P}$ is not reflected in the stable model semantics of $\mathcal{P} \cup \{p \leftarrow\sim p\}$, even though it is not related to the uncertainty about the truth-value of $p$.

The *Well-Founded Semantics* [vGRS91] avoids this problem by using a three-valued model. In contrast with three-valued logic (see [Fit85]), a three-valued interpretation of the connectives is not needed to obtain three-valued models. On the other hand, well-founded semantics has the drawback of not always inferring all atoms that one would expect to be *true* (see Apt and Bol's survey on logic programming and negation [AB94] for a comprehensive comparison between different semantics and classes of programs).

An alternative approach, relevant to the use of neural networks to implement logic programs, is to try to classify which programs will converge. To this end, consider the definition of *Acceptable Logic Programs* [AP93] (see also [Fit94]) and their significant properties, which are of importance for the neural-symbolic integration presented here.

**Definition 2.5.7.** *A level mapping for a program $\mathcal{P}$ is a function $|\,| : B_{\mathcal{P}} \rightarrow \aleph$ of atoms to natural numbers. For $A \in B_{\mathcal{P}}$, $|A|$ is the level of $A$ and $|\sim A| = |A|$.*

**Definition 2.5.8.** *Let $\mathcal{P}$ be a general logic program and $Neg_P$ be the set of atoms that occur negatively in $P$. Let $Neg_P^*$ be the set of atoms on which an atom in $Neg_P$ depends (either positively or negatively). We call $\mathcal{P}^-$ the program consisting of clauses from $\mathcal{P}$ whose head is in $Neg_P^*$.*

**Definition 2.5.9.** (Acceptable Logic Programs) *Let $\mathcal{P}$ be a general logic program, $|\,|$ a level mapping for $\mathcal{P}$, and $I$ a model of $\mathcal{P}$ whose restriction to the atoms in $Neg_P^*$ is a model of the* completion[5] *of $\mathcal{P}^-$. $\mathcal{P}$ is called acceptable w. r. t. $|\,|$ and $I$ if, for every clause $A_0 \leftarrow L_1, ..., L_n$ in $\mathcal{P}$, the following implication holds for $1 \leq i \leq n$.*

$$\text{if } I \models \bigwedge_{j=1}^{i-1} L_j \text{ then } |A_0| > |L_i|$$

*In other words, $\mathcal{P}$ is acceptable w. r. t. $|\,|$ and $I$ if*

$$|A_0| > |L_i| \text{ for } i \in [1, \overline{n}]$$

*where*

$$\overline{n} = \min(\{n\} \cup \{i \in [1, n] \mid I \not\models L_i\})$$

*$\mathcal{P}$ is called acceptable if it is acceptable w.r.t. some level mapping and a model of $\mathcal{P}$.*

**Theorem 2.5.2.** *Let $\mathcal{P}$ be an acceptable general program. Then:*

- *$T_\mathcal{P} \uparrow \omega$ is the unique fixpoint of $T_\mathcal{P}$, and*
- *$T_\mathcal{P} \uparrow \omega = T_\mathcal{P} \downarrow \omega$.*

In other words, if $\mathcal{P}$ is an acceptable program then $T_\mathcal{P}$ has a unique fixpoint, and this fixpoint is reached by iterating $T_\mathcal{P}$, starting from any valuation, after $\omega$ steps.

---

[5] The *completion* of a general (ground) logic program $\mathcal{P}$ is the logical theory obtained by the following three steps:

1. Write clauses with the same head $A_0$ equivalently as $A_0 \leftarrow B_1 \vee ... \vee B_n$, where $B_i$ is the body of the *i-th* clause with head $A_0$.
2. Change $\leftarrow$ into $\leftrightarrow$, and add the sentence $\neg A$ for any atom $A$ in $\mathcal{L}$ such that $A$ does not occur in the head of any clause in $\mathcal{P}$.
3. Add facts $\neg(c_i = c_j)$, where $c_i$ and $c_j$ are different constants in $\mathcal{L}$, and $(c_i = c_i)$, where $c_i$ is a constant in $\mathcal{L}$.

Steps (1)-(3) give the *completion* of $\mathcal{P}$ [Cla78]. The *completion* of the program of Example 2.4.1 is the set of 27 ground instances of $path(x, y) \leftrightarrow arc(x, y) \vee [arc(x, z) \wedge path(z, y)]$ together with $arc(a, b)$, $arc(b, c)$, $\neg arc(a, a)$, $\neg arc(a, c)$, $\neg arc(b, a)$, $\neg arc(b, b)$, $\neg arc(c, a)$, $\neg arc(c, b)$, $\neg arc(c, c)$, $a \neq b$, $a \neq c$, $b \neq c$, $a = a$, $b = b$, $c = c$.

**Definition 2.5.10.** *Let $\mathcal{P}$ be a program and $|\,|$ a level mapping for $\mathcal{P}$. $\mathcal{P}$ is called* acyclic *w. r. t. $|\,|$ if, for every clause $A_0 \leftarrow L_1, ..., L_n$ in $\mathcal{P}$, $|A_0| > |L_i|$ for $1 \leq i \leq n$.*

*$\mathcal{P}$ is called acyclic if it is acyclic w. r. t. some level mapping.*

**Proposition 2.5.2.** *Every acyclic program is acceptable.*

The interest in the above definitions comes from the fact that various ways of defining semantics coincide for the case of acceptable general programs (see [AP93] for proofs). More specifically, Theorem 2.5.2 will be needed in Section 3.2.

*Example 2.5.2.* Consider the programs $\mathcal{P}_1$, $\mathcal{P}_2$ and $\mathcal{P}_3$ below.

$$\mathcal{P}_1 = \{a \leftarrow c, \sim a\}$$
$$\mathcal{P}_2 = \{a \leftarrow \sim a, c\}$$
$$\mathcal{P}_3 = \{b \leftarrow a; \quad a \leftarrow \sim b; \quad b \leftarrow \sim a\}$$

In Chapter 3 it will be shown that the corresponding neural network for any acceptable program converges. However, it will become clear that the respective neural networks corresponding to programs $\mathcal{P}_1$, $\mathcal{P}_2$ and $\mathcal{P}_3$ all converge, despite only $\mathcal{P}_1$ being acceptable. Furthermore, using the standard Prolog execution rule only the first program will allow the successful computation of $a$, as indicated by acceptability.

In view of the above example, in Chapter 9 we will propose a more general definition of acceptability.

## 2.6 Belief Revision

While nonmonotonic systems allow an agent to make plausible conjectures in the absence of complete information, *Belief Revision* systems are concerned about an agent that rationally changes its knowledge base when it acquires new information. This is particularly important when the acquired information is in conflict with the agent's current knowledge. More recently, a strong relationship between nonmonotonic formalisms and belief revision has been established. In [Mak93], Makinson has shown that many properties of nonmonotonicity are shared by belief revision systems and vice versa (see also [GR94] (Chapter 6) for translating postulates for belief revision into nonmonotonic logic). In this section, we recall the basic concepts of Belief Revision, and focus on the Truth Maintenance Systems of Doyle [Doy79] and on the Compromise Revision framework of Gabbay [Gab99]. The remainder of the material presented in this section is based on [GR94] and [Gar92].

In principle, the Belief Revision framework restricts itself to modelling changes to knowledge bases that involve the addition and removal of facts only. Therefore, it does not consider the possibility of explicitly modifying individual rules, such as transforming "all birds fly" to "all birds except Tweety fly", as a primitive operation. Modifying individual rules is often seen in Machine Learning though, and can be modelled in the Belief Revision framework by observing that "all birds fly" entails "all birds except Tweety fly" (see [Ant97]). Then, removing "all birds fly" and retaining "all birds except Tweety fly" achieves the same results. Similarly, assigning a higher priority to the belief that Tweety does not fly over the one that all birds fly achieves the same result.[6]

Let us start with a simple example of revision taken from [GR94]. Suppose that we have a database which contains, among other things, the following pieces of information.

1. The bird caught in the trap is a swan ($Swan(a)$)
2. The bird caught in the trap comes from Sweden ($Swedish(a)$)
3. Sweden is part of Europe ($\forall x(Swedish(x) \rightarrow European(x))$)
4. All European swans are white ($\forall x(Swan(x) \land European(x) \rightarrow White(x))$)

If the database is coupled with a program that can compute logical inferences in a given code, the following fact can be derived

5. The bird caught in the trap is white ($White(a)$)

Now, suppose that, as a matter of fact, the bird caught turns out to be black. This means that we want to add $\neg White(a)$ in the database. But then the database becomes inconsistent. If we want to keep the database consistent, we need to revise it. In other words, some of the beliefs in the original database need to be retracted. We need to choose between retracting 1, 2, 3 or 4 above, and to provide a reason for such a choice. The problem here is that logical considerations alone do not tell us which beliefs to retain and which to retract. To make things more complicated, when we give up a belief, we also have to decide which of the logical consequences of it we should retain or retract. For instance, if we solve the inconsistency in the above example by retracting rule 4, we still need to decide whether or not to keep the following weakened version of it, which is a logical consequence of our database.

[6] In neural networks, modifications of beliefs, as opposed to additions and removals, are more likely to take place during learning. This is so because the learning process is normally done by small changes in the set of weights. Nevertheless, in Chapter 7, we will use some properties of Belief Revision, such as minimal change, to define a method for explicitly adding and/or removing beliefs from neural networks, in order to restore consistency in trained networks.

6. All European swans except the one caught in the trap are white.

When trying to handle belief revision in a computational setting, there are three main methodological questions to settle [GR94]:

1. How are the beliefs represented in the database?
2. What is the relation between the elements explicitly represented in the database and the beliefs that may be derived from these elements?
3. How are the choices made concerning what to retract?

In answering the methodological questions, a small number of basic rationality postulates (or integrity constraints) can be considered as operative:

1. (*consistency*) The beliefs in a database should be kept consistent whenever possible.
2. (*deductive closure*) If the beliefs in a database logically entail a sentence $A$, then $A$ should be included in the database.
3. (*minimal change*) The amount of information lost in a belief change should be kept minimal, and in so far as some beliefs are considered more important (or entrenched) than others, one should retract the least important ones.

In the most common case, there are two basic types of belief change: pieces of information can be inserted (*expansion*) or deleted (*contraction*) from the database. A *revision* occurs when a new piece of information that is inconsistent with the present database is inserted into it in such a way that the result is a new consistent database. In general, a belief $A$ can only be accepted or not accepted (which is not to say that the belief is rejected, i.e. that $\neg A$ is accepted) in a belief system.

The work on belief revision has followed several different perspectives, but is mainly based on two major paradigms: *coherence* and *foundations*. Basically, the difference lies in the way that the beliefs held by an agent are realised. According to the coherentist view, for a belief to be held by an agent, nothing is necessary apart from coherence of that belief with respect to the agent's other beliefs and inference system. All beliefs are equally supported as far as the agent is concerned. In the foundational view, a belief can be seen as made up of two different components: beliefs and their justifications. Beliefs are held only if they have a proper justification, and beliefs whose support is no longer accepted in the belief state are discarded. Hence there is a clear distinction between basic beliefs and derived beliefs.

In the coherentist (or logic-constrained) mode, a set of rationality postulates form the very process of belief change, so that the change operations become non-trivial. In reward, the method benefits from the use of standard, in most cases classical, propositional logic as the underlying inference mechanism. The seminal paper "On the Logic of Theory Change" [AGM85] provided the basis for the coherentist approach to theory revision. Another landmark in the area is the book "Knowledge in Flux" [Gar88].

In the foundational (or immediate) mode, one just inserts and deletes units of belief from the current database D, without bothering about any integrity constraint. The two trivial change operations are then accompanied by a sophisticated and in general paraconsistent and nonmonotonic inference operation, which tells us which beliefs are actually supported by D. The immediate mode can be instantiated, among other works, in Truth Maintenance Systems, which is the subject of the following section.

Finally, some authors argue against an outright elimination of inconsistencies from the database. Other accounts amalgamate ideas from paraconsistent logic (see [dC74]) with ideas from belief revision.

### 2.6.1 Truth Maintenance Systems

Under the name *Truth Maintenance Systems* (TMS), we find loosely a great number of systems for managing beliefs on the basis of dependency information [GR94]. In the following, however, we shall focus on the model proposed by Doyle's Justification-based TMS [Doy79] (see also [GM91]).

TMSs are quite appealing from our point of view due to a series of reasons. Firstly, TMSs are procedural by nature – they were developed as actual systems running in actual computers. Secondly, in the TMS approach one does not remove sentences from the knowledge base, but instead modifies the labels assigned to them by a labelling algorithm specific to the system. Once represented in the TMS as a node, a datum will never be erased physically but only marked as *out* (not believed) in the current belief state. Belief Revision, on this account, is relabelling of nodes. Thirdly, TMSs are based on a restricted language where nodes are objects and justifications between nodes are rules[7]. Finally, TMSs handle belief revision according to a foundational or immediate mode.

TMSs can be said to be semantic network models, but with a sophisticated technique for handling changes of belief. There are two basic types of entities in a TMS: *nodes* (representing propositional beliefs) and *justifications* (representing reasons for beliefs). A node can be *in* or *out*, which corresponds to accepting or not accepting the belief represented by it. If a node $n_1$ represents the negation of a node $n_2$, and $n_2$ is *out*, this should not entail that $n_1$ is *in*. On the other hand, as a rationality requirement, if both $n_1$ and $n_2$ are *in* then the system should start a revision of the set of nodes and their justifications in order to re-establish consistency. A justification is defined as a pair of lists: an inlist ($I$) and an outlist ($O$), together with the

[7] The *contrapositives* of such rules are not necessarily *true*, e.g. given the rule $a \rightarrow b$, the rule $\neg b \rightarrow \neg a$ is not necessarily *true*.

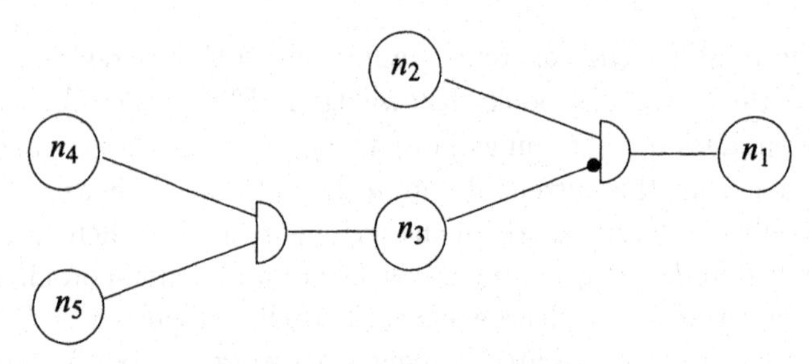


**Fig. 2.6.** An example of a TMS (taken from [GR94]).

node ($n$) that it is a justification for, and is denoted by $\langle I \mid O \rightarrow n\rangle$, where $I$ contains only nodes that are *in*, $O$ contains only nodes that are *out*, and $n$ is called the consequence of the justification. If both $I$ and $O$ are empty then $n$ is called a *premise*. The idea is that a node should be *in* if and only if it has at least one justification.

The information in a justification can be represented graphically as in Figure 2.6 (cf. [Goo87]), where nodes are represented by circles and justifications by AND-gates. The black dot means roughly negation (default negation), so that $n_3$ has to be *out* and $n_2$ has to be *in* in order for $n_1$ to be *in*, i.e. $\langle \{n_2\} \mid \{n_3\} \rightarrow n_1\rangle$.

A TMS network can be described as a triple $T = \langle N, P, J\rangle$, where $N$ is a finite set of nodes, $P \subseteq N$ is a (sometimes empty) set of premises, and $J$ is a set of justifications. A *belief state* $S$ is a subset of $N$, namely the set of nodes labelled *in*. It is the main task of the TMS to draw well-founded conclusions from a given network $T$ in order to arrive at an appropriate belief state $S$.

**Definition 2.6.1.** *A set of nodes $S$ in $T$ is* globally grounded *if and only if there is a total ordering $n_1 \prec ... \prec n_k$ on the elements of $S$ such that for each $n_i (1 \leq i \leq k)$, $n_i$ is either an element of $P$ or there is a justification of the form $\langle I \mid O \rightarrow n_i\rangle$ in $J$ such that $I \subseteq \{n_1, ..., n_{i-1}\}$ and $O \cap S = \emptyset$.*

**Definition 2.6.2.** *An* admissible belief state *with respect to $T$ is a set $S$ of nodes which is globally grounded in $T$ and closed with respect to $T$.*

A network may have one, more than one, or no admissible state.[8] If $T$ possesses more than one admissible state, we face the well-known problem of *multiple extensions* of nonmonotonic reasoning: which of the admissible states should be the current belief state sanctioned by the TMS? The *sceptical* idea is to take the intersection of all admissible states. However, this intersection need not to be an admissible state itself. For this reason – and for the sake of efficiency – TMSs follow a more *credulous* practice and remain in the first

[8] Note that the definition of admissible belief states is closely related to the stable model semantics of Logic Programming.

admissible state they happen to find (which is different from the credulous approach of taking the union of all admissible states).

An important goal of TMSs is to maintain consistency. The concept of inconsistency is introduced in the TMS framework by distinguishing a subset $N_\perp$ of $N$ as *contradiction nodes*. The definition of admissible states $S$ is then enriched by the consistency requirement $S \cap N_\perp = \emptyset$, i.e. no contradiction node is allowed to be *in*. If, for example, $n_1$ is the negation of $n_2$, the justification $\langle \{n_1, n_2\} \mid \{\emptyset\} \rightarrow n_\perp \rangle$ should make any states in which $n_1$ and $n_2$ are *in* not admissible. In order to achieve this, TMSs use what is called *dependency directed backtracking*. Neglecting many interesting details, if the TMS detects a contradiction, i.e. if an inconsistent node $n_\perp$ is labelled *in*, it traces back and looks for the set $J_\perp$ of justifications on which this labelling is currently grounded. Then it picks an element $n_j$ of the outlist of one of these justifications and creates a new justification of the form $\langle I \mid O \rightarrow n_j \rangle$, where $I$ is the union of the inlists in $J_\perp$ and $O$ is the union of the outlists of $J_\perp$ with the exception of $n_j$. So, if each element in $I$ is *in* and each element of $O$ is *out*, the newly introduced justification causes $n_j$ to be *in*. The proof of $n_\perp$ breaks down because the old justification with $n_j$ in its outlist is no longer valid. Intuitively, non-belief in $n_j$ has been identified as responsible for the contradiction [GR94].

*Example 2.6.1.* [GR94] Let $T = \langle \{a, b, n_\perp\}, \{a\}, \{\langle \{a\} \mid \{b\} \rightarrow n_\perp \rangle\} \rangle$ (i.e. having both $a$ and $\sim b$ is undesirable). Thus, $S_T = \{a, n_\perp\}$. To solve the inconsistency, a new justification $\langle \{a\} \mid \{\emptyset\} \rightarrow b \rangle$, which triggers $b$ instead of $n_\perp$, is created. If $T$ is part of a more complex network then the change in the state of $b$ should be propagated through a reconsideration of each justification with $b$ in it.

Belief change by dependency backtracking has been criticised as being insufficiently controlled. In some examples, the procedure can lead to *odd loops*, in which cases Doyle's labelling fails to terminate. Another problem is the choice of $n_j$, which is arbitrary here, so that the internally created justification with consequence $n_j$ may conflict with later extensions of the dependency network, forcing the TMS to perform *spurious belief revision* and backtracking.

In Chapter 7, we will examine the relation between neural networks and TMSs.

### 2.6.2 Compromise Revision

Consider the following scenario. Let $\Gamma$ be a consistent theory in some underlying logic $L$. Let $C$ be a consistent well-formed formula (*wff*) of $L$. We would like to put $C$ into $\Gamma$ to form a new theory, denoted by $(\Gamma + C)$. If

$\Gamma \cup \{C\}$ is consistent, it is clear that $\Gamma + C$ should be $\Gamma \cup \{C\}$. If $\Gamma \cup \{C\}$ is inconsistent, we can adopt one of the following policies:

1. The *non-insistent policy,* which rejects $C$ and lets $\Gamma + C = \Gamma$, or
2. The *insistent policy,* which must accept $C$ and restore consistency by identifying a subset $\Gamma_C$ of $\Gamma$ to be rejected such that $(\Gamma - \Gamma_C) \cup \{C\}$ is consistent.[9]

An alternative mechanism for defining $\Gamma + C$ as a consistent theory was proposed in [Gab99]. *Compromise Revision* does not necessarily either reject $C$ or throw out some $\Gamma_C$, but offers a compromise, accepting some of the logical consequences of $C$ and some of the logical consequences of $\Gamma_C$ when forming $\Gamma + C$.

Let, for example, $\Gamma = \{A, A \to B, A \wedge C \to D \wedge E, D \to \neg A\}$ and the input be $C$. Clearly, $\Gamma \cup \{C\}$ is inconsistent. We can either reject $C$ (non-insistent policy) or accept $C$ (insistent policy) and, depending on the Truth Maintenance System used, we may throw out any union of $\Gamma_{C_1} = \{A\}$, $\Gamma_{C_2} = \{A \wedge C \to D \wedge E\}$, $\Gamma_{C_3} = \{D \to \neg A\}$.

Let us give some meaning to $\Gamma$ above. Consider the context where John is flying abroad. He is booked to fly first class, but when he shows up he discovers that first class has been overbooked. Let $A$ = *John is booked first class*, $B$ = *John has double baggage allowance*, $C$ = *First class cabin is full*, $D$ = *John flies economy class*, $E$ = *John gets a letter of apology*. The database $\Gamma$ states that John is booked first class, that such passengers get double baggage allowance, and that if John is booked first class and first class is full then John flies economy class, but gets a letter of apology. Further, the airline takes pride that no passenger booked first class ever travels economy. The update is that the first class is full (this example is due to Gabbay; see [Gab99]).

In this particular example, it seems reasonable that we either end up with $\Gamma$, rejecting $C$ (in this case, John will take the next flight), or with $\Gamma_1 = \{C, A \to B, A \wedge C \to D \wedge E, D \to \neg A\}$, rejecting $\Gamma_{C_1}$ (in this case, John will fly economy). However, in the first case, if $C$ were admitted into the database, then $D$ and $E$ would be derived. Although $D$ leads to inconsistency, $E$ does not. So, we can compromise and, although we reject $C$, we can still accept $E$. Thus, a compromise approach for the non-insistent (first) case would lead to the database $\Gamma \cup \{E\}$. Similarly, for the second case, throwing out $\Gamma_{C_1}$ would block the deduction of $B$, $D$ and $E$. However, none of $\{B, D, E\}$ leads to inconsistency. Thus, a compromise approach for the insistent (second) case would end up with the database $\Gamma_1 \cup \{B, D, E\}$.

In Chapter 7, we shall consider the compromise approach to revision.

---

[9] The details of how to identify $\Gamma_C$ depend on the particular approach of the system.

# Part I

# Knowledge Refinement in Neural Networks

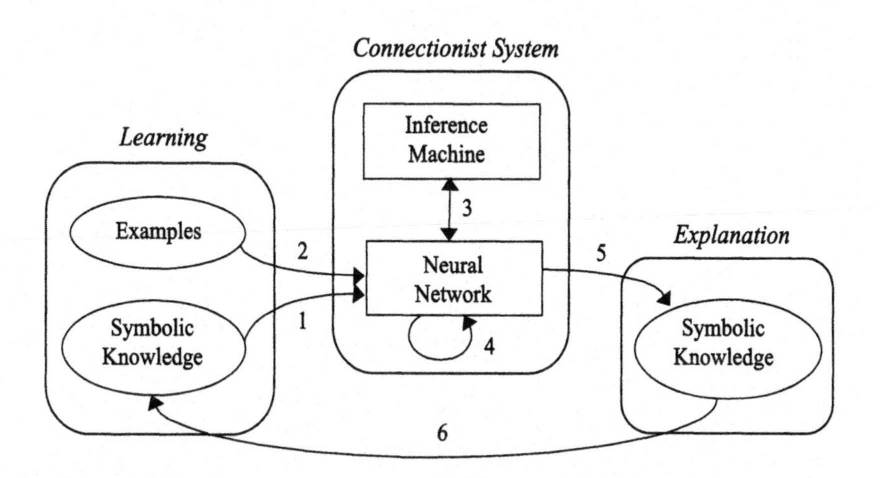


Part I of this book is about *Knowledge Representation and Learning*. It presents the theoretical and practical aspects of processes (1), (2) and (3) above. It contains a new Translation Algorithm from symbolic knowledge to neural networks, and a proof that such a translation is correct (1). It shows how the neural networks obtained could be trained with examples, using for instance the standard Backpropagation learning algorithm (2). It describes how the same neural networks could be used as massively parallel models for computing the symbolic knowledge inserted or learned with examples in the process of theory refinement (3). Part I concludes with a presentation of the empirical results of applying processes (1), (2) and (3) to real problems of DNA sequence analysis and Power Systems Fault Diagnosis.

# 3. Theory Refinement in Neural Networks

In this chapter, we present the Connectionist Inductive Learning and Logic Programming System (*C-IL²P*). *C-IL²P* is a *Neural-Symbolic Learning System* based on a feedforward neural network that integrates inductive learning from examples and background knowledge, with deductive learning from *Logic Programming*. Starting with the background knowledge represented by a propositional logic program, a translation algorithm is applied, generating a neural network that can be trained with examples. The network also computes the stable model (resp. the answer set) of the general (resp. extended) logic program inserted in it as background knowledge, or learned with the examples, thus functioning as a parallel system for Logic Programming.

In Section 3.1, we present a new translation algorithm from general logic programs ($\mathcal{P}$) to artificial neural networks ($\mathcal{N}$) with bipolar semi-linear neurons. The algorithm is based on Holldobler and Kalinke's translation algorithm from general logic programs to neural networks with binary threshold neurons [HK94]. We also present a theorem showing that $\mathcal{N}$ computes the fixed-point operator ($T_{\mathcal{P}}$) of $\mathcal{P}$. The theorem ensures that the translation algorithm is correct. In Section 3.2, we show that $\mathcal{N}$, with feedback connections, computes the stable model of $\mathcal{P}$, when $\mathcal{P}$ is an acceptable program[1]. In other words, the result obtained in [HK94] still holds, i.e. $\mathcal{N}$ is a massively parallel model for Logic Programming. However, $\mathcal{N}$ can also perform *Inductive Learning* from examples efficiently, assuming $\mathcal{P}$ as background knowledge and using, for example, the standard *Backpropagation* learning algorithm, as in [TS94a]. We outline the steps for performing inductive learning in the network in Section 3.3. Finally, in Sections 3.4 and 3.5, respectively, we extend the system to the language of extended logic programs and to cope with certain metalevel preference relations between rules. We also show that, in these cases, the network computes the answer set semantics of the program inserted in it. Section 3.6 concludes and presents some pointers to further reading.

[1] Recall that an acceptable program $\mathcal{P}$ has a unique fixpoint.

## 3.1 Inserting Background Knowledge

It has been suggested that the merging of theory (*Background Knowledge*) and data learning (learning from examples) in neural networks may provide a more effective learning system [Fu89, Fu94, TS94a]. In order to achieve this objective, one might first translate the background knowledge into a neural network's initial architecture, and then train it with examples using some neural learning algorithm, like *Backpropagation*. To do so, the *C-IL²P* system provides a translation algorithm from propositional (or grounded) general logic programs to feedforward, single hidden layer, neural networks with semi-linear neurons. A theorem then shows that the network obtained is equivalent to the original program, in the sense that what is computed by the program is computed by the network and vice versa.

The translation algorithm of *C-IL²P* is based on Holldobler and Kalinke's *Massively Parallel Model for Logic Programming*. In [HK94], Holldobler and Kalinke presented a method for computing the least fixpoint of a general logic program in a single hidden layer neural network. They have shown that for each program $\mathcal{P}$, there exists a network $\mathcal{N}$ with binary threshold units such that $\mathcal{N}$ computes $T_{\mathcal{P}}$, the program's fixpoint operator. If $\mathcal{N}$ is transformed into a *partially recurrent network* [HKP91] by linking the units in the output layer to the corresponding units in the input layer, it always settles down into a unique stable state when $\mathcal{P}$ is an acceptable program [AP93, Fit94]. This stable state is the least fixpoint of $T_{\mathcal{P}}$, which is identical to the unique stable model of $\mathcal{P}$ (see Section 2.5.1 for the stable model semantics of general logic programs).

Intuitively, the input and output layers of a feedforward neural network can represent an *interpretation* of a program $\mathcal{P}$ before and after the application of $T_{\mathcal{P}}$. Connecting the output of the network to its input yields a network with feedback that can be used to iterate $T_{\mathcal{P}}$. The neural network $\mathcal{N}$ associated with $\mathcal{P}$ can be constructed as follows [HK94]:

1. Let the input and output layers of $\mathcal{N}$ contain $k$ binary threshold neurons each, where the $i$th neuron represents the atom $A_i$, $0 \leq i \leq k$, of $\mathcal{P}$. Set the threshold of each neuron occurring in the input and output layers of $\mathcal{N}$ to 1/2.
2. For each clause of the form $A_0 \leftarrow A_1, ..., A_m, \sim A_{m+1}, ..., \sim A_n$ $(0 \leq m \leq n)$ occurring in $\mathcal{P}$, do the following:
   a) Add a binary threshold unit $c$ to the hidden layer of $\mathcal{N}$.
   b) Connect $c$ to the unit representing $A_0$ in the output layer of $\mathcal{N}$ with weight 1.
   c) For each $A_i$, $1 \leq i \leq m$, connect the unit representing $A_i$ in the input layer of $\mathcal{N}$ to $c$, and set the connection's weight to 1.

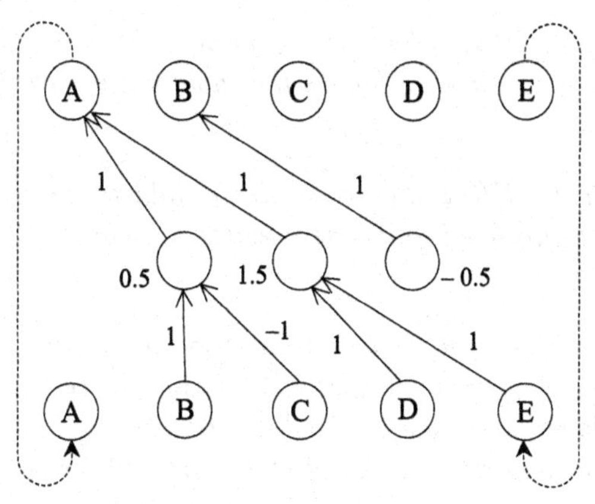


**Fig. 3.1.** The network $\mathcal{N}$ obtained by applying the above translation algorithm over program $\mathcal{P}$.

d) For each $A_i$, $m+1 \leq i \leq n$, connect the unit representing $A_i$ in the input layer of $\mathcal{N}$ to $c$, and set the connection's weight to $-1$.
e) Set the threshold $\theta_c$ of $c$ to $m - 1/2$.

*Example 3.1.1.* Consider the program $\mathcal{P} = \{A \leftarrow B \sim C; A \leftarrow DE; B \leftarrow\}$. The neural network $\mathcal{N}$ of Figure 3.1 is obtained by applying the above algorithm over $\mathcal{P}$. If, for example, $B = 1$ and $C = 0$ in the input layer of $\mathcal{N}$ then $A = 1$ in the output layer of $\mathcal{N}$, representing the rule $A \leftarrow B \sim C$ of $\mathcal{P}$. Similarly, $B = 1$ is always obtained in the output layer of $\mathcal{N}$, regardless of its input vector, indicating that $B$ is a fact. When $\mathcal{N}$ is recurrently connected as shown in Figure 3.1, its output vector feeds the input vector in an iteration of $T_\mathcal{P}$. Then, if $\mathcal{P}$ is an acceptable program, $\mathcal{N}$ eventually converges to the unique fixpoint of $T_\mathcal{P}$, for any initial input vector. In this example, $\mathcal{N}$ converges to $\{A = 1, B = 1, C = 0, D = 0, E = 0\}$, which represents the set $M_\mathcal{P} = \{A, B\}$.

One should observe that the number of units in the hidden layer of $\mathcal{N}$, as well as the number of connections between the hidden and the output layer, is equal to the number of clauses in $\mathcal{P}$. Furthermore, the number of connections between the input and the hidden layer of $\mathcal{N}$ is equal to the number of sub-goals occurring in $\mathcal{P}$, while the number of units in the input and output layers is equal to the number of atoms occurring in the program. Hence, the size of the network is bound by the size of the program and the operator $T_\mathcal{P}$ is computed in constant time, namely in two steps (one step to obtain the activations of the hidden layer and another to obtain the output of the network).

Given the above translation algorithm from $\mathcal{P}$ to $\mathcal{N}$, Holldobler and Kalinke have proved the following:

**Proposition 3.1.1.** *[HK94] For each program $\mathcal{P}$ there exists a single hidden layer feedforward network $\mathcal{N}$ of binary threshold units such that $\mathcal{N}$ computes $T_{\mathcal{P}}$.*

**Proposition 3.1.2.** *[HK94] For each single hidden layer feedforward network $\mathcal{N}$ of binary threshold units there exists a program $\mathcal{P}$ such that $T_{\mathcal{P}}$ is computed by $\mathcal{N}$.*

The proof is in the sense that each interpretation $I$ for $\mathcal{P}$ is represented by a binary vector $(i_1, ..., i_k)$ externally activating the respective units of the network's input layer, and that $T_{\mathcal{P}}(I)(A_0)$ is *true* iff the unit representing $A_0$ in the output layer becomes active two steps after $I$ is presented to the input layer of $\mathcal{N}$.

**Corollary 3.1.1.** *[HK94] Let $\mathcal{P}$ be an acceptable program. There exists a single hidden layer recurrent network such that each computation starting with an arbitrary initial input $I$ converges to a stable state and yields the unique fixpoint of $T_{\mathcal{P}}$.*

In the above corollary, note that an arbitrary initial input can be assumed because $T_{\mathcal{P}} \uparrow \omega = T_{\mathcal{P}} \downarrow \omega$ for acceptable programs (see Theorem 2.5.2 of Section 2.5).

In addition to the ability to use background knowledge, neural-symbolic learning systems must be able to perform inductive learning efficiently, one of the main attributes of neural networks. Towell and Shavlik's *Knowledge-based Artificial Neural Network* (*KBANN*) [TS94a] is one such system for rule insertion, refinement and extraction from neural networks. *KBANN* has been empirically shown to be a very efficient system for learning from examples (using *Backpropagation*) and background knowledge. This has been done by comparing the performance of *KBANN* with other hybrid, neural and purely symbolic inductive learning systems.

Briefly, the rules-to-networks algorithm of *KBANN* builds AND/OR trees. Firstly, it creates a hierarchy in the set of rules and rewrites certain rules in order to eliminate disjuncts with more than one term. For example, in this process, a set of rules $R = \{cd \rightarrow a; de \rightarrow a; a\neg f \rightarrow b\}$ becomes $R' = \{cd \rightarrow a'; a' \rightarrow a; de \rightarrow a''; a'' \rightarrow a; a\neg f \rightarrow b\}$. The translation from $R$ to $R'$ converts $cd \vee de \rightarrow a$, where $cd$ and $de$ are two-term disjuncts, into $a' \vee a'' \rightarrow a$, where $a'$ and $a''$ are one-term disjuncts. Then, *KBANN* sets weights and thresholds such that the network behaves as a set of AND/OR neurons. Figure 3.2 shows the *KBANN* network derived from the set of rules $R'$.

By extending the results of Holldobler and Kalinke in the light of the results of Towell and Shavlik, we will be able to design a system, the *C-IL²P* system, in which the theoretical results about the correctness of the

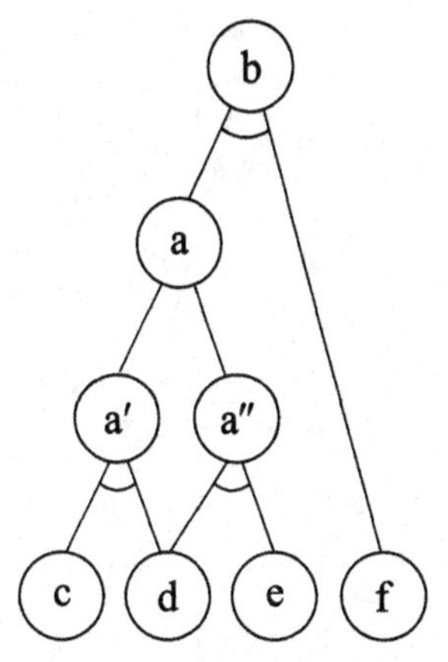


**Fig. 3.2.** A *KBANN* network. $a'$, $a''$ and $b$ are AND neurons, while $a$ is an OR neuron. Each neuron has a semi-linear activation function.

translation algorithm are verified, but the practical objectives of performing inductive learning efficiently with examples and background knowledge are also feasible [dGZC97]. In the sequel, we will show that single hidden layer feedforward networks may capture nonmonotonic inference, but also perform inductive learning efficiently. The experiments of Chapter 4 will corroborate this claim and, in fact, indicate that the performance of *C-IL²P* is at least as good as *KBANN*'s.

To insert background knowledge, described by a general logic program ($\mathcal{P}$), in the neural network ($\mathcal{N}$), we use an approach similar to Holldobler and Kalinke's [HK94]. Each general clause ($C_l$) of $\mathcal{P}$ is mapped from the input layer to the output layer of $\mathcal{N}$ through one neuron ($N_l$) in the single hidden layer of $\mathcal{N}$. Intuitively, the translation algorithm from $\mathcal{P}$ to $\mathcal{N}$ has to implement the following conditions: *(1)* The input potential of a hidden neuron ($N_l$) should only exceed $N_l$'s threshold ($\theta_l$), activating $N_l$, when all the positive antecedents of $C_l$ are assigned the truth-value *true* while all the negative antecedents of $C_l$ are assigned *false*; and *(2)* The input potential of an output neuron ($A$) should only exceed $A$'s threshold ($\theta_A$), activating $A$, when at least one hidden neuron $N_l$ that is connected to $A$ is activated.

*Example 3.1.2.* Consider the logic program $\mathcal{P} = \{A \leftarrow BC \sim D; A \leftarrow EF; B \leftarrow\}$. The translation algorithm should derive the network $\mathcal{N}$ of Figure 3.3, setting weights ($W's$) and thresholds ($\theta's$) in such a way that conditions *(1)* and *(2)* above are satisfied. Note that, if $\mathcal{N}$ ought to be fully connected, any other link (not shown in Figure 3.3) should receive weight zero initially.

Note that, in Example 3.1.2, we have labelled each input and output neuron as an atom appearing, respectively, in the body and in the head of

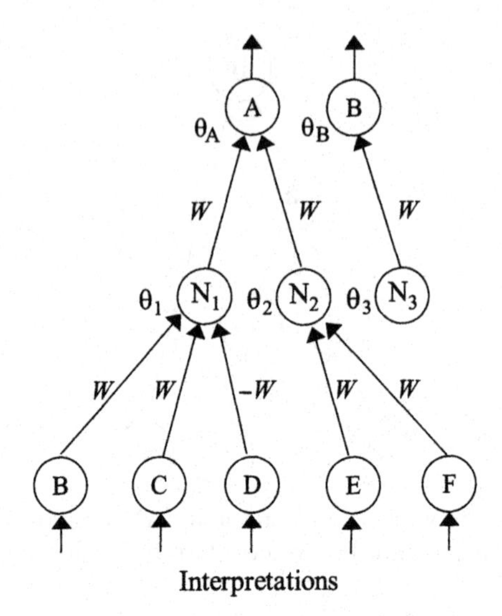


**Fig. 3.3.** Sketch of a neural network for the above logic program $\mathcal{P}$.

a clause of $\mathcal{P}$. This allows us to refer to neurons and propositional variables interchangeably and to regard each input vector $\mathbf{i} = (i_1, ..., i_m)$ of $\mathcal{N}$, where $i_{j(1 \leq j \leq m)} \in [-1, 1]$, as an interpretation for $\mathcal{P}$. If $i_j \in (A_{\min}, 1]$ then the propositional variable associated to the $j$th neuron in the network's input layer is assigned *true*, while $i_j \in [-1, -A_{\max})$ means that it is assigned *false*, where $A_{\min} \in (0, 1)$ and $A_{\max} \in (-1, 0)$ are predefined values as shown in the notation below. Note also that each hidden neuron $N_l$ corresponds to a clause $C_l$ of $\mathcal{P}$.

The following notation will be used in our translation algorithm.

Notation Given a general logic program $\mathcal{P}$, let:

1. $q$ denote the number of clauses $C_l$ $(1 \leq l \leq q)$ occurring in $\mathcal{P}$;
2. $A_{\min}$, the minimum activation for a neuron to be considered "active" (or *true*), $0 < A_{\min} < 1$;
3. $A_{\max}$, the maximum activation for a neuron to be considered "not active" (or *false*), $-1 < A_{\max} < 0$;
4. $h(x) = \frac{2}{1+e^{-\beta x}} - 1$, the bipolar semi-linear activation function, where $\beta$ is the steepness parameter (that defines the slope of $h(x)$);
5. $g(x) = x$, the identity function;
6. $W$ (resp. $-W$), the weight of connections associated with positive (resp. negative) literals;
7. $\theta_l$, the threshold of hidden neuron $N_l$ associated with clause $C_l$;
8. $\theta_A$, the threshold of output neuron $A$, where $A$ is the head of clause $C_l$;

9. $k_l$, the number of literals in the body of clause $C_l$;
10. $p_l$, the number of positive literals in the body of clause $C_l$;
11. $n_l$, the number of negative literals in the body of clause $C_l$;
12. $\mu_l$, the number of clauses in $\mathcal{P}$ with the same atom in the head for each clause $C_l$;
13. $MAX_{C_l}(k_l, \mu_l)$, the greater element among $k_l$ and $\mu_l$ for clause $C_l$;
14. $MAX_{\mathcal{P}}(k_1, ..., k_q, \mu_1, ..., \mu_q)$, the greatest element among all $k$s and $\mu$s of $\mathcal{P}$.

In addition, we use $\overrightarrow{k}$ as a shorthand for $(k_1, ..., k_q)$, and $\overrightarrow{\mu}$ as a shorthand for $(\mu_1, ..., \mu_q)$.

For example, for the program $\mathcal{P}$ of Example 3.1.2, $q = 3$, $k_1 = 3$, $k_2 = 2$, $k_3 = 0$, $p_1 = 2$, $p_2 = 2$, $p_3 = 0$, $n_1 = 1$, $n_2 = 0$, $n_3 = 0$, $\mu_1 = 2$, $\mu_2 = 2$, $\mu_3 = 1$, $MAX_{C_1}(k_1, \mu_1) = 3$, $MAX_{C_2}(k_2, \mu_2) = 2$, $MAX_{C_3}(k_3, \mu_3) = 1$, and $MAX_{\mathcal{P}}(k_1, k_2, k_3, \mu_1, \mu_2, \mu_3) = 3$.

In the following translation algorithm from a logic program $\mathcal{P}$ to a neural network $\mathcal{N}$, we define $A_{\min} = \xi_1(\overrightarrow{k}, \overrightarrow{\mu})$, $W = \xi_2(h, \overrightarrow{k}, \overrightarrow{\mu}, A_{\min})$, $\theta_l = \xi_3(k_l, A_{\min}, W)$, *and* $\theta_A = \xi_4(\mu_l, A_{\min}, W)$ for some functions $\xi_1 - \xi_4$ such that conditions *(1)* and *(2)* above are satisfied, as we will see later in the proof of Theorem 3.1.1.

– *Translation Algorithm:*

1. Given a general logic program $\mathcal{P}$, label each input neuron of a network $\mathcal{N}$ with each atom occurring in the body of the clauses of $\mathcal{P}$. Label each output neuron of $\mathcal{N}$ with each atom occurring in the head of the clauses of $\mathcal{P}$. Assume, for mathematical convenience and without loss of generality, that $A_{\max} = -A_{\min}$.
2. Calculate $MAX_{\mathcal{P}}(\overrightarrow{k}, \overrightarrow{\mu})$ of $\mathcal{P}$;
3. Calculate $A_{\min} > \frac{MAX_{\mathcal{P}}(\overrightarrow{k}, \overrightarrow{\mu})-1}{MAX_{\mathcal{P}}(\overrightarrow{k}, \overrightarrow{\mu})+1}$;
4. Calculate:

$$W \geq \frac{2}{\beta} \cdot \frac{ln\,(1 + A_{\min}) - ln\,(1 - A_{\min})}{MAX_{\mathcal{P}}(\overrightarrow{k}, \overrightarrow{\mu})\,(A_{\min} - 1) + A_{\min} + 1}; \tag{3.1}$$

5. For each clause $C_l$ of $\mathcal{P}$ of the form $A \leftarrow L_1, ..., L_k$ $(k \geq 0)$:
   a) Add a neuron $N_l$ to the hidden layer of $\mathcal{N}$;
   b) Connect each neuron $L_i$ $(1 \leq i \leq k)$ in the input layer to the neuron $N_l$ in the hidden layer. If $L_i$ is a positive literal then set the connection weight to $W$; otherwise, set the connection weight to $-W$;
   c) Connect the neuron $N_l$ in the hidden layer to the neuron $A$ in the output layer and set the connection weight to $W$;

d) Define the threshold ($\theta_l$) of the neuron $N_l$ in the hidden layer as:

$$\theta_l = \frac{(1 + A_{\min})(k_l - 1)}{2} W \tag{3.2}$$

e) Define the threshold ($\theta_A$) of the neuron $A$ in the output layer as:

$$\theta_A = \frac{(1 + A_{\min})(1 - \mu_l)}{2} W \tag{3.3}$$

6. Set $g(x)$ as the activation function of the neurons in the input layer of $\mathcal{N}$. In this way, each input vector $\mathbf{i}$ of $\mathcal{N}$ will represent an interpretation for $\mathcal{P}$.
7. Set $h(x)$ as the activation function of the neurons in the hidden and output layers of $\mathcal{N}$. In this way, a gradient descent learning algorithm, such as *Backpropagation*, can be applied on $\mathcal{N}$ efficiently.
8. If $\mathcal{N}$ ought to be fully-connected, set all other connections to *zero*.

Since $\mathcal{N}$ uses a semi-linear (differentiable) activation function ($h(x)$), the activation values of its output neurons are real numbers in the range $[-1, 1]$. Therefore, we say that an output within the range $(A_{\min}, 1]$ represents the truth-value *true*, while an output within $[-1, -A_{\min})$ represents $false$. We will see later in the proof of Theorem 3.1.1 that the above defined weights and thresholds do not allow the network to present activation values in the range $[-A_{\min}, A_{\min}]$.

Note that the translation of facts of $\mathcal{P}$ into $\mathcal{N}$ (for instance $B \leftarrow$ in Example 3.1.2), is done by simply taking $k_l = 0$ in the above algorithm. Alternatively, each fact of the form $A \leftarrow$ may be converted to a rule of the form $A \leftarrow T$, as in Figure 3.4, where $T$ denotes *true* and is an extra neuron that is always active in the input layer of $\mathcal{N}$, i.e. $T$ has input data fixed at "1". From the point of view of the computation of $T_{\mathcal{P}}$ by $\mathcal{N}$, there is absolutely no difference between the above two ways of inserting facts of $\mathcal{P}$ into $\mathcal{N}$.

Note also that in the subsequent process of inductive learning, in which $\mathcal{P}$ is regarded as background knowledge, the set of examples to be learned could defeat some of the rules of $\mathcal{P}$ by changing weights and/or establishing new connections in $\mathcal{N}$. If, however, certain weights and thresholds are fixed, or some inputs and outputs are clamped, then the respective rules of $\mathcal{P}$ can not be defeated by the set of examples. Hence, defeasible and nondefeasible knowledge could be inserted in the network by defining variable and fixed weights, respectively. From now on, we assume that the background knowledge is defeasible, unless stated otherwise.

The above translation algorithm is based upon the one presented in [HK94], where $\mathcal{N}$ is defined with binary threshold neurons. It is known that such networks have limited ability to learn (see [DS96]). Here, in order to perform inductive learning efficiently, $\mathcal{N}$ is defined using $h(x)$ as activation function. An immediate result is that $\mathcal{N}$ can also perform inductive learning from examples and background knowledge as in *KBANN* [TS94a]. Moreover, the restriction imposed over $W$ in [HK94], where it is shown that $\mathcal{N}$ computes $T_{\mathcal{P}}$ for $W = 1$, is weakened here, since the weights must be able to change during training.

Nevertheless, in [TS94a], and more clearly in [Tow92], the background knowledge of *KBANN* must have a *sufficiently small* number of rules as well as a *sufficiently small* number of antecedents in each rule, in order to be accurately encoded in the neural network. The *sufficiently small* restrictions are given by equations $\mu_l A_{\max} \leq 0.5$ and $k_l A_{\max} \leq 0.5$, respectively, where $A_{\max} > 0$, since $i_{j(1 \leq j \leq m)} \in [0, 1]$ in *KBANN* (see [Tow92]). Unfortunately, these restrictions become quite strong or even unfeasible if, for instance, $A_{\max} = 0.5$ as in [TS94a] (Section 5: Empirical Tests of *KBANN*). Consequently, an interpretation that does not satisfy a clause can wrongly activate a neuron in the output layer of $\mathcal{N}$. This results from the use of the standard (unipolar) semi-linear activation function, where each neuron's activation is in the range $[0, 1]$. Hence, in [TS94a] both $false$ and $true$ are represented by *positive* numbers in the ranges $[0, A_{\max})$ and $(A_{\min}, 1]$, respectively. For example, if $A_{\max} = 0.3$ and $k_l = 2$, $k_l A_{\max} \leq 0.5$ is violated and an interpretation that assigns $false$ to positive literals in the input layer of $\mathcal{N}$ can generate a *positive* input potential greater than the hidden neuron's threshold, wrongly activating the neuron in the output layer of $\mathcal{N}$.

In order to solve this problem, we use bipolar activation functions, where each neuron's activation is in the range $[-1, 1]$. Now, an interpretation that does not satisfy a clause contributes *negatively* to the hidden neuron's input potential, since $false$ is represented by a number in $[-1, -A_{\min})$, while an interpretation that does satisfy a clause contributes *positively* to the input potential, because $true$ is in $(A_{\min}, 1]$. Theorem 3.1.1 will show that the choice of a bipolar activation function is sufficient to solve the above problem. Furthermore, during training, the choice of $-1$ instead of $zero$ to represent $false$ will lead to faster convergence in almost all cases. The reason for this is that the update of a weight connected to an input variable will be $zero$ when the corresponding variable is $zero$ in the training pattern (see [BL96, HKP91]).

Making use of a bipolar semi-linear activation function $h(x)$, let us see how we have obtained the values for the hidden and output neurons' thresholds ($\theta_l$ and $\theta_A$). To confer symmetric mathematical results, we have assumed that $A_{\max} = -A_{\min}$. Consider a clause $C_l = A \leftarrow L_1, ..., L_k$. From the input to the hidden layer of $\mathcal{N}$, i.e. from $L_1, ..., L_k$ to $N_l$, if an interpretation satisfies $L_1, ..., L_k$ then the contribution of $L_1, ..., L_k$ to the input potential of $N_l$ is greater than $I_+ = n_l(-A_{\min})(-W) + p_l A_{\min} W = k_l A_{\min} W$. If, conversely, an interpretation does not satisfy $L_1, ..., L_k$ then the contribution of $L_1, ..., L_k$

to the input potential of $N_l$ is smaller than $I_- = (p_l - 1)W - A_{\min}W + n_lW$. Therefore, we define $\theta_l = \frac{I_+ + I_-}{2} = \frac{(1+A_{\min})(k_l-1)}{2}W$ (Translation Algorithm, step 5d). Now, from the hidden to the output layer of $\mathcal{N}$, i.e. from $N_l$ to $A$, if an interpretation satisfies $L_1, ..., L_k$ then the contribution of $N_l$ to the input potential of $A$ is greater than $I_+ = A_{\min}W - (\mu_l - 1)W$. If, conversely, an interpretation does not satisfy $L_1, ..., L_k$ then the contribution of $N_l$ to the input potential of $A$ is smaller than $I_- = \mu_l(-A_{\min})W$. Similarly, we define $\theta_A = \frac{I_+ + I_-}{2} = \frac{(1+A_{\min})(1-\mu_l)}{2}W$ (Translation Algorithm, step 5e). Obviously, $I_+ > I_-$ should be satisfied in both cases above. Therefore, $A_{\min} > \frac{k_l-1}{k_l+1}$ and $A_{\min} > \frac{\mu_l-1}{\mu_l+1}$ must be verified and, more generally, the condition imposed over $A_{\min}$ in the Translation Algorithm (step 3). Finally, given $A_{\min}$, the value of $W$ (Translation Algorithm (step 4)) results from the proof of Theorem 3.1.1 below.

In what follows, we show that the network $\mathcal{N}$, with semi-linear neurons, computes the fixpoint operator $T_\mathcal{P}$ of the program $\mathcal{P}$. The following theorem ensures that the above *Translation Algorithm* is sound. Recall that the function $T_\mathcal{P}$ mapping interpretations to interpretations is defined as follows: Let $\mathbf{i}$ be an interpretation and $A$ an atom. $T_\mathcal{P}(\mathbf{i})(A) = true$ *iff there exists* $A \leftarrow L_1, \ldots, L_k$ *in* $\mathcal{P}$ *s.t.* $\bigwedge_{i=1}^{k} \mathbf{i}(L_i) = true$.

**Theorem 3.1.1.** *For each propositional general logic program $\mathcal{P}$, there exists a feedforward artificial neural network $\mathcal{N}$ with exactly one hidden layer and semi-linear neurons such that $\mathcal{N}$ computes $T_\mathcal{P}$.*

*Proof. We have to show that there exists $W > 0$ such that the network obtained by the above* Translation Algorithm *computes $T_\mathcal{P}$. In order to do so, we need to prove that given an input vector $\mathbf{i}$, each neuron $A$ in the output layer of $\mathcal{N}$ is "active" if and only if there exists a clause of $\mathcal{P}$ of the form $A \leftarrow L_1, ..., L_k$ s.t. $L_1, ..., L_k$ are satisfied by interpretation $\mathbf{i}$. The proof takes advantage of the monotonically non-decreasing property of the bipolar semi-linear activation function $h(x)$, which allows the analysis to focus on the boundary cases. As before, we assume that $A_{\max} = -A_{\min}$ without loss of generality.*

*($\leftarrow$) "$A \geq A_{\min}$ if $L_1, ..., L_k$ is satisfied by $\mathbf{i}$". Assume that the $p_l$ positive literals in $L_1, ..., L_k$ are true, while the $n_l$ negative literals in $L_1, ..., L_k$ are false. Consider the mapping from the input layer to the hidden layer of $\mathcal{N}$. The input potential ($I_l$) of $N_l$ is minimum when all the neurons associated with a positive literal in $L_1, ..., L_k$ are at $A_{\min}$, while all the neurons associated with a negative literal in $L_1, ..., L_k$ are at $-A_{\min}$. Thus, $I_l \geq p_lA_{\min}W + n_lA_{\min}W - \theta_l$, and assuming $\theta_l = \frac{(1+A_{\min})(k_l-1)}{2}W$, $I_l \geq p_lA_{\min}W + n_lA_{\min}W - \frac{(1+A_{\min})(k_l-1)}{2}W$.*

*If* $h(I_l) \geq A_{\min}$*, i.e.* $I_l \geq -\frac{1}{\beta} ln\left(\frac{1-A_{\min}}{1+A_{\min}}\right)$*, then* $N_l$ *is active. Therefore, Equation 3.4 must be satisfied.*[2]

$$p_l A_{\min} W + n_l A_{\min} W - \frac{(1+A_{\min})(k_l-1)}{2} W \geq -\frac{1}{\beta} ln\left(\frac{1-A_{\min}}{1+A_{\min}}\right) \quad (3.4)$$

*Solving Equation 3.4 for the connection weight (W) yields Equations 3.5 and 3.6, given that* $W > 0$*.*

$$W \geq -\frac{2}{\beta} \cdot \frac{ln(1-A_{\min}) - ln(1+A_{\min})}{k_l(A_{\min}-1) + A_{\min} + 1} \quad (3.5)$$

$$A_{\min} > \frac{k_l - 1}{k_l + 1} \quad (3.6)$$

*Consider now the mapping from the hidden layer to the output layer of* $\mathcal{N}$*. By Equations 3.5 and 3.6, at least one neuron* $N_l$ *that is connected to A is "active". The input potential* $(I_l)$ *of A is minimum when* $N_l$ *is at* $A_{\min}$*, while the other* $\mu_l - 1$ *neurons connected to A are at* $-1$*. Thus,* $I_l \geq A_{\min} W - (\mu_l - 1) W - \theta_l$*, and assuming* $\theta_l = \frac{(1+A_{\min})(1-\mu_l)}{2} W$*,* $I_l \geq A_{\min} W - (\mu_l - 1) W - \frac{(1+A_{\min})(1-\mu_l)}{2} W$*.*

*If* $h(I_l) \geq A_{\min}$*, i.e.* $I_l \geq -\frac{1}{\beta} ln\left(\frac{1-A_{\min}}{1+A_{\min}}\right)$*, then A is active. Therefore, Equation 3.7 must be satisfied.*

$$A_{\min} W - (\mu_l - 1) W - \frac{(1+A_{\min})(1-\mu_l)}{2} W \geq -\frac{1}{\beta} ln\left(\frac{1-A_{\min}}{1+A_{\min}}\right) \quad (3.7)$$

*Solving Equation 3.7 for the connection weight W yields Equations 3.8 and 3.9, given that* $W > 0$*.*

$$W \geq -\frac{2}{\beta} \cdot \frac{ln(1-A_{\min}) - ln(1+A_{\min})}{\mu_l(A_{\min}-1) + A_{\min} + 1} \quad (3.8)$$

$$A_{\min} > \frac{\mu_l - 1}{\mu_l + 1} \quad (3.9)$$

*(→) "*$A \leq -A_{\min}$ *if* $L_1, ..., L_k$ *is not satisfied by* $\mathbf{i}$*". Assume that at least one of the* $p_l$ *positive literals in* $L_1, ..., L_k$ *is false or one of the* $n_l$ *negative literals in* $L_1, ..., L_k$ *is true. Consider the mapping from the input layer to the hidden layer of* $\mathcal{N}$*. The input potential* $(I_l)$ *of* $N_l$ *is maximum when only one neuron associated to a positive literal in* $L_1, ..., L_k$ *is at* $-A_{\min}$ *or when only one neuron associated to a negative literal in* $L_1, ..., L_k$ *is at* $A_{\min}$*. Thus, either* $I_l \leq (p_l - 1) W - A_{\min} W + n_l W - \theta_l$ *or* $I_l \leq (n_l - 1) W - A_{\min} W +$

[2] Throughout, we use "*Equation*", as in "*Equation 3.4*", even though "*Equation 3.4*" is an *Inequality*.

*$p_l W - \theta_l$, and assuming $\theta_l = \frac{(1+A_{\min})(k_l-1)}{2}W$, $I_l \leq (k_l - 1 - A_{\min})W - \frac{(1+A_{\min})(k_l-1)}{2}W$.*

*If $-A_{\min} \geq h(I_l)$, i.e. $-A_{\min} \geq \frac{2}{1+e^{-\beta(I_l)}} - 1$, then $I_l \leq -\frac{1}{\beta}ln\left(\frac{1+A_{\min}}{1-A_{\min}}\right)$, and so $N_l$ is not active. Therefore, Equation 3.10 must be satisfied.*

$$(k_l - 1 - A_{\min})W - \frac{(1+A_{\min})(k_l-1)}{2}W \leq -\frac{1}{\beta}ln\left(\frac{1-A_{\min}}{1+A_{\min}}\right) \quad (3.10)$$

*Solving Equation 3.10 for the connection weight $W$ yields Equations 3.11 and 3.12, given that $W > 0$.*

$$W \geq \frac{2}{\beta} \cdot \frac{ln(1+A_{\min}) - ln(1-A_{\min})}{k_l(A_{\min}-1) + A_{\min} + 1} \quad (3.11)$$

$$A_{\min} > \frac{k_l - 1}{k_l + 1} \quad (3.12)$$

*Consider now the mapping from the hidden layer to the output layer of $\mathcal{N}$. By Equations 3.11 and 3.12, all neurons $N_l$ that are connected to $A$ are "not active". The input potential ($I_l$) of $A$ is maximum when all the neurons connected to $A$ are at $-A_{\min}$. Thus, $I_l \leq -\mu_l A_{\min} W - \theta_l$, and assuming $\theta_l = \frac{(1+A_{\min})(1-\mu_l)}{2}W$, $I_l \leq -\mu_l A_{\min} W - \frac{(1+A_{\min})(1-\mu_l)}{2}W$.*

*If $-A_{\min} \geq h(I_l)$, i.e. $-A_{\min} \geq \frac{2}{1+e^{-\beta(I_l)}} - 1$, then $I_l \leq -\frac{1}{\beta}ln\left(\frac{1+A_{\min}}{1-A_{\min}}\right)$, and so $A$ is not active. Therefore, Equation 3.13 must be satisfied.*

$$-\mu_l A_{\min} W - \frac{(1+A_{\min})(1-\mu_l)}{2}W \leq -\frac{1}{\beta}ln\left(\frac{1+A_{\min}}{1-A_{\min}}\right) \quad (3.13)$$

*Solving Equation 3.13 for the connection weight $W$ yields Equations 3.14 and 3.15, given that $W > 0$.*

$$W \geq \frac{2}{\beta} \cdot \frac{ln(1+A_{\min}) - ln(1-A_{\min})}{\mu_l(A_{\min}-1) + A_{\min} + 1} \quad (3.14)$$

$$A_{\min} > \frac{\mu_l - 1}{\mu_l + 1} \quad (3.15)$$

*Notice that Equations 3.5 and 3.8 are equivalent to Equations 3.11 and 3.14, respectively. Hence, the theorem holds if, for each clause $C_l$ in $\mathcal{P}$, Equations 3.5 and 3.6 are satisfied by $W$ and $A_{\min}$ from the input to the hidden layer of $\mathcal{N}$, while Equations 3.8 and 3.9 are satisfied by $W$ and $A_{\min}$ from the hidden to the output layer of $\mathcal{N}$.*

*In order to unify the weights of $\mathcal{N}$ for each clause $C_l$ of $\mathcal{P}$, given the definition of $MAX_{C_l}(k_l, \mu_l)$, it is sufficient that Equations 3.16 and 3.17 below are satisfied by $W$ and $A_{\min}$, respectively.*

$$W \geq \frac{2}{\beta} \cdot \frac{ln\,(1 + A_{\min}) - ln\,(1 - A_{\min})}{MAX_{C_l}(k_l, \mu_l)(A_{\min} - 1) + A_{\min} + 1} \tag{3.16}$$

$$A_{\min} > \frac{MAX_{C_l}(k_l, \mu_l) - 1}{MAX_{C_l}(k_l, \mu_l) + 1} \tag{3.17}$$

*Finally, in order to unify all the weights of $\mathcal{N}$ for a program $\mathcal{P}$, given the definition of $MAX_{\mathcal{P}}(k_1, ..., k_q, \mu_1, ..., \mu_q)$, it is sufficient that Equations 3.18 and 3.19 are satisfied by $W$ and $A_{\min}$, respectively.*

$$W \geq \frac{2}{\beta} \frac{ln\,(1 + A_{min}) - ln\,(1 - A_{min})}{MAX_{\mathcal{P}}(k_1, ..., k_q, \mu_1, ..., \mu_q)(A_{min}-1)+A_{min}+1} \tag{3.18}$$

$$A_{\min} > \frac{MAX_{\mathcal{P}}(k_1, ..., k_q, \mu_1, ..., \mu_q) - 1}{MAX_{\mathcal{P}}(k_1, ..., k_q, \mu_1, ..., \mu_q) + 1} \tag{3.19}$$

*As a result, if Equations 3.18 and 3.19 are satisfied by $W$ and $A_{\min}$, respectively, then $\mathcal{N}$ computes $T_{\mathcal{P}}$.* □

*Example 3.1.3.* Consider the program $\mathcal{P} = \{A \leftarrow BC \sim D;\ A \leftarrow EF;\ B \leftarrow\}$. Converting fact $B \leftarrow$ to rule $B \leftarrow T$ and applying the *Translation Algorithm*, we obtain the neural network $\mathcal{N}$ of Figure 3.4. Firstly, we calculate $MAX_{\mathcal{P}}(\overrightarrow{k}, \overrightarrow{\mu}) = 3$ (step 2), and $A_{\min} > 0.5$ (step 3). Then, suppose $A_{\min} = 0.6$, we obtain $W \geq 6.931/\beta$ (step 4). Alternatively, suppose $A_{\min} = 0.7$, then $W \geq 4.336/\beta$. Let us take $A_{\min} = 0.7$ and $h(x)$ as the standard bipolar semi-linear activation function ($\beta = 1$), then if $W = 4.5$, $\mathcal{N}$ computes the operator $T_{\mathcal{P}}$ of $\mathcal{P}$.[3]

In the above example, the neuron $B$ appears at both the input and the output layers of $\mathcal{N}$. This indicates that there are at least two clauses of $\mathcal{P}$ that are linked through $B$ (in the example: $A \leftarrow BC \sim D$ and $B \leftarrow$), defining a *dependency chain* [AB94]. We represent that chain in the network using the recurrent connection $W_r = 1$ to denote that the output of $B$ must feed the input of $B$ in the next learning or recall step. In this way, regardless of the length of the dependency chains in $\mathcal{P}$, $\mathcal{N}$ always contains a single hidden layer, thus obtaining a better learning performance.[4] We will explain in detail the use of recurrent connections in Sections 3.2 and 3.3. In Chapter 4, we will compare the learning results of *C-IL²P* with *KBANN*'s, where the number of hidden layers is equal to the length of the greatest dependency chain in the background knowledge.

[3] Note that a sound translation from $\mathcal{P}$ to $\mathcal{N}$ does not require all the weights in $\mathcal{N}$ to have the same absolute value. We unify the weights ($|W|$) for the sake of simplicity of the translation algorithm and to comply with previous work.

[4] It is known that an increase in the number of hidden layers in a neural network results in a corresponding degradation in learning performance.

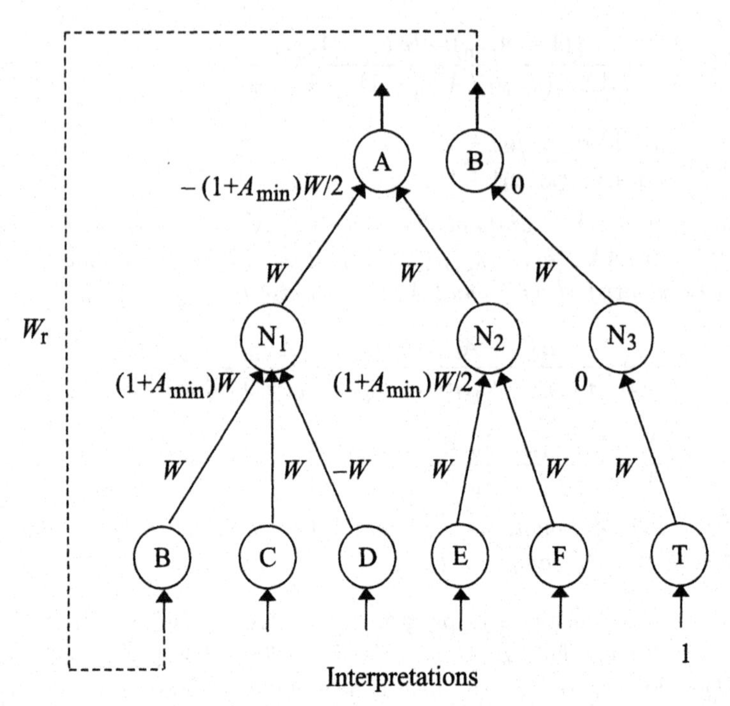


**Fig. 3.4.** The neural network $\mathcal{N}$ obtained by the translation over $\mathcal{P}$. Connections with weight *zero* are not shown.

## 3.2 Massively Parallel Deduction

The neural network $\mathcal{N}$ can perform deduction and induction. In order to perform deduction, $\mathcal{N}$ is transformed into a *partially recurrent network* $\mathcal{N}_r$ by connecting each neuron in the output layer to its correspondent neuron in the input layer with weight $W_r = 1$, as shown in Figure 3.5. In this way, $\mathcal{N}_r$ is used to iterate $T_{\mathcal{P}}$ in parallel, because its output vector becomes its input vector in the next computation of $T_{\mathcal{P}}$.

Let us now show that, as in [HK94], if $\mathcal{P}$ is an acceptable program then $\mathcal{N}_r$ always settles down in a stable state that yields the unique fixpoint of $T_{\mathcal{P}}$, since $\mathcal{N}_r$ computes the upward powers ($T_{\mathcal{P}} \uparrow m$ (**i**)) of $T_{\mathcal{P}}$. A similar result could also be easily proved for the class of locally stratified programs.

**Theorem 3.2.1.** *[Fit94] (Theorem 2.5.2 rewritten) For each acceptable general program $\mathcal{P}$, the function $T_{\mathcal{P}}$ has a unique fixpoint. The sequence of all $T_{\mathcal{P}} \uparrow m$ (**i**), $m \in \aleph$, converges to this fixpoint $T_{\mathcal{P}} \uparrow \omega$ (**i**) (which is identical to the stable model of $\mathcal{P}$ [GL88]), for each $\mathbf{i} \subseteq B_{\mathcal{P}}$.*

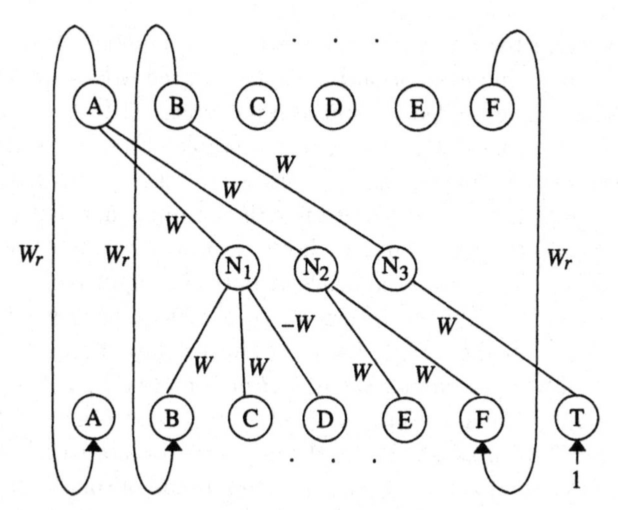


**Fig. 3.5.** The recurrent neural network $\mathcal{N}_r$.

Recall that, since $\mathcal{N}_r$ has semi-linear neurons, for each real value $o_i$ in the output vector ($\mathbf{o}$) of $\mathcal{N}_r$, if $o_i \geq A_{\min}$ then the corresponding $i$th atom in $\mathcal{P}$ is assigned *true*, while $o_i \leq A_{\max}$ means that it is assigned $false$.

**Corollary 3.2.1.** *Let $\mathcal{P}$ be an acceptable general program. There exists a recurrent neural network $\mathcal{N}_r$ with semi-linear neurons such that, starting from an arbitrary initial input, $\mathcal{N}_r$ converges to a stable state and yields the unique fixpoint ($T_\mathcal{P} \uparrow \omega$ ($\mathbf{i}$)) of $T_\mathcal{P}$, which is identical to the stable model of $\mathcal{P}$.*

*Proof. Assume that $\mathcal{P}$ is an acceptable program. By Theorem 3.1.1, $\mathcal{N}_r$ computes $T_\mathcal{P}$. Recurrently connected, $\mathcal{N}_r$ computes the upwards powers ($T_\mathcal{P} \uparrow m$ ($\mathbf{i}$)) of $T_\mathcal{P}$. By Theorem 3.2.1, $\mathcal{N}_r$ computes the unique stable model of $\mathcal{P}$ ($T_\mathcal{P} \uparrow \omega$ ($\mathbf{i}$)).* □

Hence, in order to use $\mathcal{N}$ as a massively parallel model for Logic Programming, we simply have to follow two steps: (1) add neurons to the input and output layers of $\mathcal{N}$, allowing it to be partially recurrently connected; and (2) add the correspondent recurrent links with fixed weight $W_r = 1$.

*Example 3.2.1.* (Example 3.1.3 continued): Given any initial activation in the input layer of $\mathcal{N}_r$ (Figure 3.5), it always converges to the following stable state: $A$ = *false*, $B$ = *true*, $C$ = *false*, $D$ = *false*, $E$ = *false* and $F$ = *false*, which represents the unique stable model of $\mathcal{P}$: $M_\mathcal{P} = \{B\}$.

*Remark 3.2.1.* Following [HK94], in this section we have completed the input and output layers of $\mathcal{N}$ so that both contain the same number of neurons (see Figure 3.5). This allows $\mathcal{N}$ to compute the upper powers of $T_\mathcal{P}$ starting from any input vector.[5] However, even when the number of input neurons is different from the number of output neurons, a network can compute the upper powers of $T_{\mathcal{P}}$, provided the initial input vector is $\{-1, -1, ..., -1\}$. We need to redefine, though, the concept of a stable state, previously defined as "input vector" = "output vector". Now, we say that a network $\mathcal{N}$ is stable when "each literal appearing in both the input and output layers of $\mathcal{N}$ presents the same truth value in both neurons after at least one recall step". As a result, if no literal appears in both the input and output layers of $\mathcal{N}$, a single recall step is necessary for $\mathcal{N}$ to reach stability. The program $P = \{B \leftarrow A\}$ is a good example. Let $N$ be the network obtained from $P$ such that $N$ has $A$ only in its input layer and $B$ only in its output layer. Since we want to compute minimal models, $A \in [-1, -A_{\min})$ and $B \in [-1, -A_{\min})$, i.e. $A = false$ and $B = false$, should be the unique stable state of $N$. Differently from when both $A$ and $B$ are present in the input and output layers of $N$, had we now started from $A = 1$, we would have obtained $B \in (A_{\min}, 1]$, that is, $A = true$ and $B = true$. This is the reason why, when the input and output layers of $N$ are not completed, one needs to constrain the initial input vector of $N$ to be $\{-1, -1, ..., -1\}$. Starting from $A = -1$, the unique stable state of $N$, according to the above definition, will correspond to the unique minimal model of $P$, as expected. Analogously to Corollary 3.2.1, it is not difficult to show that, if $\mathcal{P}$ is an acceptable program, there exists a neural network $\mathcal{N}$ with different sets of input and output neurons such that, starting from $\{-1, -1, ..., -1\}$, $\mathcal{N}$ will eventually converge and settle down in the unique fixpoint of $\mathcal{P}$, according to the above definition of a stable state.

## 3.3 Performing Inductive Learning

One of the main features of artificial neural networks is their learning capability. The program $\mathcal{P}$, viewed as background knowledge, may now be refined with examples in a neural training process on $\mathcal{N}_r$.

Hornik *et al.* [HSW89] have proved that standard feedforward neural networks with only a single hidden layer are capable of approximating any (Borel measurable) function from one finite dimensional space to another, to any desired degree of accuracy, provided sufficiently many hidden units are available. Hence we can train single hidden layer neural networks to approximate the operator $T_{\mathcal{P}}$ associated with a logic program $\mathcal{P}$. Powerful neural learning algorithms have been established theoretically and applied extensively in practice. These algorithms may be used to learn the operator $T_{\mathcal{P}'}$ of a previously unknown program $\mathcal{P}'$, and therefore to learn the program $\mathcal{P}'$ itself.

[5] Note that one does not need to physically add neurons into $\mathcal{N}$. It suffices to complete the input and output vectors of $\mathcal{N}$ with $-1$s for any unknown (input or output) value.

Moreover, DasGupta and Schinitger [DS96] have proved that neural networks with continuously differentiable activation functions are capable of computing a certain family of boolean functions with constant size ($n$), while networks composed of binary threshold functions require at least $O(log(n))$ size. Hence analogue neural networks have more computational power than discrete neural networks, even when computing boolean functions.

The network's recurrent connections contain fixed weights $W_r = 1$, with the sole purpose of ensuring that the output feed the input in the next learning or recall process. As $\mathcal{N}_r$ does not learn in its recurrent connections[6], the standard *Backpropagation* learning algorithm can be applied directly [HKP91] (see also [Jor86] and [Hay99a]). Hence, in order to perform inductive learning with examples on $\mathcal{N}_r$, four simple steps should be followed: (1) add neurons to the input and output layers of $\mathcal{N}_r$, according to the training set (the training set may contain concepts not represented in the background knowledge and vice versa); (2) add neurons to the hidden layer of $\mathcal{N}_r$, if it is so required for the convergence of the learning algorithm; (3) add connections with weight zero, in which $\mathcal{N}_r$ will learn new concepts; (4) perturb the connections by adding small random numbers to its weights in order to avoid learning problems caused by symmetry[7]. The implementation of steps (1) to (4) will become clearer in Chapter 4, where we describe some applications of the *C-IL²P* system using *Backpropagation*.

## 3.4 Adding Classical Negation

According to Lifschitz and McCarthy, commonsense knowledge can be represented more easily when *classical negation* ($\neg$), sometimes called *explicit negation,* is available. In [GL91], Gelfond and Lifschitz have extended the notion of stable models to programs with classical negation. *Extended Logic Programs* can be viewed as a fragment of Default theories (see [MT93a]), and thus are of interest with respect to the relation between Logic Programming and nonmonotonic formalisms. In this section, we extend *C-IL²P* to incorporate classical negation. The extended *C-IL²P* system computes the *Answer Set Semantics* [GL91] of extended logic programs. As a result, it can be applied in a broader range of domains. For example, the application of the system in power system fault diagnosis, which we describe in Chapter 4, uses the extended *C-IL²P*.

General logic programs provide negative information implicitly, by the closed-world assumption, while extended programs include explicit negation,

---

[6] The recurrent connections represent an external process between output and input.

[7] The perturbation should be small enough not to have any effects on the computation of the background knowledge.

allowing the presence of incomplete information in the database. "In the language of extended programs, we can distinguish between a query which fails in the sense that it does not succeed, and a query which fails in the stronger sense that its negation succeeds" [GL91]. The following example, due to John McCarthy, illustrates such a difference: a school bus may cross railway tracks unless there is an approaching train. This would be expressed in a general logic program by the rule $cross \leftarrow \sim train$, in which case the absence of *train* in the database is interpreted as the absence of an approaching train, i.e. using the closed-world assumption. Such an assumption is unacceptable if one reasons with incomplete information. However, if we use classical negation and represent the above knowledge as the extended program: $cross \leftarrow \neg train$, then *cross* will not be derived until the fact $\neg train$ is added to the database.

Therefore, it is essential to differentiate between $\neg A$ and $\sim A$ in a logic program whenever the closed-world assumption is not applicable to $A$. Nevertheless, the closed-world assumption can be explicitly included in extended programs by adding rules of the form $\neg A \leftarrow \sim A$, whenever the information about $A$ in the database is assumed to be complete. Moreover, for some literals, the opposite assumption $A \leftarrow \sim \neg A$ may be appropriate.

The semantics of extended programs, called the *Answer Set Semantics*, is an extension of the stable model semantics for general logic programs. "A 'well-behaved' general program has exactly one stable model, and the answer that it returns for a ground query ($A$) is *yes* or *no*, depending on whether $A$ belongs or not to the stable model of the program. A 'well behaved' extended program has exactly one answer set, and this set is consistent. The answer that an extended program returns for a ground query ($A$) is *yes*, *no* or *unknown*, depending on whether its answer set contains $A$, $\neg A$ or neither" [GL91]. If a program does not contain classical negation, then its answer sets are exactly the same as its stable models.

**Definition 3.4.1.** *[GL91] An* extended logic program *is a finite set of clauses of the form* $L_0 \leftarrow L_1, ..., L_m, \sim L_{m+1}, ..., \sim L_n$, *where* $L_i$ $(0 \le i \le n)$ *is a literal (an atom or the classical negation of an atom, denoted by* $\neg$*).*

**Definition 3.4.2.** *[GL91] Let* $\mathcal{P}$ *be an extended program. By Lit we denote the set of ground literals in the language of* $\mathcal{P}$. *For any set* $\mathcal{S} \subset Lit$, *let* $\mathcal{P}^+$ *be the extended program obtained from* $\mathcal{P}$ *by deleting* (1) *each clause that has a formula* $\sim L$ *in its body when* $L \in S$, *and* (2) *all formulas of the form* $\sim L$ *present in the bodies of the remaining clauses.*

Following [BE99], we say that $\mathcal{P}^+ = \mathbf{R}_\mathcal{S}(\mathcal{P})$, which should be read as "$\mathcal{P}^+$ is the *Gelfond–Lifschitz reduction* of $\mathcal{P}$ w.r.t. $\mathcal{S}$" (after its inventors).

By the above definition, $\mathcal{P}^+$ does not contain default negation ($\sim$), and its answer set can be defined as follows.

**Definition 3.4.3.** *[GL91] The* answer set *of $\mathcal{P}^+$ is the smallest subset $\mathcal{S}^+$ of Lit such that (1) for any rule $L_0 \leftarrow L_1, ..., L_m$ of $\mathcal{P}^+$, if $L_1, ..., L_m \in \mathcal{S}^+$ then $L_0 \in \mathcal{S}^+$, and (2) if $\mathcal{S}^+$ contains a pair of complementary literals then $\mathcal{S}^+ = Lit$.*[8]

Finally, the answer set of an extended program $\mathcal{P}$ that contains default negation ($\sim$) can be defined as follows.

**Definition 3.4.4.** *[GL91] Let $\mathcal{P}$ be an extended program and $\mathcal{S} \subset Lit$. Let $\mathcal{P}^+ = \mathbf{R}_\mathcal{S}(\mathcal{P})$ and $\mathcal{S}^+$ be the answer set of $\mathcal{P}^+$. $\mathcal{S}$ is the* answer set *of $\mathcal{P}$ iff $\mathcal{S} = \mathcal{S}^+$.*

For example, the program $\mathcal{P} = \{r \leftarrow\sim p; \neg q \leftarrow r\}$ has $\{r, \neg q\}$ as its only answer set, since no other subset of the literals in $\mathcal{P}$ has the same fixpoint property. The answers that $\mathcal{P}$ gives to the queries $p$, $q$ and $r$ are, respectively, *unknown*, $false$ and $true$. Note that the answer set semantics assigns different meanings to the rules $\neg q \leftarrow r$ and $\neg r \leftarrow q$, i.e. it is not contrapositive w. r. t. $\leftarrow$ and $\neg$. For instance, the answer set of $\mathcal{P}' = \{r \leftarrow\sim p; \neg r \leftarrow q\}$ is $\{r\}$, differently from the answer set of (the classically equivalent) $\mathcal{P}$.

If $\mathcal{P}$ does not contain classical negation ($\neg$) then its answer sets do not contain negative literals. As a result, the answer sets of a general logic program are identical to its stable models. However, the absence of an atom $A$ in the stable model of a general program means that $A$ is $false$ (by default), while the absence of $A$ (and $\neg A$) in the answer set of an extended program means that nothing is known about $A$.[9]

An extended logic program ($\mathcal{P}$) that has a consistent answer set can be reduced to a general logic program ($\mathcal{P}^*$) as follows. For any negative literal $\neg A$ occurring in $\mathcal{P}$, let $A'$ be a positive literal that does not occur in $\mathcal{P}$. $A'$ is called the *positive form* of $\neg A$. $\mathcal{P}^*$ is obtained from $\mathcal{P}$ by replacing all the negative literals of each rule of $\mathcal{P}$ by its positive form. $\mathcal{P}^*$ is called the positive form of $\mathcal{P}$. For example, the program $\mathcal{P} = \{a \leftarrow b, \neg c; c \leftarrow\}$ can be reduced to its positive form $\mathcal{P}^* = \{a \leftarrow b, c'; c \leftarrow\}$.

**Definition 3.4.5.** *[GL91] For any set $\mathcal{S} \subset Lit$, let $\mathcal{S}^*$ denote the set of the positive forms of the elements of $\mathcal{S}$.*

---

[8] $\mathcal{S}^+ = Lit$ works as though the schema $X \leftarrow L, \sim L$ were present for all $L$ and $X$ in $\mathcal{P}^+_\mathcal{S}$.

[9] Gelfond and Lifschitz think of answer sets as incomplete theories rather than three-valued models. Intuitively, the answer sets of a program $\mathcal{P}$ are possible sets of beliefs that a rational agent may hold on the basis of the incomplete information expressed by $\mathcal{P}$. "When a program has several answer sets, it is incomplete also in another sense – it has several different interpretations, and the answer to a query may depend on the interpretation" [GL91].

**Proposition 3.4.1.** *[GL91] A consistent set $\mathcal{S} \subset Lit$ is an answer set of $\mathcal{P}$ if and only if $\mathcal{S}^*$ is a stable model of $\mathcal{P}^*$.*

The mapping from $\mathcal{P}$ to $\mathcal{P}^*$ reduces extended programs to general programs, although $\mathcal{P}^*$ alone does not indicate that $A'$ represents the negation of $A$.

By Proposition 3.4.1, in order to translate an extended program ($\mathcal{P}$) into a neural network ($\mathcal{N}$), we can use the same approach as the one for general programs (see Section 3.1 for the description of the Translation Algorithm), with the only difference that input and output neurons should be labelled as literals, instead of atoms. In the case of general logic programs, a concept $A$ is represented by a neuron, and its weights indicate whether $A$ is a positive or a negative literal in the sense of default negation; that is, the weights differentiate $A$ from $\sim A$. In the case of extended logic programs, we must also be able to represent the concept $\neg A$ in the network. We do so by explicitly labelling input and output neurons as $\neg A$, while the weights differentiate $\neg A$ from $\sim \neg A$. Note that, in this case, both neurons $A$ and $\neg A$ might be present in the same network.

Analogously to Theorem 3.1.1, the following proposition ensures that the translation of extended programs into a neural network is sound.

**Proposition 3.4.2.** *For each extended logic program $\mathcal{P}$, there exists a feedforward artificial neural network $\mathcal{N}$ with exactly one hidden layer and semilinear neurons such that $\mathcal{N}$ computes $T_{\mathcal{P}^*}$, where $\mathcal{P}^*$ is the positive form of $\mathcal{P}$.*

*Proof. By Definition 3.4.4, we simply need to rename each negative literal $\neg A_i$ $(0 \leq i \leq n)$ in $\mathcal{P}$ by $A'_i$, and label the corresponding neuron in $\mathcal{N}$ as $\neg A_i$. Then, by Theorem 3.1.1, $\mathcal{N}$ computes $T_{\mathcal{P}^*}$.* □

*Example 3.4.1.* Consider the extended program $\mathcal{P} = \{A \leftarrow B, \neg C; \neg C \leftarrow B, \sim \neg E; B \leftarrow \sim D\}$. *C-IL²P's* Translation Algorithm over the positive form $\mathcal{P}^*$ of $\mathcal{P}$ obtains the network $\mathcal{N}$ of Figure 3.6 such that $\mathcal{N}$ computes the fixpoint operator $T_{\mathcal{P}^*}$ of $\mathcal{P}^*$.

As before, the network of Figure 3.6 can be transformed into a partially recurrent network by connecting neurons in the output layer (e.g. $B$) to its correspondent neuron in the input layer, with weight $W_r = 1$, so that $\mathcal{N}$ computes the upward powers of $T_{\mathcal{P}^*}$.

**Definition 3.4.6.** *An extended logic program $\mathcal{P}$ is called an acceptable program if its positive form $\mathcal{P}^*$ is an acceptable program.*

**Corollary 3.4.1.** *Let $\mathcal{P}$ be a consistent acceptable extended program. Let $\mathcal{P}^*$ be the positive form of $\mathcal{P}$. There exists a recurrent neural network $\mathcal{N}$ with*

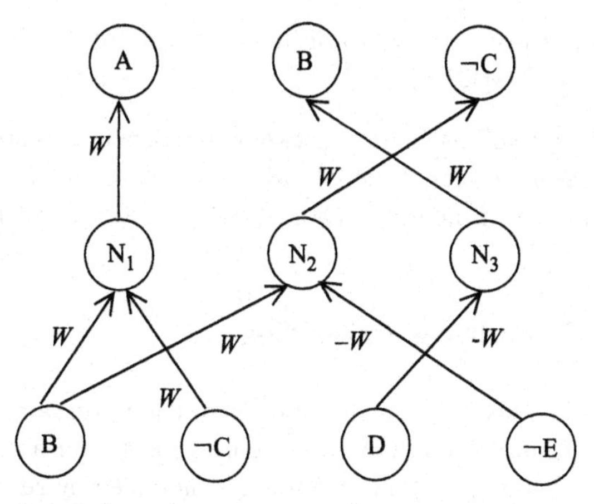


**Fig. 3.6.** From extended programs to neural networks.

*semi-linear neurons such that, starting from an arbitrary input, $\mathcal{N}$ converges to the unique fixpoint of $\mathcal{P}^*$ ($T_{\mathcal{P}^*} \uparrow \omega$ (**i**)), which is identical to the unique answer set of $\mathcal{P}$.*

*Proof.* *By Proposition 3.4.2, $\mathcal{N}$ computes $T_{\mathcal{P}^*}$. Assume that $\mathcal{P}$ is an acceptable program. By Corollary 3.2.1, when recurrently connected, $\mathcal{N}$ computes the unique stable model of $\mathcal{P}^*$. By Proposition 3.4.1, a consistent set $\mathcal{S} \subset Lit$ is an answer set of $\mathcal{P}$ if and only if $\mathcal{S}^*$ is a stable model of $\mathcal{P}^*$. Thus, $\mathcal{N}$ computes the unique answer set of $\mathcal{P}$.* □

*Example 3.4.2.* (Example 3.4.1 continued) Given any initial activation in the input layer of $\mathcal{N}$ (Figure 3.6), it always converges to the following stable state: $A = true, B = true, \neg C = true, D = false, \neg E = false$, which represents the answer set of $\mathcal{P}$, $\mathcal{S}_{\mathcal{P}} = \{A, B, \neg C\}$.

Let us now briefly discuss the case of inconsistent extended programs. Consider, for example, the contradictory program $\mathcal{P} = \{B \leftarrow A; \neg B \leftarrow A; A \leftarrow\}$. As it is an acceptable program, its associated network always converges to the stable state that represents the set $\{A, B, \neg B\}$. At this point, we have to make a choice; either we adopt Gelfond and Lifschitz's definition of answer set semantics, and assume that the answer set of $\mathcal{P}$ is the set of all literals in the language ($Lit$), or we use a *paraconsistent* approach (see [BS89]). We believe that the second option is more appropriate, due to the following argument:

> "Inconsistencies can be read as signals to take external actions, such as 'ask the user', or as signals for internal actions that activate some rules and deactivate other rules. There is a need to develop a framework in which inconsistency can be viewed according to context, as

a trigger for actions, for learning, and as a source of directions in argumentation." [GH91]

In Chapter 7, we will use neural networks to solve inconsistencies through learning. Some of the results of the following section, about the representation of metalevel priorities in neural networks, will be vital for such a task.

## 3.5 Adding Metalevel Priorities

"Recently there has been increasing interest in logic programming-based default reasoning approaches which are not using negation as failure in their object language. Instead, default reasoning is modelled by rules with explicit negation and a metalevel priority relation between rules" [AMB00].

So far, we have seen that a single hidden layer network can represent either a general or an extended logic program. In both cases, the network does not contain negative connections from the hidden layer to the output. What would then be the meaning of negative weights from the hidden to the output layer of the network? In this section, we are interested in answering this question. Throughout, we use $r_j \succ r_i$ to indicate that rule $r_j$ has higher priority than rule $r_i$. We start with an example.

*Example 3.5.1.* Consider program: $P = \{r_1 : fingertips,\ r_2 : alibi,\ r_3 : fingertips \rightarrow guilty,\ r_4 : alibi \rightarrow \neg guilty\}$, and the priority relation $r_3 \succ r_4$, stating that $fingertips \rightarrow guilty$ is stronger evidence than $alibi \rightarrow \neg guilty$. A neural network that encodes the program $P$ but not the relation $r_3 \succ r_4$ will compute the inconsistent answer set $\{fingertips, alibi, guilty, \neg guilty\}$. Alternatively, the metalevel priority relation could be incorporated in the object-level by changing the rule $alibi \rightarrow \neg guilty$ to $alibi, \sim fingertips \rightarrow \neg guilty$. The new program would compute the answer set $\{fingertips, alibi, guilty\}$, which contains the intended answers for $P$ complemented by $r_3 \succ r_4$.

How could we represent the above priority explicitly in the neural network? In the same way that negative weights from input to hidden neurons are interpreted as negation by default because they contribute to blocking the activation of the hidden neurons, negative weights from hidden to output neurons could be seen as the implementation of metalevel priorities. [dGBG00] A negative weight from hidden neuron $r_3$ to output neuron $\neg guilty$ could implement $r_3 \succ r_4$, provided that whenever $r_3$ is activated then $r_3$ *blocks* (or *inhibits*) the activation of $\neg guilty$, which is the conclusion of $r_4$. Figure 3.7 illustrates the idea.

In the above example, $r_3 \succ r_4$ means that, whenever the conclusion of $r_3$ holds, the conclusion of $r_4$ does not hold. Hence, $\succ (r_i, r_j)$ defines priorities

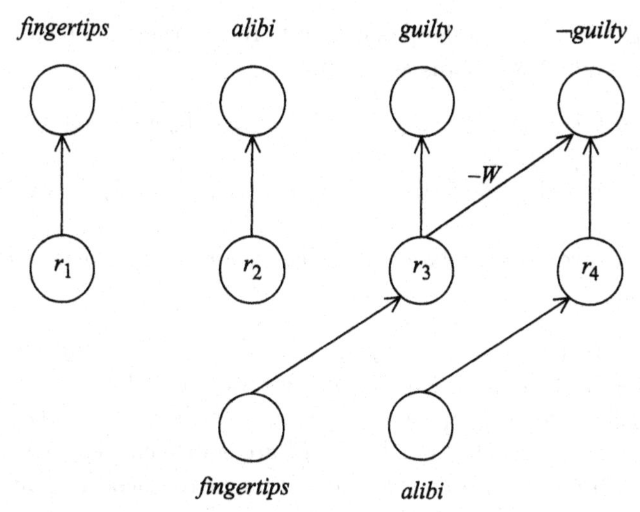


**Fig. 3.7.** Adding metalevel priorities ($r_3 \succ r_4$).

among rules that can override the conclusions one of another. It is thus similar to the *superiority relation* defined by Nute in [Nut87, Nut94] and later investigated by Antoniou in [ABM98].

In the following, we recall some of the basic definitions of Nute's superiority relation. A superiority (binary) relation $\succ$ is a strict partial order, i.e. an irreflexive and transitive relation on a set. Rule $r_1$ is said to be superior to rule $r_2$ if $r_1 \succ r_2$. When the antecedents of two rules are derivable, the superiority relation is used to adjudicate the conflict. If either rule is superior to the other then the superior rule is applied. In other words, $\succ$ provides information about the relative strength of the rules. Following [ABM98], we will define superiority relations over rules with contradictory conclusions only. More precisely, for all $r_i, r_j$, if $\succ (r_i, r_j)$ then $r_i$ and $r_j$ have complementary literals ($x$ and $\neg x$) as consequents.[10] A cycle in the superiority relation (e.g. $r_1 \succ r_2$ and $r_2 \succ r_1$) is counter-intuitive from the knowledge representation perspective, and thus $\succ$ is also required to be an acyclic relation, i.e. we assume that the transitive closure of $\succ$ is irreflexive.

It has been proved in [ABM98] that the above superiority relation does not add to the expressive power of Defeasible Logic [Nut94]. In fact, (object level)

[10] In [ABM98], Antoniou argues that "It turns out that we only need to define the superiority relation over rules with contradictory conclusions". In [PS97], however, Prakken and Sartor present examples in which priorities have to be given between non-contradictory information, which propagate in order to solve an inconsistency. Otherwise, they claim, the result obtained is counter-intuitive. For the sake of simplicity, we restrict $\succ (r_i, r_j)$ to conflicting rules $r_i$ and $r_j$.

default negation and (metalevel) superiority relations are interchangeable (see [KMD94]). The translation is as follows:

1. Replace the clause $L_1, ..., L_p, \sim L_{p+1}, ..., \sim L_q \rightarrow L_{q+1}$ by the clause $r : L_1, ..., L_p \rightarrow L_{q+1}$;
2. Add to $\mathcal{P}$, $j$ clauses of the form $r_{i(1 \leq i \leq j)} : L_j \rightarrow \neg L_{q+1}$, where $p+1 \leq j \leq q$;
3. Define a preference relation between $r_i$ and $r$. In general, $r_i \succ r$ for $1 \leq i \leq j$.

*Example 3.5.2.* Let $P = \{\rightarrow p;\ p \rightarrow b;\ b \sim p \rightarrow f\}$. $P$ can be translated to the following logic program without negation as failure: $P' = \{r_1 : \rightarrow p;\ r_2 : p \rightarrow b;\ r_3 : b \rightarrow f;\ r_4 : p \rightarrow \neg f\}$, $r_4 \succ r_3$, such that $\mathcal{S}(P') \approx \mathcal{S}(P)$, i.e. the answer set semantics of $P$ and $P'$ are equivalent. In a trained neural network, both representations ($P$ and $P'$) might be encoded simultaneously, so that the network is more robust.

Nute's superiority relation adds, however, to the epistemic power of *Defeasible Logic* because it allows one to represent information in a more natural way. For example, the *Nixon Diamond* Problem can be expressed as follows: $r_1 : quaker(Nixon) \leadsto pacifist(Nixon)$ and $r_2 : republican(Nixon) \leadsto \neg pacifist(Nixon)$, where $\leadsto$ should be read as *normally implies.* The definition of an adequate priority relation between $r_1$ and $r_2$ would then solve the inconsistency regarding the pacifism of Nixon, when both $quaker(Nixon)$ and $republican(Nixon)$ are *true.* Also for epistemological reasons, in many cases it is useful to have both default negation and metalevel priorities. This facilitates the expression of (object-level) priorities in the sense of default reasoning and (metalevel) priorities in the sense that a given conclusion should be overridden by another with higher priority.

Of course, preference relations can represent much more than a superiority relation between conflicting consequents. Brewka's preferred subtheories [Bre91] and Prakken and Sartor's argument-based extended logic programming with defeasible priorities [PS97] are only two examples in which a preference relation establishes a partial ordering between rules that describes the relative degree of belief in some beliefs in general, and not only between conflicting conclusions. In a more philosophical aspect, the *AGM* Theory [AGM85] defines a priority relation as the representative of the *epistemic entrenchment* of a set of beliefs, in which formulas are analysed closed under logical consequence. In [Rod97], a preference relation is defined as a pre-order, and in this case, one can differentiate between two incomparable beliefs and two beliefs with the same priority.

In this section, however, we are interested in finding out what is the preference relation that easily fits into a single hidden layer neural network. For this reason, we stick to Nute's superiority relation $\succ$. It would be interesting, though, to try to implement some of the above, more sophisticated, preference relations. Due to the limitations of $\succ$, it might be more appropriate to treat it as *Inconsistency Handling* rather than *Preference Handling*.[11]

Hence the superiority relation we discuss here essentially makes explicit the priorities encoded in the object level. As a result, a network $N_{\succ}$ encoding a program $P_{\succ}$ with explicit priorities is expected to behave exactly as the network $N$ that encodes the equivalent extended program $P$ without priorities. The following definition clarifies what we mean by the equivalence between $P_{\succ}$ and $P$ in terms of $T_{\mathcal{P}}$.

**Definition 3.5.1.** *Let $P_{\succ} = \{r_1, r_2, ..., r_n\}$ be an extended program with an explicit superiority relation $\succ$. Let $P$ be the same extended program $P_{\succ}$ without the superiority relation $\succ$. For any two rules $r_i, r_j$ in $P_{\succ}$ such that $r_j \succ r_i$, let $P' = P - r_i$, i.e. $P'$ is the program $P$ without rule $r_i$. We define $T_{P_{\succ}} = T_{P'}$ if $r_j$ fires, and $T_{P_{\succ}} = T_P$ otherwise.*

The following example illustrates how we can encode metalevel priorities in a neural network, in some special cases, by adding a very simple and straightforward step to the Translation Algorithm.

*Example 3.5.3.* Consider the following *labelled* logic program $P = \{r_1 : ab \sim c \rightarrow x,\ r_2 : de \rightarrow \neg x\}$. $P$ can be encoded in a neural network by using the Translation Algorithm of Section 3.1. If we have also the information that, say, $r_1 \succ r_2$, then we know that the consequent of $r_1$ ($x$) is preferred over the consequent of $r_2$ ($\neg x$). We can represent this in the network by ensuring that whenever $x$ is activated, the activation of $\neg x$ is blocked. We might do so by simply connecting the hidden neuron $r_1$ (remember that each hidden neuron represents a rule of the background knowledge) to the output neuron $\neg x$ with a negative weight (see Figure 3.8). The idea is that, from the Translation Algorithm: (a) when $x$ is activated then $r_1$ also is; and in this case $r_1$ should inhibit $\neg x$ regardless of the activation of $r_2$. Hence, we need to set the weight $-\delta W$ ($\delta, W \in \Re^+$) from $r_1$ to $\neg x$ accordingly. We also need to guarantee that (b) the addition of such a new weight does not change the behaviour of the network when $r_1$ is not activated. In this example, to satisfy (a) above, if the weight from $r_2$ to $\neg x$ is $W$ then $h(W - A_{\min}\delta W - \theta_{\neg x}) < -A_{\min}$ has to be satisfied. To satisfy (b), we need to guarantee that, when $r_1$ is not activated, $\neg x$ is activated if and only if $r_2$ is activated. Therefore, $h(A_{\min}W + \delta W - \theta_{\neg x}) > A_{\min}$ and $h(-A_{\min}W + A_{\min}\delta W - \theta_{\neg x}) < -A_{\min}$

[11] In [BE99], Brewka and Eiter differentiate between the use of priorities to resolve conflicts that emerge from rules with opposite conclusions, and the use of priorities for choosing a rule out of a set of (not necessarily conflicting) rules for application. While Brewka and Eiter's preferred answer set semantics uses the latter, this section offers a neural implementation of the former.

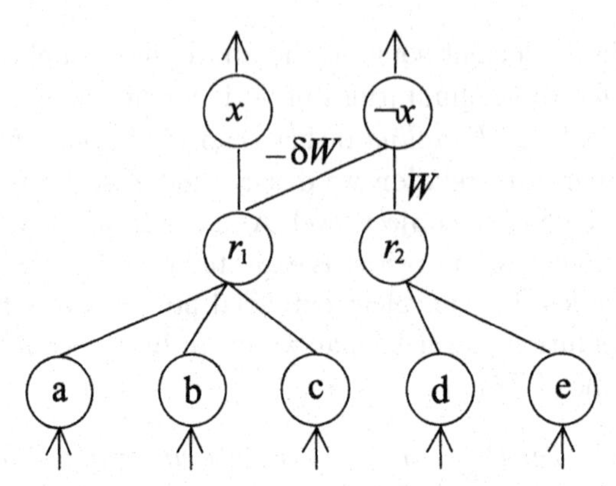


**Fig. 3.8.** Adding metalevel priorities ($r_1 \succ r_2$).

have to be satisfied. This imposes a constraint on $\theta_{\neg x}$.[12] We will see in the sequel that, if for example $W = 6.0$ and $A_{\min} = 0.8$, then $3.4 < \theta_{\neg x} < 86.0$ and, if we take $\theta_{\neg x} = 5$ then $0.666 < \delta < 1.267$ is such that the intended meaning of $r_1 \succ r_2$ is obtained.

Firstly, let us consider the case where a single rule $r$ with consequent $x$ has higher priority than $n$ rules $r_j$ with consequent $\neg x$. The easiest way to extend the *Translation Algorithm* in order to accommodate such a priority relation ($r \succ r_j$) is as follows:

*Metalevel Priorities Algorithm 1:*[13]

1. Consider a Logic Program $P = \{r_1, ..., r_q\}$ and its equivalent neural network $N$, obtained by applying the *Translation Algorithm* of Section 3.1. If there is a single rule $r$ in $P$ with consequent $x$ and $r \succ r_j$ for $1 \leq j \leq n$, do:
   a) Set the threshold $\theta_{\neg x}$ of output neuron $\neg x$ such that:[14]

$$h^{-1}(A_{\min}) - nW\left(\frac{A_{\min}^2 - 1}{1 + A_{\min}}\right) < \theta_{\neg x} < \frac{-2h^{-1}(A_{\min}) + W(1 + A_{\min})(n+1)}{1 - A_{\min}};$$

   b) Add a connection $c$ from the hidden neuron $r$ to the output neuron $\neg x$;
   c) If $(1 + A_{\min})W > 2\theta_{\neg x}$, calculate:

$$\frac{h^{-1}(A\min) + nW - \theta_{\neg x}}{A\min W} < \delta < \frac{-h^{-1}(A_{\min}) + nA_{\min}W + \theta_{\neg x}}{W};$$

[12] Recall that, originally (by Equation 3.3) $\theta_{\neg x} = 0$ for this example.
[13] The value of $\delta$ results from the proof of Proposition 3.5.1 below.
[14] Note that $\neg x$ is the consequent of rules $r_j$.

d) Else calculate:

$$\frac{h^{-1}(A_{\min}) - A_{\min}W + (n-1)W + \theta_{\neg x}}{A_{\min}W} < \delta < \frac{-h^{-1}(A_{\min}) + nA_{\min}W + \theta_{\neg x}}{W};$$

e) Set the weight of $c$ as $-\delta W$.

**Proposition 3.5.1.** *Let $r, r_j \in P$, $r$ the only rule in $P$ with consequent $x$, and $r \succ r_j$ for $1 \leq j \leq n$. Let $N$ be the network obtained by using the* Translation Algorithm *of Section 3.1 over $P$. If $N$ is modified by using the* Metalevel Priorities Algorithm 1 *then $N$ computes $T_{P_\succ}$, where $P_\succ$ is the program $P$ together with the preference relation $r \succ r_j$.*

*Proof.* *We need to show that whenever the hidden neuron $r$ is activated in $N$, the consequent of rules $r_j$ ($\neg x$) is not activated in $N$. By setting the weight of $c$ as $-\delta W$, we want the activation of $\neg x$ to be smaller than $-A_{\min}$ when the activation of $r$ is, at least, $A_{\min}$, and the activation of all hidden neurons $r_j$ ($1 \leq j \leq n$) connected to $\neg x$ is, at most, 1. So we want $h(nW - A_{\min}\delta W - \theta_{\neg x}) < -A_{\min}$, where $n$ is the number of connections to output neuron $\neg x$ prior to the addition of connection $c$. Thus, $\delta > \frac{h^{-1}(A\min)+nW-\theta_{\neg x}}{A\min W}$ needs to be satisfied. However, when the hidden neuron $r$ is not activated in $N$, we still want output $\neg x$ to be activated if and only if at least one of $r_j$ ($1 \leq j \leq n$) is activated. Considering the worst cases, similarly to the proof of Theorem 3.1.1, we derive two constraints: $h(-nA_{\min}W + \delta W - \theta_{\neg x}) < -A_{\min}$ (indicating that $\neg x$ must not be activated when $r_1, ..., r_n$ are not activated, even when the contribution of $r$ is maximum) and $h(A_{\min}W - (n-1)W + A_{\min}\delta W - \theta_{\neg x}) > A_{\min}$ (indicating that $\neg x$ must be activated when one of $r_1, ..., r_n$ is activated, even when the contribution of $r$ is minimum). These yield equation 3.20 below:*

$$\frac{h^{-1}(A_{\min}) - A_{\min}W + (n-1)W + \theta_{\neg x}}{A_{\min}W} < \delta$$

$$< \frac{-h^{-1}(A_{\min}) + nA_{\min}W + \theta_{\neg x}}{W} \tag{3.20}$$

*Equation 3.20 and $\delta > \frac{h^{-1}(A\min)+nW-\theta_{\neg x}}{A\min W}$ impose the following constraint on the threshold $\theta_{\neg x}$ of output neuron $\neg x$ :*

$$h^{-1}(A_{\min}) - nW(\frac{A_{\min}^2 - 1}{1 + A_{\min}}) < \theta_{\neg x}$$

$$< \frac{-2h^{-1}(A_{\min}) + W(1 + A_{\min})(n+1)}{1 - A_{\min}} \tag{3.21}$$

*which is clearly satisfied by* Metalevel Priorities Algorithm 1. *Finally, it is not difficult to see that if $(1 + A_{\min})W > 2\theta_{\neg x}$ then $\frac{h^{-1}(A\min)+nW-\theta_{\neg x}}{A\min W} >$*

$\frac{h^{-1}(A_{\min})-A_{\min}W+(n-1)W+\theta_{\neg x}}{A_{\min}W}$ *and vice versa. As a result, the above constraints on* $\delta$ *are satisfied by* Metalevel Priorities Algorithm 1. *This completes the proof.* □

Let us now consider the case in which more than one rule with consequent $x$ has higher priority than a single rule $r$ with consequent $\neg x$, i.e. $r_i \succ r$ for $1 < i \leq m$. The following example shows that the threshold $\theta_{\neg x}$ of any output neuron $\neg x$ needs to be changed when more than one hidden neuron has to inhibit the activation of the same output neuron.

*Example 3.5.4.* Let $P = \{r_1 : ab \rightarrow \neg x; r_2 : \sim a \sim b \rightarrow \neg x; r_3 : c \rightarrow x\}$, $r_1 \succ r_3$ and $r_2 \succ r_3$, i.e. either $r_1$ or $r_2$ should block the activation of $x$. Let $A_{r_1}$ and $A_{r_2}$ denote, respectively, the activations of hidden neurons $r_1$ and $r_2$. Let $\delta_1$ and $\delta_2$ denote, respectively, the weights from hidden neurons $r_1$ and $r_2$ to output neuron $x$. Thus, $h(W + \delta_1 A_{r_1} + \delta_2 A_{r_2} - \theta_x) < -A_{\min}$ should be satisfied either when $A_{r_1} \in (A_{\min}, 1]$ and $A_{r_2} \in [-1, 1]$, or when $A_{r_1} \in [-1, 1]$ and $A_{r_2} \in (A_{\min}, 1]$. Taking $A_{r_1} = 1$ and $A_{r_2} = -1$, we obtain $\delta_1 - \delta_2 < h^{-1}(-A_{\min}) - W + \theta_x$. Taking $A_{r_1} = -1$ and $A_{r_2} = 1$, we obtain $-\delta_1 + \delta_2 < h^{-1}(-A_{\min}) - W + \theta_x$. Thus, $h^{-1}(-A_{\min}) - W + \theta_x > 0$ should be satisfied. However, since $h^{-1}(-A_{\min}) < 0, W > 0$ and $\theta_x \leq 0$, we have $h^{-1}(-A_{\min}) - W + \theta_x < 0$; a contradiction! As a result, even certain very simple preference relations such as $\{r_1 \succ r_3, r_2 \succ r_3\}$ require changing the threshold of output neuron $\theta_x$.

It turns out that when $m$ rules ($m > 1$) have higher priority than $n$ rules ($n \geq 1$), the threshold of the output neuron in question, and possibly the weights of its connections, have to be changed.

*Example 3.5.5.* Let $P = \{r_1, r_2, r_3\}, r_1 \succ r_3, r_2 \succ r_3$. Let the consequents of $r_1$ and $r_2$ be $x$ and the consequent of $r_3$ be $\neg x$. We want to add negative weights $-\delta_1$ from $r_1$ to $\neg x$ and $-\delta_2$ from $r_2$ to $\neg x$ such that (1) the activation of $\neg x$ is greater than $A_{\min}$ when $r_3$ is activated, provided $r_1$ and $r_2$ are not activated, and (2) the activation of $\neg x$ is smaller than $-A_{\min}$ when either $r_1$ or $r_2$ are activated, regardless of the activation of $r_3$. Assume that the weight from $r_3$ to $\neg x$ is $W$ and take, for the sake of simplicity, $\delta_1 = \delta_2 = W$.

In Case (1), the minimum activation of $r_3$ is $A_{\min}$ while, in the worst case, the activation of $r_1$ and $r_2$ are both $-A_{\min}$. Thus, we have:

$$A_{\min}W + A_{\min}W + A_{\min}W - \delta' W > h^{-1}(A_{\min}) \tag{3.22}$$

In Case (2), either $r_1$ presents activation $A_{\min}$, while, in the worst case, $r_2$ is at $-1$ and $r_3$ is at 1, or $r_2$ presents activation $A_{\min}$ while, again in the worst case, $r_1$ is at $-1$ and $r_3$ is at 1. Since we have taken $-\delta_1 = -\delta_2 = -W$, both cases yield the same inequality, and we have:

$$W + W - A_{\min}W - \delta' W < h^{-1}(-A_{\min}) \tag{3.23}$$

From Equation 3.22, we obtain $W > \frac{h^{-1}(A_{\min})}{3A_{\min}-\delta'}$ and the constraint $3A_{\min} - \delta' > 0$. From Equation 3.23, we obtain $W > \frac{h^{-1}(-A_{\min})}{2-A_{\min}-\delta'}$ and the constraint $2 - A_{\min} - \delta' < 0$. Therefore, we need to satisfy $2 - A_{\min} < \delta' < 3A_{\min}$, that is, $A_{\min} > 0.5$. Taking $A_{\min} = 0.6$ and $\delta' = 1.5$, if $W = 15$, and therefore $\theta_{\neg x} = 22.5$, the network will encode $r_1 \succ r_3$ and $r_2 \succ r_3$. Note that, we might also need to change the original value of $W$ in $N$, obtained from the *Translation Algorithm* of Section 3.1 (step 4), in order to satisfy $W > \frac{h^{-1}(A_{\min})}{3A_{\min}-\delta'}$ and $W > \frac{h^{-1}(-A_{\min})}{2-A_{\min}-\delta'}$.

In light of the above examples, when $r_i \succ r_j$ for a number $m$ $(m > 1)$ of rules $r_i$ with consequent $\neg x$ and a number $n$ $(n \geq 1)$ of rules $r_j$ with consequent $x$ in $\mathcal{P}$, we extend the *Translation Algorithm* of Section 3.1 as follows to accommodate $r_i \succ r_j$:

*Metalevel Priorities Algorithm 2:*[15]

1. Consider a Logic Program $P = \{r_1, ..., r_q\}$ and its equivalent neural network $N$, obtained by applying the *Translation Algorithm* of Section 3.1. Let $r_1, ..., r_n$ be the rules in $\mathcal{P}$ with consequent $x$, and $r_{n+1}, ..., r_o$, the rules in $\mathcal{P}$ with consequent $\neg x$. Let $m = o - n$. In other words, there are $m$ rules with consequent $\neg x$ and $n$ rules with consequent $x$ in $\mathcal{P}$. In what follows, $m_p$ and $n_p$ will denote, respectively, the values $m$ and $n$ of each pair $p$ of output neurons $x$ and $\neg x$ in $\mathcal{P}$. Let $r_i \succ r_j$ for $1 \leq j \leq n$ and $n + 1 \leq i \leq o$.
2. For each pair $p$ $(1 \leq p \leq q')$ of output neurons of the form $(x, \neg x)$ in $\mathcal{P}$, do:

   Calculate

   $$A^p_{\min} = \frac{n_p(m_p + 1) - 1}{n_p(m_p + 1) + 1};$$

3. Calculate:

   $A_{\min} > MAX\left(\frac{MAX_{\mathcal{P}}(\vec{k},\vec{\mu})-1}{MAX_{\mathcal{P}}(\vec{k},\vec{\mu})+1}, MAX\left(A^1_{\min}, A^2_{\min}, ..., A^{q'}_{\min}\right)\right)$, where $MAX(f_1, ..., f_s)$ returns the greatest number among $f_1, ..., f_s$;

4. For each pair $p$ $(1 \leq p \leq q')$ of output neurons of the form $(x, \neg x)$:
   a) Add a connection $c_i$ from each hidden neuron $r_i$ to the output neuron $x$;

[15] The values of $A_{\min}$, $W$, $\delta$ and $\delta'$ result from the proof of Proposition 3.5.2 below.

b) Calculate:

$$\frac{1-2n_p+A_{\min}}{m_p-1-(m_p+1)A_{\min}} < \delta < \frac{1-n_p+(n_p+1)A_{\min}}{m_p(1-A_{\min})};$$

c) Calculate:

$$MAX\,(n_p+\delta(m_p-1-A_{\min}), -n_pA_{\min}+m_p\delta) < \delta'$$

$$< 1-n_p+A_{\min}+m_pA_{\min}\delta;$$

d) Calculate:

$$W' > MAX\left(W, \frac{h^{-1}(-A_{\min})}{n_p+\delta(m_p-1-A_{\min})-\delta'},\right.$$

$$\left.\frac{h^{-1}(A_{\min})}{1-n_p+A_{\min}+m_pA_{\min}\delta-\delta'}, \frac{h^{-1}(-A_{\min})}{-n_pA_{\min}+m_p\delta-\delta'}\right);$$

e) Change the weights of connections $c_j$ from each hidden neuron $r_j$ to the output neuron $x$ to $W'$;
f) Set the weights of connections $c_i$ from each hidden neuron $r_i$ to the output neuron $x$ as $-\delta W'$;
g) Change the threshold $\theta_x$ of output neuron $x$ to $\delta' W'$.

**Proposition 3.5.2.** *Let $r_1, ..., r_n$ be the rules in $P$ with consequent $x$, and $r_{n+1}, ..., r_o$, the rules in $P$ with consequent $\neg x$. Let $r_i \succ r_j$ for $1 \leq j \leq n$ and $n+1 \leq i \leq o$, and $m = o - n$. Let $N$ be the network obtained by using the* Translation Algorithm *of Section 3.1 over $P$. If $N$ is modified by using the* Metalevel Priorities Algorithm 2 *then $N$ computes $T_{P_\succ}$, where $P_\succ$ is the program $P$ together with the preference relation $r_i \succ r_j$.*

*Proof.* *We need to show that when any hidden neuron $r_i$ is activated in $N$, the consequent of rules $r_j$ ($x$) is not activated in $N$ (*Case 1*). We also need to guarantee, since we are changing the threshold of $x$, that when any hidden neuron $r_j$ is activated in $N$, provided no hidden neuron $r_i$ is activated, $x$ is also activated in $N$ (*Case 2i*), and that when no hidden neuron $r_j$ is activated in $N$, provided no hidden neuron $r_i$ is activated, $x$ is not activated in $N$ (*Case 2ii*).*

Case 1: *In the worst case, $n$ neurons $r_j$ present activation 1, one neuron $r_i$ is at $A_{\min}$ and $m-1$ neurons $r_i$ are at $-1$. Let $W > 0$ be the weight from* *neurons $r_j$ to $x$. Let $-\delta W$ ($\delta \in \Re^+$) be the weight from neurons $r_i$ to $x$, and $\delta' W$ ($\delta' \in \Re$) be the threshold of $x$. Thus, Equation 3.24 has to be satisfied.*[16]

$$nW - A_{\min}\delta W + (m-1)\delta W - \delta' W < h^{-1}(-A_{\min}) \tag{3.24}$$

*Solving Equation 3.24 for the connection weight $W$ yields Equations 3.25 and 3.26.*

$$W > \frac{h^{-1}(-A_{\min})}{n + \delta(m - 1 - A_{\min}) - \delta'} \tag{3.25}$$

$$n + \delta(m - 1 - A_{\min}) - \delta' < 0 \tag{3.26}$$

Case 2i: *Again in the worst case scenario, when one neuron $r_j$ presents activation $A_{\min}$ and $n-1$ neurons $r_j$ are at $-1$, provided $m$ neurons $r_i$ are at $-A_{\min}$, $x$ should be activated. Equation 3.27 has to be satisfied.*

$$A_{\min}W - (n-1)W + mA_{\min}\delta W - \delta' W > h^{-1}(A_{\min}) \tag{3.27}$$

*Solving Equation 3.27 for the connection weight $W$ yields Equations 3.28 and 3.29.*

$$W > \frac{h^{-1}(A_{\min})}{1 - n + A_{\min} + mA_{\min}\delta - \delta'} \tag{3.28}$$

$$1 - n + A_{\min} + mA_{\min}\delta - \delta' > 0 \tag{3.29}$$

Case 2ii: *Finally, when all neurons $r_j$ are at $-A_{\min}$ and all neurons $r_i$ are at $-1$, the output neuron $x$ should not be activated.*

$$-nA_{\min}W + m\delta W - \delta' W < h^{-1}(-A_{\min}) \tag{3.30}$$

*Solving Equation 3.30 yields Equations 3.31 and 3.32 below.*

$$W > \frac{h^{-1}(-A_{\min})}{-nA_{\min} + m\delta - \delta'} \tag{3.31}$$

$$-nA_{\min} + m\delta - \delta' < 0 \tag{3.32}$$

*From Equations 3.26, 3.29 and 3.32, we derive the following constraints on $\delta'$ :*

$$n + \delta(m - 1 - A_{\min}) < \delta' < 1 - n + A_{\min} + mA_{\min}\delta \tag{3.33}$$

$$-nA_{\min} + m\delta < \delta' < 1 - n + A_{\min} + mA_{\min}\delta \tag{3.34}$$

*From Equations 3.33 and 3.34, and assuming $m - 1 - (m+1)A_{\min} < 0$, we obtain the following constraint on $\delta$ :*

[16] Recall that $0 < A_{\min} < 1$ and $h^{-1}(A_{\min}) = -\frac{1}{\beta} ln\left(\frac{1-A_{\min}}{1+A_{\min}}\right)$.

$$\frac{1-2n+A_{\min}}{m-1-(m+1)A_{\min}} < \delta < \frac{1-n+(n+1)A_{\min}}{m(1-A_{\min})} \quad (3.35)$$

*From Equation 3.35 we derive Equation 3.36, and solving Equation 3.36 for $A_{\min}$, we obtain Equation 3.37.*

$$(n(m+1)+1)\,A_{\min}{}^2 + 2A_{\min} - n(m+1) + 1 > 0 \quad (3.36)$$

$$A_{\min} > \frac{n(m+1)-1}{n(m+1)+1} \quad (3.37)$$

*Since we have assumed that $m-1-(m+1)A_{\min} < 0$ then $A_{\min} > \frac{m-1}{m+1}$ must hold. However, $\frac{n(m+1)-1}{n(m+1)+1} > \frac{m-1}{m+1}$ for $m > 1$ and $n \geq 1$, and thus Equation 3.37 suffices.*

*From Equation 3.37 and the* Translation Algorithm *(step 3) of Section 3.1, we finally obtain the new constraint on $A_{\min}$, given by Equation 3.38 below.*

$$A_{\min} > MAX\left(\frac{MAX_{\mathcal{P}}(\overrightarrow{k},\overrightarrow{\mu})-1}{MAX_{\mathcal{P}}(\overrightarrow{k},\overrightarrow{\mu})+1}, MAX\left(A^1_{\min}, A^2_{\min}, \ldots, A^{q'}_{\min}\right)\right) \quad (3.38)$$

*where $A^p_{\min}(1 \leq p \leq q') = \frac{n_p(m_p+1)-1}{n_p(m_p+1)+1}$, for each pair $p$ of output neurons of the form $(x, \neg x)$.*[17]

*If the value of $A_{\min}$ satisfies Equation 3.38, the value of $\delta$ should be calculated using Equation 3.35 and the value of $\delta'$ should be calculated using Equations 3.33 and 3.34. Finally, the value of $W$ should satisfy Equations 3.25, 3.28, and 3.31. Since the* Metalevel Priorities Algorithm 2 *complies with the above constraints, this completes the proof.* □

Let us briefly analyse the main constraint in the above algorithm, namely, the one on $A_{\min}$ (see Equation 3.38). The value of $A_{\min}$ now depends also on $m$ and $n$. When either $m$ or $n$ are large numbers, we might need to change the value of $A_{\min}$ previously defined in Section 3.1, in order to satisfy Equation 3.37 as well. If, however, we allow the use of different values for $A_{\min}$ in $N$, Equation 3.37 could be used to define the new $A_{\min}$ for a particular output neuron $x$. Notwithstanding, for consistency, we will maintain a single value for $A_{\min}$ throughout $N$, by using Equation 3.38. Summarising, we proceed as follows: when $m = 1$, use *Metalevel Priorities Algorithm 1.* Otherwise, in addition to thresholds, certain weights might need to be changed in $N$. In this case, use *Metalevel Priorities Algorithm 2.*

[17] As before, we want to define a single value $A_{\min}$ for $N$. Hence the constraint $A_{\min} > \frac{n(m+1)-1}{n(m+1)+1}$ for each pair $(x, \neg x)$ of output neurons becomes $A_{\min} > MAX\left(A^1_{\min}, A^2_{\min}, \ldots, A^{q'}_{\min}\right)$ for the set of $q'$ pairs of complementary output neurons.

So far, we have investigated the cases in which a superiority relation between rules defines a preference relation between *models*. This is so because, when $r_j$ are the rules in $P$ with consequent $x$, and $r_i$ are the rules in $P$ with consequent $\neg x$, $r_i \succ r_j$ actually states that any model of $P$ containing $\neg x$ is preferred over any model of $P$ containing $x$. This does not cater for the case of exceptions to exceptions, i.e. when a preference relation overrides a conclusion that is derived based on another preference relation. Take, for instance, the following example. Let $P = \{r_1 : birds \rightarrow fly,\ r_2 : penguins \rightarrow birds,\ r_3 : penguins \rightarrow \neg fly,\ r_4 : superpenguins \rightarrow fly\}$ and $r_4 \succ r_3 \succ r_1$. In this example, $r_3$ should block $r_1$ but not $r_4$. In other words, the conclusion of $fly$ from $r_4$ is more reliable than from $r_1$, and thus the weights from $r_1$ and $r_4$ to $fly$ should be different now. Intuitively, we should have $W_{fly,r_1}, W_{fly,r_4} > 0$ while $W_{fly,r_3} < 0$. In fact, $W_{fly,r_1} = 1.94, W_{fly,r_3} = -1.93, W_{fly,r_4} = 3.97$ and $\theta_{fly} = -1.93$ implements $r_4 \succ r_3 \succ r_1$. These values were obtained by actually training a network to encode $r_4 \succ r_3 \succ r_1$ using fixed weights from the input to the hidden layer (given by the Translation Algorithm of Section 3.1 over $P$).

Following [Gab99], we will consider the case of finite linearly ordered sets. Assume $P = \{r_1, ..., r_q\}$ and $r_q \succ ... \succ r_2 \succ r_1$. As a result, any subset of $P$ is also linearly ordered. We are interested in the subset of $P$ with complementary conclusions $x$ and $\neg x$. We represent it as a list $(r_1, r_2, ..., r_j)$ where $r_j \succ ... \succ r_2 \succ r_1$. Assume $r_1$ has consequent $x$. We need to assign values to weights $W_{xr_1}, W_{xr_2}, W_{xr_3}, ..., W_{xr_j}, \theta_x$ and $W_{\neg xr_2}, W_{\neg xr_3}, ..., W_{\neg xr_j}, \theta_{\neg x}$ such that $r_j \succ ... \succ r_2 \succ r_1$ holds. Let:

$$\begin{cases} W_{xr_1} = W, \\ W_{xr_2} = -W + \varepsilon, \\ W_{xr_n} = W_{xr_{n-2}} - W_{xr_{n-1}} + \varepsilon, \text{if} n \in \{3, 5, 7...\} \text{or} \\ W_{xr_n} = W_{xr_{n-2}} - W_{xr_{n-1}} - \varepsilon, \text{if} n \in \{4, 6, 8...\}. \end{cases}$$

where $W > 0$ and $\varepsilon$ is a small positive number such that $W >> \varepsilon$ (typically $\varepsilon = 0.01$). Similarly, let $W_{\neg xr_2} = W, W_{\neg xr_3} = -W + \varepsilon$, and $W_{\neg xr_n} = W_{\neg xr_{n-2}} - W_{\neg xr_{n-1}} + \varepsilon$ for $n = 5, 7, 9...$ while $W_{\neg xr_n} = W_{\neg xr_{n-2}} - W_{\neg xr_{n-1}} - \varepsilon$ for $n = 4, 6, 8...$[18]

*Metalevel Priorities Algorithm 3 (linear ordering):*[19]

1. Consider a Logic Program $P = \{r_1, ..., r_q\}$ and its equivalent neural network $N$, obtained by applying the *Translation Algorithm* of Section 3.1.

[18] Note that $W_n = W_{n-2} - W_{n-1} \pm \varepsilon$ is responsible for assigning the weights such that: if $j$ is an even number, $W_{xr_1}, W_{xr_3}, ..., W_{xr_{j-1}} > 0$ and $W_{xr_2}, W_{xr_4}, ..., W_{xr_j} < 0$; if $j$ is an odd number, $W_{xr_1}, W_{xr_3}, ..., W_{xr_j} > 0$ and $W_{xr_2}, W_{xr_4}, ..., W_{xr_{j-1}} < 0$.

[19] The values of $\theta_x$ and $\theta_{\neg x}$ are obtained from the proof of Proposition 3.5.3.

For each subset $(r_1, r_2, ..., r_j)$ of rules in $P$ with consequents $x$ and $\neg x$, do:

a) Calculate $\theta_1$ such that:
$(1 - A_{\min})W + \varepsilon A_{\min} - A_{\min} \sum_{k=3,5...}^{j} (W_{xr_k}) - \sum_{k=4,6...}^{j-1} (W_{xr_k}) - h^{-1}(-A_{\min}) < \theta_1 < A_{\min} W - A_{\min} \sum_{k=2,4...}^{j-1} (W_{xr_k}) - \sum_{k=3,5...}^{j} (W_{xr_k}) - h^{-1}(A_{\min})$.

b) Calculate $\theta_2$ such that:
$(1 - A_{\min})W + \varepsilon A_{\min} - A_{\min} \sum_{k=4,6...}^{j-1} (W_{\neg xr_k}) - \sum_{k=5,7...}^{j} (W_{\neg xr_k}) - h^{-1}(-A_{\min}) < \theta_2 < A_{\min} W - A_{\min} \sum_{k=3,5...}^{j} (W_{\neg xr_k}) - \sum_{k=4,6...}^{j-1} (W_{\neg xr_k}) - h^{-1}(A_{\min})$.

c) Add a connection from each hidden neuron $r_i$ $(1 \leq i \leq j)$ to the output neuron $x$, and set the connection weight to $W_{xr_i}$;

d) Add a connection from each hidden neuron $r_i$ $(2 \leq i \leq j)$ to the output neuron $\neg x$, and set the connection weight to $W_{\neg xr_i}$;

e) If $j$ is an *odd* number:
   i. Set the threshold $\theta_x$ of output neuron $x$ to $\theta_1$,
   ii. Set the threshold $\theta_{\neg x}$ of output neuron $\neg x$ to $\theta_2$.

f) If $j$ is an *even* number:
   i. Set the threshold $\theta_x$ of output neuron $x$ to $\theta_2$,
   ii. Set the threshold $\theta_{\neg x}$ of output neuron $\neg x$ to $\theta_1$.

**Proposition 3.5.3.** *Let $P = \{r_1, r_2, ..., r_q\}$ and $r_q \succ ... \succ r_2 \succ r_1$. Let $N$ be the network obtained by using the* Translation Algorithm *of Section 3.1 over $P$. If $N$ is modified by using the* Metalevel Priorities Algorithm 3 *then $N$ computes $T_{P_\succ}$, where $P_\succ$ is the program $P$ together with the preference relation $r_q \succ ... \succ r_2 \succ r_1$.*

*Proof. By induction.*

*We consider the subset $R = (r_1, r_2, ..., r_j)$ of $P$ containing the rules with conclusions $x$ and $\neg x$. Let $r_{i(1 \leq i \leq j)} \in R$ be the last neuron from left to right in $(r_1, r_2, ..., r_j)$ to be activated in $N$. Assume $r_1$ has consequent $x$. From the definition of $W_{xr_i}$, the activation of $x$, $Act(x) = h(Wr_1 - (W - \varepsilon)r_2 + 2Wr_3 - 3Wr_4 + (5W + \varepsilon)r_5 - (8W + 2\varepsilon)r_6 + ... - \theta_x)$*

*We distinguish several cases:*

1. *If $j$ is an odd number and $i$ is an odd number, show that $x$ is activated: **Basis:** If $i = 1$ then $x$ is activated. In the worst case, $r_3, r_5, ..., r_j = -1$ and $r_2, r_4, ..., r_{j-1} = -A_{\min}$, while $r_1 = A_{\min}$. We need to satisfy:*

$$A_{\min} W - A_{\min} \sum_{k=2,4...}^{j-1} W_{xr_k} - \sum_{k=3,5...}^{j} W_{xr_k} - \theta_x > h^{-1}(A_{\min}) \quad (3.39)$$

*which yields:*

$$\theta_x < A_{\min} W - A_{\min} \sum_{k=2,4...}^{j-1} W_{xr_k} - \sum_{k=3,5...}^{j} W_{xr_k} - h^{-1}(A_{\min}) \quad (3.40)$$

*and, from the* Metalevel Priorities Algorithm 3 (steps 1(a) and 1(e)), *Equation 3.40 is clearly satisfied.*
*Inductive Step: if $x$ is activated for $i = n$ then $x$ is activated for $i = n+2$. For $i = n$, in the worst case, $r_1, r_3, ..., r_{n-2} = -1$ and $r_2, r_4, ..., r_{n-1} = 1$, $r_n = A_{\min}$ and $r_{n+1}, ..., r_j$ are not activated, i.e. $r_{n+1}, r_{n+3}, ..., r_{j-1} = -A_{\min}$ and $r_{n+2}, r_{n+4}, ..., r_j = -1$. If $x$ is activated then*

$$- \sum_{k=1,3...}^{n-2} W_{xr_k} + \sum_{k=2,4...}^{n-1} W_{xr_k} + A_{\min} W_{xr_n} -$$

$$A_{\min} \sum_{k=n+1,n+3...}^{j-1} W_{xr_k} - \sum_{k=n+2,n+4...}^{j} W_{xr_k} > h^{-1}(A_{\min}) + \theta_x \quad (3.41)$$

*holds.*
*For $i = n+2$, we have $r_1, r_3, ..., r_n = -1$ and $r_2, r_4, ..., r_{n+1} = 1$, $r_{n+2} = A_{\min}$ and $r_{n+3}, ..., r_j$ not activated, i.e. $r_{n+3}, r_{n+5}, ..., r_{j-1} = -A_{\min}$ and $r_{n+4}, r_{n+6}, ..., r_j = -1$. We need to show that*

$$- \sum_{k=1,3...}^{n} W_{xr_k} + \sum_{k=2,4...}^{n+1} W_{xr_k} + A_{\min} W_{xr_{n+2}} -$$

$$A_{\min} \sum_{k=n+3,n+5...}^{j-1} W_{xr_k} - \sum_{k=n+4,n+6...}^{j} W_{xr_k} > h^{-1}(A_{\min}) + \theta_x \quad (3.42)$$

*also holds.*
*It is sufficient to show that:*

$$- \sum_{k=1,3,5...}^{n} W_{xr_k} + \sum_{k=2,4,6...}^{n+1} W_{xr_k} + A_{\min} W_{xr_{n+2}} - A_{\min} \sum_{k=n+3,n+5...}^{j-1} W_{xr_k} -$$

$$\sum_{k=n+4,n+6...}^{j} W_{xr_k} > - \sum_{k=1,3...}^{n-2} W_{xr_k} + \sum_{k=2,4...}^{n-1} W_{xr_k} + A_{\min} W_{xr_n} -$$

$$A_{\min} \sum_{k=n+1,n+3...}^{j-1} W_{xr_k} - \sum_{k=n+2,n+4...}^{j} W_{xr_k} \quad (3.43)$$

*and, simplifying Equation 3.43, we obtain:*

$$(1 + A_{\min})W_{xr_{n+2}} + (1 + A_{\min})W_{xr_{n+1}} > (1 + A_{\min})W_{xr_n} \quad (3.44)$$

*Thus, if* $W_{xr_{n+2}} + W_{xr_{n+1}} > W_{xr_n}$ *then* $x$ *is activated for* $i = n + 2$. *Since* $W_{xr_{n+2}} = W_{xr_n} - W_{xr_{n+1}} + \varepsilon$ *and* $\varepsilon > 0$, $W_{xr_n} + \varepsilon > W_{xr_n}$. *This completes the proof of the inductive step.*

2. *If* $j$ *is an odd number and* $i$ *is an even number, show that* $x$ *is not activated:*
*Basis: If* $i = 2$ *then* $x$ *is not activated. In the worst case,* $r_1 = 1, r_2 = A_{\min}, r_3, r_5, ..., r_j = -A_{\min}$ *and* $r_4, r_6, ..., r_{j-1} = -1$. *We need to satisfy:*

$$W - A_{\min}(W - \varepsilon) - A_{\min} \sum_{k=3,5...}^{j} W_{xr_k} - \sum_{k=4,6...}^{j-1} W_{xr_k} - \theta_x < h^{-1}(-A_{\min}) \quad (3.45)$$

*which yields:*

$$\theta_x > (1 - A_{\min})W + \varepsilon A_{\min} - A_{\min} \sum_{k=3,5...}^{j} W_{xr_k} - \sum_{k=4,6...}^{j-1} W_{xr_k} - h^{-1}(-A_{\min}) \quad (3.46)$$

*and, from the* Metalevel Priorities Algorithm 3 (steps 1(a) and 1(e)), *Equation 3.46 is clearly satisfied.*
*Inductive Step: if* $x$ *is not activated for* $i = n$ *then* $x$ *is not activated for* $i = n + 2$. *For* $i = n$, *in the worst case,* $r_1, r_3, ..., r_{n-1} = 1$ *and* $r_2, r_4, ..., r_{n-2} = -1$, $r_n = A_{\min}$ *and* $r_{n+1}, ..., r_j$ *are not activated, i.e.* $r_{n+1}, r_{n+3}, ..., r_j = -A_{\min}$ *and* $r_{n+2}, r_{n+4}, ..., r_{j-1} = -1$. *If* $x$ *is not activated then*

$$\sum_{k=1,3...}^{n-1} W_{xr_k} - \sum_{k=2,4...}^{n-2} W_{xr_k} + A_{\min} W_{xr_n} - A_{\min} \sum_{k=n+1,n+3...}^{j} W_{xr_k} - \sum_{k=n+2,n+4...}^{j-1} W_{xr_k} < h^{-1}(-A_{\min}) + \theta_x \quad (3.47)$$

*holds.*
*For* $i = n + 2$ *we have* $r_1, r_3, ..., r_{n+1} = 1$ *and* $r_2, r_4, ..., r_n = -1$, $r_{n+2} = A_{\min}$ *and* $r_{n+3}, ..., r_j$ *not activated, i.e.* $r_{n+3}, r_{n+5}, ..., r_j = -A_{\min}$ *and* $r_{n+4}, r_{n+6}, ..., r_{j-1} = -1$. *We need to show that*

$$\sum_{k=1,3\ldots}^{n+1} W_{xr_k} - \sum_{k=2,4\ldots}^{n} W_{xr_k} + A_{\min} W_{xr_{n+2}} -$$

$$A_{\min} \sum_{k=n+3,n+5\ldots}^{j} W_{xr_k} - \sum_{k=n+4,n+6\ldots}^{j-1} W_{xr_k} < h^{-1}(-A_{\min}) + \theta_x \quad (3.48)$$

*also holds.*
*It is sufficient to show that:*

$$\sum_{k=1,3\ldots}^{n+1} W_{xr_k} - \sum_{k=2,4\ldots}^{n} W_{xr_k} + A_{\min} W_{xr_{n+2}} - A_{\min} \sum_{k=n+3,n+5\ldots}^{j} W_{xr_k} -$$

$$\sum_{k=n+4,n+6\ldots}^{j-1} W_{xr_k} < \sum_{k=1,3\ldots}^{n-1} W_{xr_k} - \sum_{k=2,4\ldots}^{n-2} W_{xr_k} + A_{\min} W_{xr_n} -$$

$$A_{\min} \sum_{k=n+1,n+3\ldots}^{j} W_{xr_k} - \sum_{k=n+2,n+4\ldots}^{j-1} W_{xr_k} \quad (3.49)$$

*and, simplifying Equation 3.49, we obtain:*

$$(1 + A_{\min}) W_{xr_{n+2}} + (1 + A_{\min}) W_{xr_{n+1}} < (1 + A_{\min}) W_{xr_n} \quad (3.50)$$

*Thus, if* $W_{xr_{n+2}} + W_{xr_{n+1}} < W_{xr_n}$ *then* $x$ *is not activated for* $i = n + 2$. *Now,* $W_{xr_{n+2}} = W_{xr_n} - W_{xr_{n+1}} - \varepsilon$, *and since* $\varepsilon > 0$, $W_{xr_n} - \varepsilon < W_{xr_n}$. *This completes the proof of the inductive step.*

3. *If* $j$ *is an even number and* $i$ *is an odd number, show that* $x$ *is activated:* *Basis: If* $i = 1$ *then* $x$ *is activated. In the worst case,* $r_3, r_5, \ldots, r_{j-1} = -1$ *and* $r_2, r_4, \ldots, r_j = -A_{\min}$, *while* $r_1 = A_{\min}$. *We need to satisfy:*

$$A_{\min} W - A_{\min} \sum_{k=2,4\ldots}^{j} W_{xr_k} - \sum_{k=3,5\ldots}^{j-1} W_{xr_k} - \theta_x > h^{-1}(A_{\min}) \quad (3.51)$$

*which yields:*

$$\theta_x < A_{\min} W - A_{\min} \sum_{k=2,4\ldots}^{j} W_{xr_k} - \sum_{k=3,5\ldots}^{j-1} W_{xr_k} - h^{-1}(A_{\min}) \quad (3.52)$$

*and, from the* Metalevel Priorities Algorithm 3 (steps 1(b) and 1(f)), *Equation 3.52 is clearly satisfied. Note that* $W_{\neg xr_k} = W_{xr_{k-1}}$ *for* $2 \leq k \leq j$.
*Inductive Step: Identical to Case 1.*

4. *If $j$ is an even number and $i$ is an even number, show that $x$ is not activated:*
*Basis: If $i = 2$ then $x$ is not activated. In the worst case, $r_1 = 1, r_2 = A_{\min}, r_3, r_5, ..., r_{j-1} = -A_{\min}$ and $r_4, r_6, ..., r_j = -1$. We need to satisfy:*

$$W - A_{\min}(W - \varepsilon) - A_{\min} \sum_{k=3,5...}^{j-1} W_{xr_k} -$$

$$\sum_{k=4,6...}^{j} W_{xr_k} - \theta_x < h^{-1}(-A_{\min}) \quad (3.53)$$

*which yields:*

$$\theta_x > (1 - A_{\min})W + \varepsilon A_{\min} - A_{\min} \sum_{k=3,5...}^{j-1} W_{xr_k} -$$

$$\sum_{k=4,6...}^{j} W_{xr_k} - h^{-1}(-A_{\min}) \quad (3.54)$$

*and, from the* Metalevel Priorities Algorithm 3 (steps 1(b) and 1(f)), *Equation 3.54 is clearly satisfied. Again, $W_{\neg xr_k} = W_{xr_{k-1}}$ for $2 \leq k \leq j$.*
*Inductive Step: Identical to Case 2.*
5. *Finally, we consider the special case in which no neuron in $R$ is activated: If $r_1, r_2, ..., r_j$ are not activated then $x$ is not activated. If $j$ is an odd number, in the worst case, $r_1, r_3, ..., r_j = -A_{\min}$ and $r_2, r_4, ..., r_{j-1} = -1$. Otherwise, $r_1, r_3, ..., r_{j-1} = -A_{\min}$ and $r_2, r_4, ..., r_j = -1$. This yields, respectively, Equations 3.55 and 3.56.*

$$\theta_x > -A_{\min} \sum_{k=1,3...}^{j} W_{xr_k} - \sum_{k=2,4...}^{j-1} W_{xr_k} - h^{-1}(-A_{\min}) \quad (3.55)$$

$$\theta_x > -A_{\min} \sum_{k=1,3...}^{j-1} W_{xr_k} - \sum_{k=2,4...}^{j} W_{xr_k} - h^{-1}(-A_{\min}) \quad (3.56)$$

*But if $\theta_x$ satisfies Equation 3.46 then $\theta_x$ also satisfies Equation 3.55, and if $\theta_x$ satisfies Equation 3.54, it also satisfies Equation 3.56. This is so because $(1 - A_{\min})W + \varepsilon A_{\min} > -A_{\min}W + W - \varepsilon$. This completes the proof.*

*The proof for output neuron $\neg x$ is analogous.* □

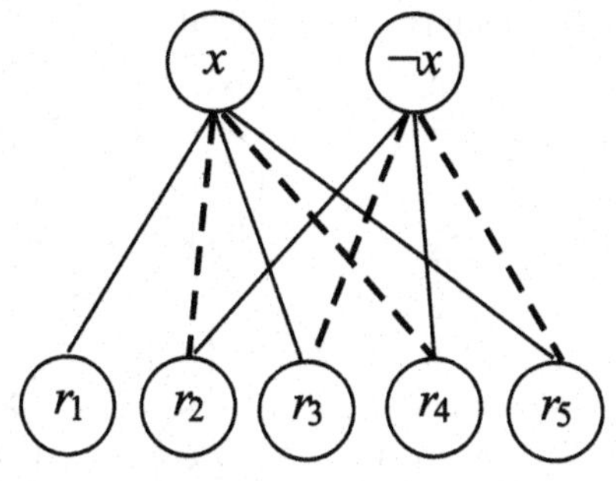


**Fig. 3.9.** Implementing a linear ordering in a neural network. Dotted lines indicate negative weights.

*Example 3.5.6.* Let $r_5 \succ ... \succ r_2 \succ r_1$ and assume that the consequent of $r_1$ is $x$, the consequent of $r_2$ is $\neg x$, the consequent of $r_3$ is $x$, and so on. Taking $A_{\min} = 0.9$, $W = 20$ and $\varepsilon = 0.1$, we calculate $-61.06 < \theta_1 < -53.13$ and $29.03 < \theta_2 < 46.97$. $W_{xr_1} = W_{\neg xr_2} = 20$, $W_{xr_2} = W_{\neg xr_3} = -19.9$, $W_{xr_3} = W_{\neg xr_4} = 40$, $W_{xr_4} = W_{\neg xr_5} = -60$, and $W_{xr_5} = 100.1$. Taking $\theta_x = -55$ and $\theta_{\neg x} = 45$, the network of Figure 3.9 will implement $r_5 \succ ... \succ r_2 \succ r_1$. Note that, if a smaller value for $A_{\min}$ is desired, a larger value for $W$ may be necessary in order to satisfy the constraints on $\theta_x$ and $\theta_{\neg x}$.

We now investigate several well-known examples, extracted from [ABM98], in order to identify to which extent neural networks can represent metalevel priorities.[20]

*Example 3.5.7.* **(The Nixon Diamond)**

$quaker(nixon)$
$republican(nixon)$
$r_1 : quaker(x) \rightarrow \neg militarist(x)$
$r_2 : republican(x) \rightarrow militarist(x)$

If the superiority relation is empty then we derive $militarist\ (nixon)$ and $\neg militarist(nixon)$. But if we add $r_2 \succ r_1$ then, by using *Metalevel Priorities Algorithm 1*, we can conclude $militarist(nixon)$.

*Example 3.5.8.* **(The birds example)**

$bird(tweety)$
$penguin(opus)$
$r_1 : penguin(x) \rightarrow bird(x)$
$r_2 : bird(x) \rightarrow flies(x)$
$r_3 : penguin(x) \rightarrow \neg flies(x)$

[20] Rules with free variables are interpreted as rule schemas, that is, as the set of all ground instances.

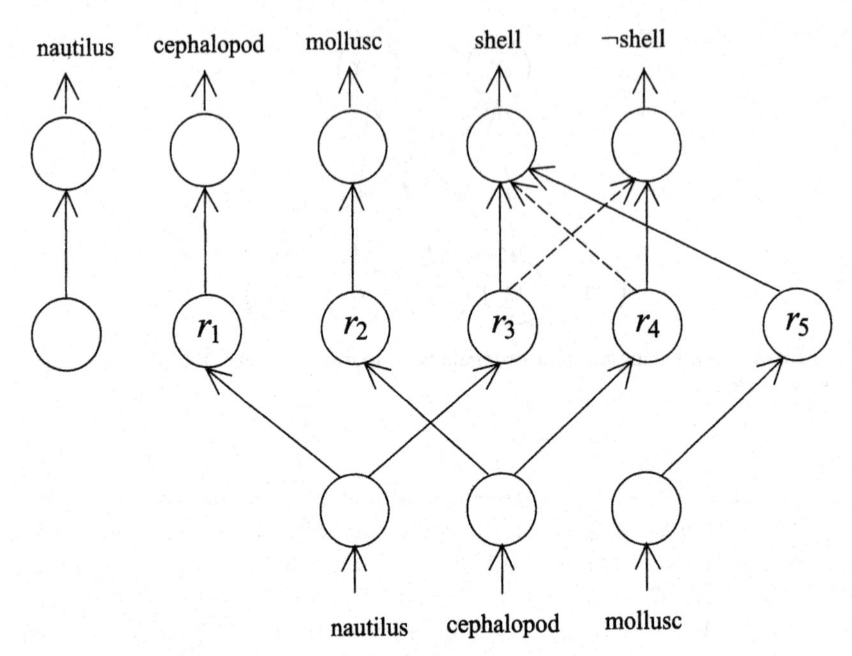


**Fig. 3.10.** The case of exceptions of exceptions.

$r_3 \succ r_2$

Here we can prove $flies(tweety)$ and $\neg flies(opus)$, by applying *Metalevel Priorities Algorithm 1* twice (for tweety and opus).

*Example 3.5.9.* **(Hierarchies with exceptions)**

$nautilus(nancy)$
$r_1 : nautilus(x) \rightarrow cephalopod(x)$
$r_2 : cephalopod(x) \rightarrow mollusc(x)$
$r_3 : nautilus(x) \rightarrow shell(x)$
$r_4 : cephalopod(x) \rightarrow \neg shell(x)$
$r_5 : mollusc(x) \rightarrow shell(x)$
$r_4 \succ r_5$
$r_3 \succ r_4$

Here we can derive $shell(nancy)$ using *Metalevel Priorities Algorithm 3* (see Figure 3.10).

*Example 3.5.10.* **(No propagation of ambiguity)**

$quaker(nixon)$
$republican(nixon)$
$r_1 : quaker(x) \rightarrow pacifist(x)$
$r_2 : republican(x) \rightarrow \neg pacifist(x)$

$r_3 : republican(x) \rightarrow footballfan(x)$
$r_4 : pacifist(x) \rightarrow antimilitary(x)$
$r_5 : footballfan(x) \rightarrow \neg antimilitary(x)$
$r_5 \succ r_4$

Here we can prove $\neg antimilitary(nixon)$, despite the conflict regarding Nixon's pacifism. We use *Metalevel Priorities Algorithm 1.* Note that neither propagation of ambiguity nor trivialisation of theory occur in this example. If, on the other hand, the superiority relation were empty, an ambiguity about Nixon's militarism would exist.

*Example 3.5.11.* **(Partial ordering)**
$laysEggs(platypus)$
$hasFur(platypus)$
$monotreme(platypus)$
$hasBill(platypus)$
$r_1 : monotreme(x) \rightarrow mammal(x)$
$r_2 : hasFur(x) \rightarrow mammal(x)$
$r_3 : layEggs(x) \rightarrow \neg mammal(x)$
$r_4 : hasBill(x) \rightarrow \neg mammal(x)$
$r_1 \succ r_3$
$r_2 \succ r_4$

Intuitively, we should be able to derive $mammal(platypus)$ because for every reason against $mammal(platypus)$, namely $r_3$ and $r_4$, there is a stronger reason for $mammal(platypus)$, respectively, $r_1$ and $r_2$. However, in order to achieve such a result in the network, one needs to explicitly state that $r_1 \succ r_3, r_1 \succ r_4$ and $r_2 \succ r_3, r_2 \succ r_4$. This is so because, as we will see in Section 5.4, $(a \wedge b) \vee (c \wedge d) \rightarrow x$ cannot be represented in a network without hidden layers. Since $r_1 \succ r_3$ and $r_2 \succ r_4$ can be rewritten as $(r_3 \wedge \sim r_1) \vee (r_4 \wedge \sim r_2) \rightarrow \neg mammal(x)$, and we work with hidden to output connections in order to encode priorities, trying to represent $r_1 \succ r_3$ and $r_2 \succ r_4$ in the network is equivalent to trying to encode $(a \wedge b) \vee (c \wedge d) \rightarrow x$ in a network without hidden layers. On the other hand, $r_1 \succ r_3, r_1 \succ r_4$ and $r_2 \succ r_3, r_2 \succ r_4$ can be encoded in the network using *Metalevel Priorities Algorithm 2* such that $mammal(platypus)$ is derived.

In this section we have seen that negative weights from the hidden layer to the output of a single hidden layer neural network may be regarded as the implementation of metalevel priorities, which define a superiority relation between conflicting rules of an extended logic program. In this way, the implementation of such a superiority relation in the network is straightforward. More complex and elaborated preference relations, for example when one is allowed to define priorities among literals that are not complementary, or even among literals that are not in the head of the rules, or when the preference relation is any partial ordering, cannot be represented in the network as easily, and would require changes in the basic neural structure investigated here. Such more sophisticated preference relations are left as future work.

## 3.6 Summary and Further Reading

In this chapter, we have presented the Connectionist Inductive Learning and Logic Programming System ($C\text{-}IL^2P$) – a massively parallel computational model, based on artificial neural networks, that integrates inductive learning from examples and background knowledge, with deductive learning from Logic Programming.

We have also presented an extension of $C\text{-}IL^2P$ that defines a massively parallel model for extended logic programming. Consequently, some facts of commonsense knowledge can be represented more easily, since classical negation is available.

In a rather satisfying way, the Translation Algorithm of $\mathcal{P}$ into $\mathcal{N}$ associates each hidden neuron of $\mathcal{N}$ with a rule of $\mathcal{P}$. Such a representation led to yet another extension of $C\text{-}IL^2P$, in which negative connections from hidden to output layers are used to inhibit the presence of a neuron in the answer set of $\mathcal{P}$, thereby acting as metalevel priorities.

Summarising, we have shown that single hidden layer neural networks can represent, in a very natural way, the class of extended logic programs augmented with metalevel priorities given by a superiority relation. The question whether this is the language that single hidden layer networks can learn remains without an answer, and is an interesting theoretical result to pursue.

Both kinds of Intelligent Computational Systems, Symbolic and Connectionist, have virtues and deficiencies. Research into the integration of the two has important implications [Hil95], in that one is able to benefit from the advantages that each confers. We believe that our approach contributes to this area of research.

In the next chapter, we will present the experimental results obtained by applying $C\text{-}IL^2P$ to two real-world problems in the domain of molecular biology, and to a simplified version of a real power plant fault diagnosis system.

Many different systems that combine background knowledge and learning from examples – ranging from *Bayesian Networks* [Mit97] to *Neuro-Fuzzy Systems* [CZ00] – have been proposed in the last decade. Among those that combine *Mathematical Logic* [End72] and *Connectionist Models* [Hay99a], the work of Pinkas [Pin91, Pin95] and Holldobler [Hol90, HK92, Hol93] are particularly relevant. In an attempt to show the capabilities and limitations of neural networks at performing logical inference, Pinkas defined a bi-directional mapping between symmetric neural networks[21] and mathematical logic. He presented a theorem showing the equivalence between the problem of satisfiability of propositional logic and minimising the energy function associated with symmetric networks: for every *well-formed formula (wff),* a quadratic energy function can efficiently be found, and for every energy function there exists a *wff* (inefficiently found) such that the global minima of the function are exactly equal to the satisfying models of the formula. Holldobler presented a parallel unification algorithm and an automated reasoning system for first-order Horn clauses, implemented in a feedforward neural network, called the *Connectionist Horn Clause Logic (CHCL) System.*

The relations between neural networks and nonmonotonicity can be traced back to [BG91], in which Balkenius and Gardenfors have shown that symmetric neural networks can compute a form of nonmonotonic inference relation, satisfying some of the most fundamental postulates for Nonmonotonic Logic. Gardenfors went on to produce his theory of conceptual representations [Gar00], according to which a bridge between the symbolic and connectionist approaches to cognitive science could be built from geometric structures based on a number of quality dimensions such as temperature, weight and brightness, as well as the spatial dimensions. Nonmonotonic Neural-Symbolic Learning Systems are, to the best of our knowledge, restricted to the *C-IL²P* system and, more recently, Boutsinas and Vrahatis' *Artificial Nonmonotonic Neural Networks* (*ANNN*) [BV01], which use a multiple inheritance scheme with exceptions to learn inheritance networks. Although Pinkas' more recent work [Pin95] deals with the representation of *Penalty Logic*, a kind of nonmonotonic logic, in symmetric neural networks, the task of learning (penalties) from examples in such networks has not been investigated to the same extent as the task of representation. Finally, *Neural Networks* have been combined with *Explanation-based Learning (EBL)* systems [MT93b] into *Explanation-based Neural Networks (EBNN)* [Thr96], *EBL* has been combined with *Inductive Logic Programming (ILP)* [MZ94] (see [LD94] and [MR94] for introductions to *ILP*), and the relations between *ILP* and *Neural Networks* start to be unravelled [BZB01].

[21] A neural network is called *symmetric* if $W_{ij} = W_{ji}$ for all neurons $i, j$ (e.g. Hopfield networks [Hop82]).

# 4. Experiments on Theory Refinement

In this chapter, we apply the *C-IL²P* system to two problems of DNA classification, which have become benchmark data sets for testing the accuracy of machine learning systems. We compare the results obtained by different neural, symbolic and hybrid inductive learning systems. For example, the test-set performance of *C-IL²P* is at least as good as those of *KBANN* and *Backpropagation*, while *C-IL²P's* training-set performance is considerably superior to *KBANN* and *Backpropagation*. We also apply *C-IL²P* to fault diagnosis, using a simplified version of a real power generator plant. In this application, we use the system extended with classical negation. We then compare *C-IL²P* with *Backpropagation*, using three different architectures. The results corroborate the importance of the background knowledge for learning in the presence of noisy data sets.

## 4.1 DNA Sequence Analysis

Molecular Biology is an area of increasing interest for computational learning systems analysis and application. Specifically, *DNA* sequence analysis problems have recently become a benchmark for learning systems' performance comparison. We have applied the *C-IL²P* system to two real-world problems in the domain of Molecular Biology, in particular the *Promoter Recognition* and *splice junction Determination* problems of DNA sequence analysis[1]. In this section, we compare the experimental results obtained by *C-IL²P* with a variety of learning systems.

In what follows we briefly introduce the problems in question from a computational application perspective (see [WHR+87] for a fuller treatment on the subject). A DNA molecule contains two strands that are linear sequences of nucleotides. The DNA is composed from four different nucleotides – *adenine*, *guanine*, *thymine*, *cytosine* – which are abbreviated by $a, g, t, c$, respec-

[1] These are the same problems that were investigated in [TS94a] for the evaluation of *KBANN*. We have followed as much as possible the methodology used by Towell and Shavlik, and we have used the same background knowledge and set of examples as *KBANN*.

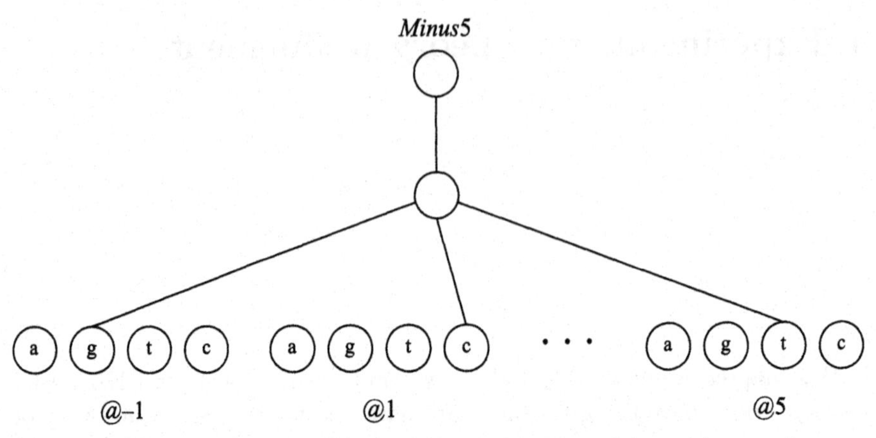


**Fig. 4.1.** Inserting rule $Minus5 \leftarrow @-1`gc', @5`t'$ into the neural network.

tively. Some sequences of the DNA strand, called genes, serve as a blueprint for the synthesis of proteins. Interspersed among the genes are segments, called non-coding regions, that do not encode proteins.

Following [TS94a], we use a special notation to identify the location of nucleotides in a DNA sequence. Each nucleotide is numbered with respect to a fixed, biologically meaningful, reference point. Rule antecedents of the form "*@3 atcg*" state the location relative to the reference point in the DNA, followed by the sequence of symbols that must occur. For example, "*@3 atcg*" means that an $a$ must appear three nucleotides to the right of the reference point, followed by a $t$ four nucleotides to the right of the reference point, and so on. By convention, location zero is not used, while '$*$' means that any nucleotide will suffice in a particular location. In this way, a rule of the form $Minus35 \leftarrow @-36`ttg_*ca'$ is a short representation for $Minus35 \leftarrow @-36`t'$, $@-35`t'$, $@-34`g'$, $@-32`c'$, $@-31`a'$. Each location is encoded in the network by four input neurons, representing nucleotides $a$, $g$, $t$ and $c$, in this order. The rules are, therefore, inserted in the network as depicted in Figure 4.1 for the hypothetical rule $Minus5 \leftarrow @-1`gc'$, $@5`t'$.

In addition to the reference point notation, Table 4.1 specifies a standard notation for referring to all possible combinations of nucleotides using a single letter. This notation is compatible with the EMBL, GenBank and PIR libraries – three major collections of molecular biology data.

The first application in which we test *C-IL²P* is the prokaryotic[2] promoter recognition. Promoters are short DNA sequences that precede the beginning of genes. The aim of *promoter recognition* is to identify the starting location of

[2] Prokaryotes are single-celled organisms that do not have a nucleus, e.g. *E. Coli*.

genes in long sequences of DNA. Table 4.2 contains the background knowledge for this task.[3]

**Table 4.1.** Single-letter codes for expressing uncertain DNA sequence.

| Code | Meaning | Code | Meaning | Code | Meaning |
|---|---|---|---|---|---|
| *m* | *a or c* | *r* | *a or g* | *W* | *a or t* |
| *s* | *c or g* | *y* | *c or t* | *K* | *g or t* |
| *v* | *a or c or g* | *h* | *a or c or t* | *D* | *a or g or t* |
| *b* | *c or g or t* | *x* | *a or g or c or t* | | |

The background knowledge of Table 4.2 is translated by *C-IL²P's* Translation Algorithm to the neural network of Figure 4.2. In addition, two hidden neurons are added in order to facilitate the learning of new concepts from examples. Note that the network is fully connected, but low-weighted links are not shown in the figure. The network's input vector for this task contains 57 consecutive DNA nucleotides. The training examples consist of 53 promoter and 53 nonpromoter DNA sequences.

**Table 4.2.** Background knowledge for promoter recognition.

$Promoter \leftarrow Contact, Conformation$
$Contact \leftarrow Minus10, Minus35$
$Minus10 \leftarrow @ - 14`tataat\text{'}$; $Minus35 \leftarrow @ - 37`cttgac\text{'}$
$Minus10 \leftarrow @ - 13`tataat\text{'}$; $Minus35 \leftarrow @ - 36`ttgaca\text{'}$
$Minus10 \leftarrow @ - 13`ta_*a_*t\text{'}$; $Minus35 \leftarrow @ - 36`ttgac\text{'}$
$Minus10 \leftarrow @ - 12`ta_{***}t\text{'}$; $Minus35 \leftarrow @ - 36`ttg_*ca\text{'}$
$Conformation \leftarrow @ - 45`aa_{**}a\text{'}$
$Conformation \leftarrow @ - 45`a_{***}a\text{'}, @ - 28`t_{***}t_*aa_{**}t\text{'}, @ - 4`t\text{'}$
$Conformation \leftarrow @ - 49`a_{****}t\text{'}, @ - 27`t_{****}a_{**}t_*tg\text{'}, @ - 1`a\text{'}$
$Conformation \leftarrow @ - 47`caa_*tt_*ac\text{'}, @ - 22`g_{***}t_*c\text{'}, @ - 8`gcgcc_*cc\text{'}$

The second application that we use to test *C-IL²P* is eukaryotic[4] splice junction determination. Splice junctions are points on a DNA sequence at which the non-coding regions are removed during the process of protein synthesis. The aim of *splice junction determination* is to recognise the boundaries between the parts of the DNA that are retained – called exons – and the parts that are removed – called introns. The task consists therefore of recognising

[3] Rules obtained from [TS94a], and derived from the biological literature [O'N89] from Noordewier [TSN90].

[4] Unlike prokaryotic cells, eukaryotic cells contain a nucleus, and so are higher up the evolutionary scale.

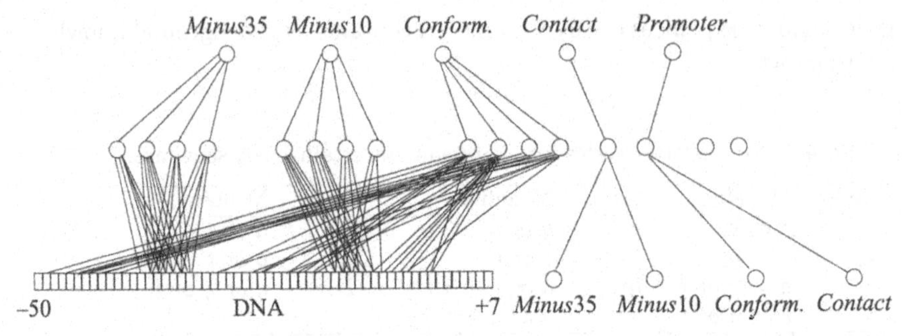


**Fig. 4.2.** Initial neural network for promoter recognition. Each box at the input layer represents one sequence location that is encoded by four input neurons $\{a, g, t, c\}$.

exon/intron (E/I) boundaries and intron/exon (I/E) boundaries. Table 4.3 contains the background knowledge for splice junction determination[5].

**Table 4.3.** Background knowledge for splice junction.

| | | |
|---|---|---|
| $EI \leftarrow @-3$'*maggtragt*', $\sim EI\_Stop$ | | |
| $EI\_Stop \Leftarrow @-3$'*taa*'; | $EI\_Stop \Leftarrow @-4$'*taa*'; | $EI\_Stop \Leftarrow @-5$'*taa*' |
| $EI\_Stop \Leftarrow @-3$'*tag*'; | $EI\_Stop \Leftarrow @-4$'*tag*'; | $EI\_Stop \Leftarrow @-5$'*tag*' |
| $EI\_Stop \Leftarrow @-3$'*tga*'; | $EI\_Stop \Leftarrow @-4$'*tga*'; | $EI\_Stop \Leftarrow @-5$'*tga*' |
| $IE \leftarrow pyramidine\_rich, @-3$'*yagg*', $\sim IE\_Stop$ | | |
| $pyramidine\_rich \leftarrow 6\ of\ (@-15$'*yyyyyyyyyy*') | | |
| $IE\_Stop \Leftarrow @1$'*taa*'; | $IE\_Stop \Leftarrow @2$'*taa*'; | $IE\_Stop \Leftarrow @3$'*taa*' |
| $IE\_Stop \Leftarrow @1$'*tag*'; | $IE\_Stop \Leftarrow @2$'*tag*'; | $IE\_Stop \Leftarrow @3$'*tag*' |
| $IE\_Stop \Leftarrow @1$'*tga*'; | $IE\_Stop \Leftarrow @2$'*tga*'; | $IE\_Stop \Leftarrow @3$'*tga*' |

The background knowledge of Table 4.3 is translated by *C-IL²P* to the neural network of Figure 4.3. Rules containing symbols other than the original $(a, g, t, c)$ are split into a number of equivalent rules containing only the original symbols, according to Table 4.1. For example, since $y \equiv c \vee t$, the rule $IE \leftarrow pyramidine\_rich, @-3$'*yagg*', $\sim IE\_Stop$ is encoded in the network as $IE \leftarrow pyramidine\_rich, @-3$'*cagg*', $\sim IE\_Stop$ and $IE \leftarrow pyramidine\_rich, @-3$'*tagg*', $\sim IE\text{-}Stop$. In Table 4.3, $\Leftarrow$ indicates nondefeasible rules, which cannot be altered during training. Therefore, the weights and thresholds set in the network by these rules are fixed, and learning is disabled. Rule $pyramidine\_rich \leftarrow 6\ of\ (@-15$'*yyyyyyyyyy*') indicates that a DNA sequence is pyramidine-rich if 6 or more nucleotides $y$ appear from location $-15$ to $-6$. Generically, rules of the form "$x \leftarrow m\ of\ (y_1, ..., y_n)$",

[5] Rules obtained from [TS94a] and derived from the biological literature from Noordewier [NTS91].

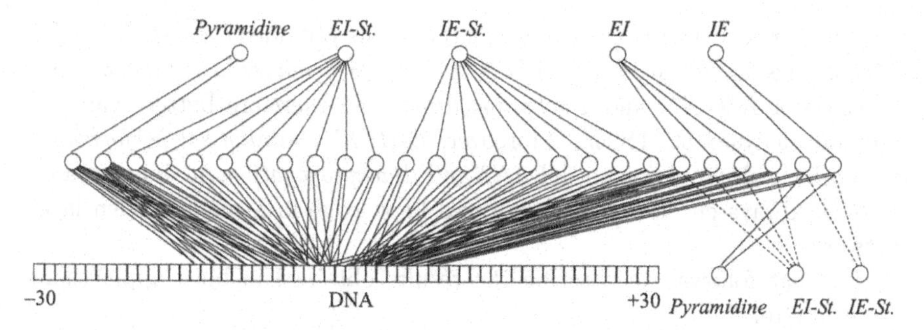


**Fig. 4.3.** Initial neural network for splice junction determination. Each box at the input layer of the network represents one sequence location, which is encoded by four input neurons $\{a, g, t, c\}$.

called $m\ of\ n$ rules, are satisfied if at least $m$ of $(y_1, ..., y_n)$ are *true*. The translation of $m\ of\ n$ rules to the network is done by simply making $k_l = m$ in *C-IL²P*'s Translation Algorithm.[6]

The training set for the task of splice junction determination contains 3190 examples, with approximately 25% of I/E boundaries, 25% of E/I boundaries and 50% of neither. The third category (neither E/I nor I/E) is considered true when neither output neuron I/E nor output neuron E/I is active. Each training example is a DNA sequence with 60 nucleotides, where the centre is the reference point. Remember that the network of Figure 4.3 is fully connected, but that low-weighted links are not shown. Dotted lines indicate links with negative weights.

In both applications, unless stated otherwise, the background knowledge is assumed defeasible, i.e. the weights are allowed to change during the learning process. Hence, some of the background knowledge may be revised by the training examples. Note, however, that *C-IL²P*'s recurrent connections are responsible for reinforcing the background knowledge during training. For instance, in the network of Figure 4.3, the concepts *Pyramidine*, *EI-St.* and *IE-St.*, called intermediate concepts, have their input values calculated by the network in action, according to the background knowledge and to the DNA sequence input vector.

[6] $m\ of\ (y_1, ..., y_n)$ is a compact representation of the disjunct $c_1 \vee c_2 \vee ... \vee c_j$, where each $c_i$ is a conjunction of $m$ distinct terms chosen from $(y_1, ..., y_n)$, and $j$ is the number of combinations of $m$ distinct objects chosen from $n$ distinct objects, namely $C_n^m$. For example, $2(a, b, c)$ represents $ab \vee ac \vee bc$.

When inserting $x \leftarrow 2(a, b, c)$ into a neural network, instead of creating three hidden neurons, one for each rule $x \leftarrow ab$, $x \leftarrow ac$ and $x \leftarrow bc$, one may simply connect input neurons $a, b, c$ to a single hidden neuron $N_l$, and use $k_l = 2$ in the Translation Algorithm. In this way, the threshold of $N_l$ will be such that any two of $\{a, b, c\}$ activates $N_l$ – the intended meaning of $2(a, b, c)$.

Let us now describe the experimental results obtained by *C-IL²P* in the applications above. Briefly, *C-IL²P*'s *test set performance* is at least as good as those of *KBANN* and *Backpropagation*, and therefore better than any method analysed in [TS94a]. Moreover, *C-IL²P*'s *training set performance* is considerably superior to *KBANN* and *Backpropagation*, mainly because *C-IL²P* always encodes the background knowledge in a single hidden layer network.

In what follows, we describe the results obtained in both applications according to:

- Test set performance (percentage of test set examples correctly classified);
- Test set error rate given smaller training sets (the training set is partitioned into sets ($s$) of increasing sizes and, for each $s$, the percentage of test set examples wrongly classified is calculated); and
- Training set performance (we use the root mean square (RMS) error rate decay[7] to measure how well and how fast the network has learned the training set examples).

Firstly, let us consider *C-IL²P*'s test-set performance, i.e. its ability to generalise over examples not seen during training. We compare the results obtained by *C-IL²P* in both applications with some of the main inductive learning systems from examples: Backpropagation [RHW86], Perceptron [Ros62] (neural systems), ID3 [Qui86] and Cobweb [Fis87] (symbolic systems). We also compare the results in the promoter recognition problem with a method suggested by biologists [Sto90]. In addition, we compare *C-IL²P* with systems that learn from both examples and background knowledge: Either [OM94], Labyrinth-K [TLI91], FOCL [PK92] (symbolic systems), and *KBANN* [TS94a] (hybrid system).[8]

As in [TS94a], we evaluate the systems using *cross-validation,* a methodology that divides the set of examples into $n$ sets so that one division is used for testing and the remaining $n-1$ divisions are used for training. This is repeated $n$ times, allowing every division to be used once for testing. For the 106-examples promoter data set, we use leaving-one-out cross-validation, in which each example is successively left out of the training set. Hence, it requires 106 training phases, in which the training set has 105 examples and the testing set has 1 example. Since leaving-one-out becomes prohibitive as the number of available examples grows, following [TS94a], we use 10-fold cross-validation for the splice junction determination data set. As in [TS94a],

[7] The RMS error computes the average deviation between target vectors $\mathbf{t}^i$ $(1 \leq i \leq N)$ and output vectors $\mathbf{o}^i$ such that: $RMS\_Error = \sqrt{\frac{1}{N}\sum_{i=1}^{N} \|\mathbf{t}^i - \mathbf{o}^i\|^2}$.

[8] Towell and Shavlik compared *KBANN* with other hybrid systems, such as [Fu89] and [Kat89], obtaining better results.

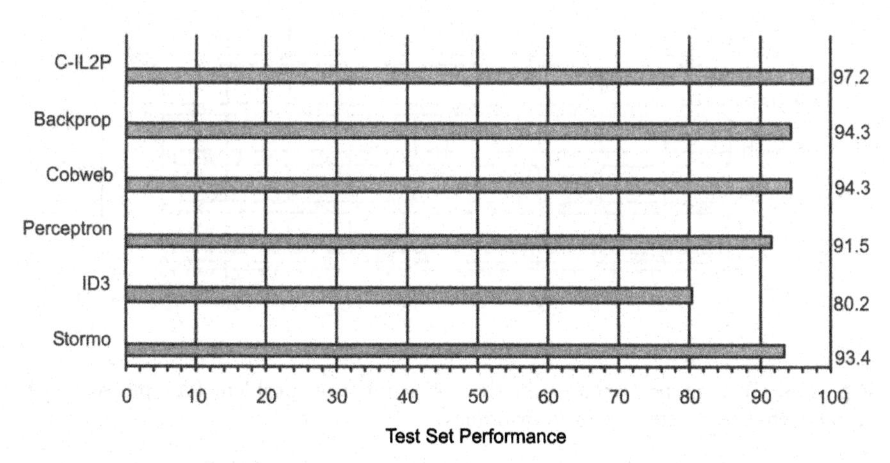


**Fig. 4.4.** Test-set performance in the promoter problem (comparison with systems that learn strictly from examples).

we randomly select 1000 examples from the splice junction data set and perform 10 distinct training phases. In each phase, the weights that are not predefined by *C-IL²P*'s *Translation Algorithm* are initialised randomly in the range $[-1, 1]$. In addition, we use a *batch* updating procedure during training, that is, the weights are updated only after each *epoch* (one presentation of the complete training set).

Again as in [TS94a], the learning systems that are based on neural networks have been trained until one of the following three *stopping criteria* was satisfied: (1) on 99% of the training examples, the activation of every output unit is within 0.10 of correctness; (2) every training example is presented to the network 100 times, i.e. the network has been trained for 100 *epochs*; (3) the network classifies at least 90% of the training examples correctly, but has not improved its classification ability for 5 epochs. We used the standard backpropagation learning algorithm with momentum to train *C-IL²P* networks. The momentum constant was set to 0.9, and the learning rate was kept small, typically 0.1. In both applications, stopping criterion (1) was satisfied ahead of (2) and (3), i.e. training time was not a problem.

*C-IL²P* generalises better than any empirical learning system (see Figures 4.4 and 4.5) and better than any system that learns from examples and background knowledge (see Figures 4.6 and 4.7) tested on both applications. However, *C-IL²P* is only marginally better than *Backpropagation* in the splice junction problem. This is because both systems are based upon the same learning algorithm and have been trained using the same architecture. The superior performance of *C-IL²P* in the promoter problem must thus be due to the presence of the background knowledge.

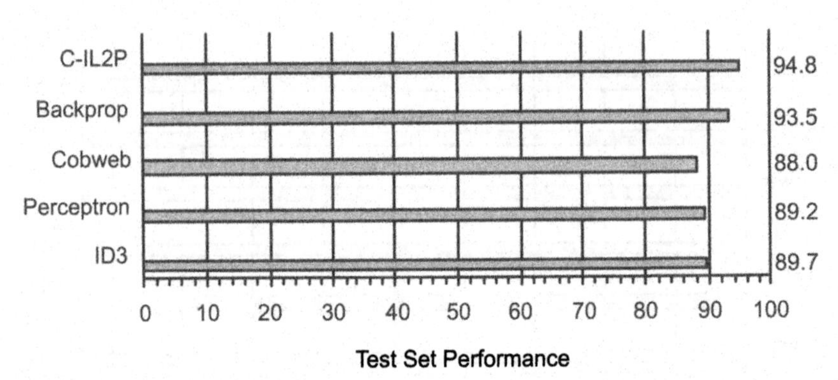


**Fig. 4.5.** Test-set performance in the splice junction problem (comparison with systems that learn strictly from examples).

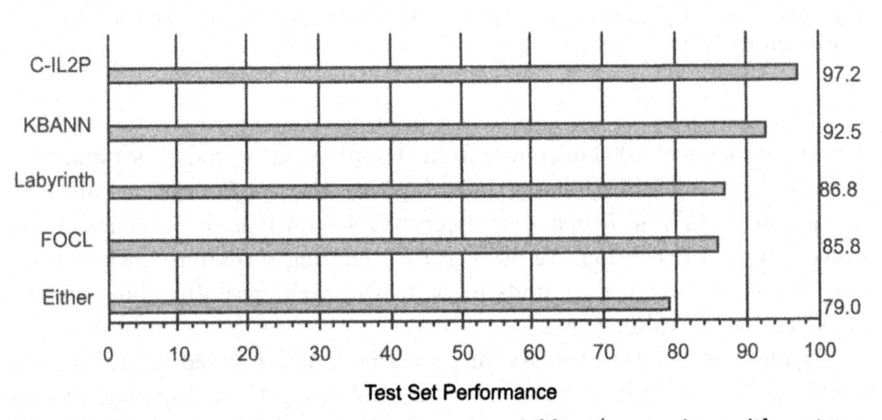


**Fig. 4.6.** Test-set performance in the promoter problem (comparison with systems that learn both from examples and theory).

Background knowledge should help a learning system to produce good generalisations from smaller sets of examples. In fact, theory and data learning systems normally require fewer training examples than systems that learn only from data. This is quite important since, in general, it is not easy to obtain large and accurate training sets. Hence let us now analyse *C-IL²P*'s test-set performance given smaller sets of examples. The following tests will compare the performance of *C-IL²P* with *KBANN* and *Backpropagation* only, because these systems have shown to be the most effective ones in the previous tests (see Figures 4.4, 4.5, 4.6 and 4.7).

Following [TS94a], the generalisation ability over small sets of examples is analysed by splitting the examples into two subsets: the testing set containing approximately 25% of the examples, and the training set containing the remaining examples. The training set $S$ is then partitioned into subsets

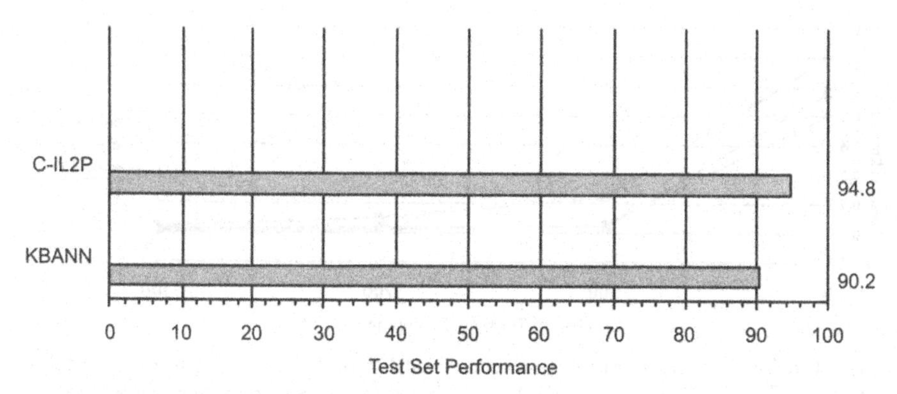


**Fig. 4.7.** Test-set performance in the splice junction problem (comparison with systems that learn both from examples and theory).

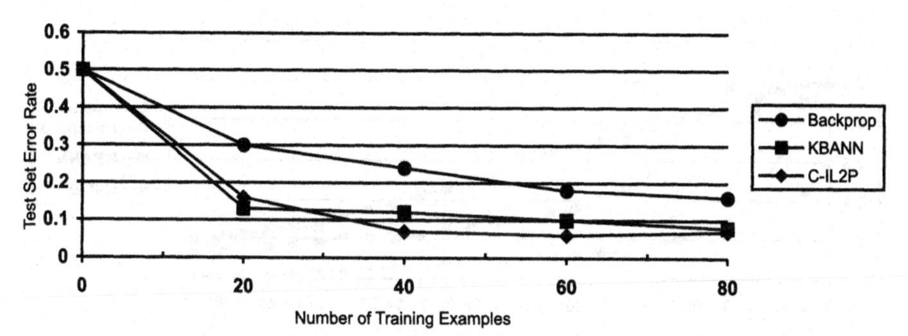


**Fig. 4.8.** Test-set error rate in the promoter problem (26 examples reserved for testing). For *C-IL²P* and *KBANN*, zero training examples represents the accuracy of the background knowledge w.r.t. the test set. For *Backpropagation*, zero training examples represents random guess.

$S_1, S_2, ..., S$ of increasing sizes ($S_1 \subset S_2 \subset ... \subset S$) and the networks are trained using each subset at a time.

Figures 4.8 and 4.9 show that, in both applications, *C-IL²P* generalises over small sets of examples better than *Backpropagation*. The results empirically show that the initial set of weights of the network, defined by the background knowledge, gives it a better generalisation capability in the presence of fewer training examples (compare, e.g. the results obtained by *C-IL²P* and *Backpropagation* in Figure 4.8 for sets of 20 and 40 training examples). Note that the results obtained by *C-IL²P* and *KBANN* are very similar, since both systems use the same background knowledge. Nevertheless, in the splice junction problem, the initial *C-IL²P* network (before any training takes place) has an accuracy superior to that of the initial *KBANN* network (see

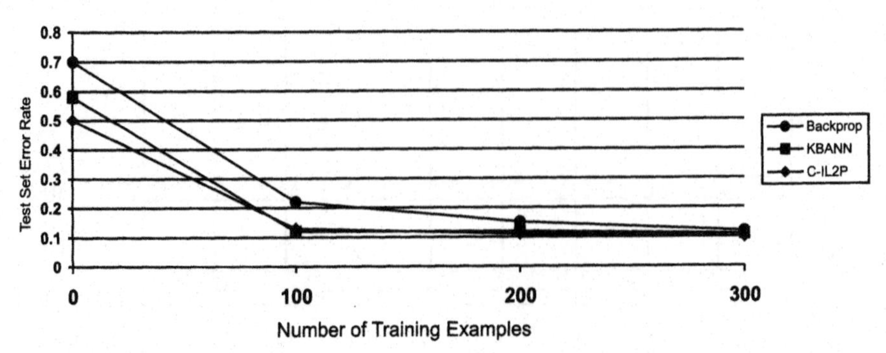


**Fig. 4.9.** Test-set error rate in the splice junction problem (798 examples reserved for testing). For *C-IL²P* and *KBANN*, zero training examples represents the accuracy of the background knowledge w.r.t. the test set. For *Backpropagation*, zero training examples represents random guess.

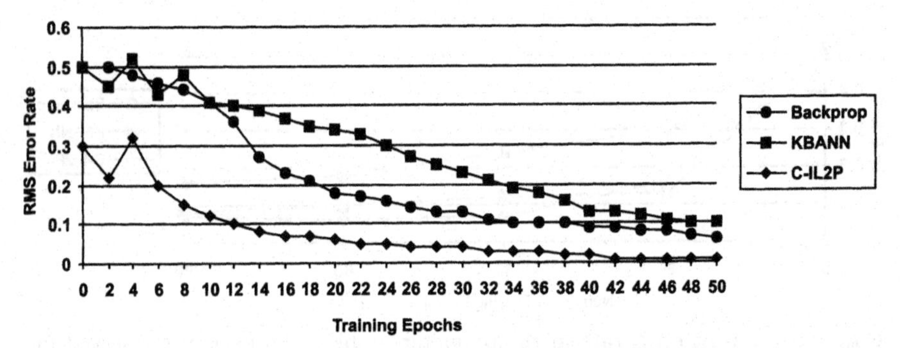


**Fig. 4.10.** Training-set RMS error decay during learning the promoter problem. For *C-IL²P* and *KBANN*, zero training epochs represents the accuracy of the background knowledge w.r.t. the training set. For *Backpropagation*, zero training epochs represents random guess.

Figure 4.9). This is probably due to the unsoundness of *KBANN*'s translation algorithm, as discussed in Section 3.1.

To conclude the tests, we check the training-set performance of *C-IL²P* in comparison again with *KBANN* and *Backpropagation*. Figures 4.10 and 4.11 describe the training-set RMS error rate decay obtained by each system during learning, respectively, in each application. The RMS parameter indicates how fast a neural network learns a set of examples w.r.t. training epochs. The learning performance of Neural networks is a major concern, since it can become prohibitive in certain applications, usually as a result of the problem of local minima.

Figures 4.10 and 4.11 show that *C-IL²P's* learning performance is considerably better than *KBANN*'s. The results suggest that *C-IL²P*'s *Translation Algorithm* from symbolic knowledge to neural networks has advantages

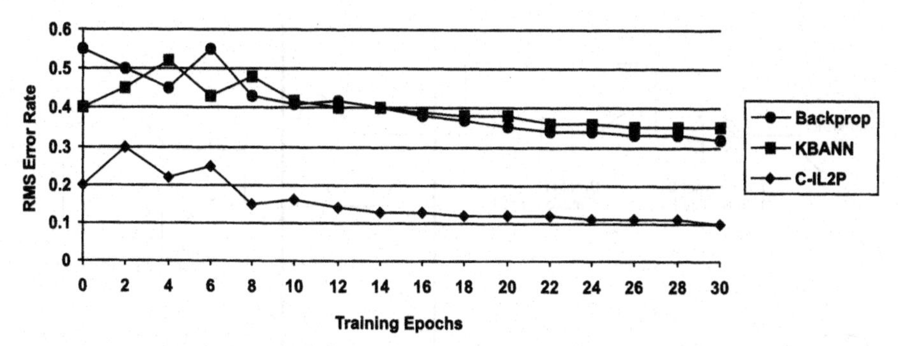


**Fig. 4.11.** Training-set RMS error decay during learning the splice junction problem. For *C-IL²P* and *KBANN*, zero training epochs represents the accuracy of the background knowledge w.r.t. the training set. For *Backpropagation*, zero training epochs represents random guess.

over the algorithm presented in [TS94a]. *C-IL²P* always encodes background knowledge in a single hidden-layer neural network. Differently, *KBANN*'s translation algorithm generates networks with as many hidden layers as the number of dependencies that exist in the background knowledge. For example, if $P = \{B \leftarrow A;\ C \leftarrow B;\ D \leftarrow C;\ E \leftarrow D\}$, *KBANN* generates a network with three hidden-layers, in which literals $B$, $C$ and $D$ should be represented. Obviously, this creates a respective degradation in learning performance. Towell and Shavlik have tried to overcome this problem with a symbolic pre-processor of rules for *KBANN* [TS94b]. However, this introduces another preliminary phase to their translation process[9]. In our opinion, the problem lies in *KBANN*'s translation algorithm, and can be straightforwardly solved by an accurate translation mechanism.

## 4.2 Power Systems Fault Diagnosis

In this section, we apply *C-IL²P* to fault diagnosis using a simplified power generator plant (see [dGZdS97, SZS95]). This application illustrates the use of rules with classical negation (¬) because the background knowledge, which we have translated into an extended logic program, had been originally described using Poole's Default Logic system Theorist[10] [Poo87, Poo91]. The results corroborate the claims about the importance of the background knowledge for learning.

[9] *KBANN* already contains a preliminary phase of rules hierarchying, that rewrites the rules before translating them.

[10] The background knowledge and the data sets were obtained from [dGZdS97], and were originally derived by Vitor Navarro da Silva [SZS95].

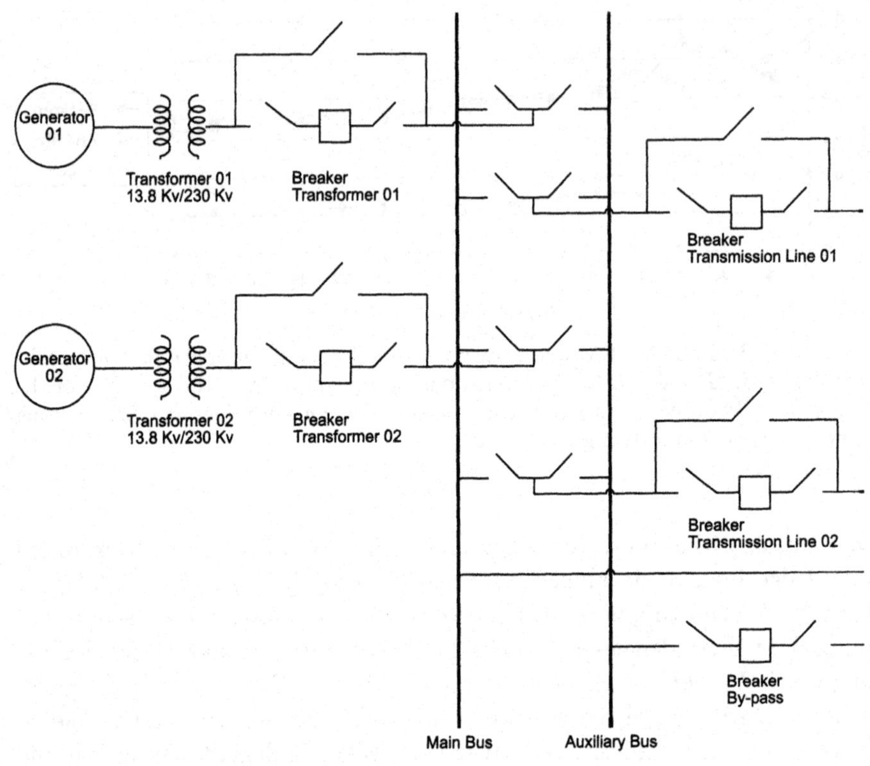


**Fig. 4.12.** Configuration of a simplified power system generation plant.

The aim of using *C-IL²P* for power systems fault diagnosis is twofold: to allow the use of background knowledge in the neural learning process, and to explain the reasons for the activated sets of alarms of the power plant. In this section, we investigate the former. The latter will be explored in Chapter 6 (Experiments on Knowledge Extraction).

In short, the task is to quickly identify possible faults in the power plant when a given set of alarms is activated. The neural network should map the set of alarms represented in its input layer to the set of faults represented in its output layer. However, in the case of a system fault, alarms might fail to activate due to equipment failure. This degree of uncertainty motivated the use of artificial neural networks for fault diagnosis, since they normally present good generalisation performance in noisy situations (in this case, alarms that might fail to activate).

Figure 4.12 shows a simplified version of a real power plant. The system has two generators, two transformers with their respective circuit breakers, two buses (main and auxiliary), and two transmission lines and a bypass circuit, also with their respective circuit breakers.

Each transmission line has six associated alarms: *breaker* status (indicates that the breaker is open), *phase* over-current (shows that there was an over-current in the phase line), *ground* over-current (shows that there was an over-current in the ground line), *timer* (shows that there was a fault distant from the power plant generator), *instantaneous* (shows that there was a fault close up to the power plant generator), and *auxiliary* (indicates that the transmission line is connected to the auxiliary bus). In addition, each transformer has three associated alarms: *breaker* status (indicates that the breaker is open), *overload* (shows that there was a transformer overload) and *auxiliary* (indicates that the transformer is connected to the auxiliary bus). Finally, there are five alarms associated with the bypass circuit breaker: *breaker* status, *phase* over-current, *ground* over-current, *timer* and *instantaneous*. We describe the set of alarms by the predicate

$$Alarm(type, local)$$

where *type* $\in$ {*breaker_open, phase, ground, timer, instantaneous, auxiliary, overload*} and *local* $\in$ {*line01, line02, transformer01, transformer02, bypass*}.

Certain combinations of alarms indicate certain faults in the system. They indicate the *component* affected by the fault (whether a transmission line or a transformer), the *reason* for the fault (if it is a phase or ground over-current, a problem in the main or auxiliary bus, or a transformer overload), and the *proximity* of the fault to the power plant generator. In addition, some faults should indicate whether the system had been using the *bypass* circuit when an alarm was activated. We describe the set of faults by the predicate

$$Fault(occurrence, proximity, component, bypass\_use)$$

such that *occurrence* $\in$ {*phase, ground, main_bus, auxiliary_bus, overload*}, *proximity* $\in$ {*close-up, distant*}, *component* $\in$ {*line01, line02, transformer01, transformer02*}, and *bypass_use* $\in$ {*no_bypass, bypass*}.

In the above electric system, it is known that if the Instantaneous alarm of Transmission Line 01 is activated then there is a fault at Transmission Line 01 close up to the power plant generator. This relation can be described as follows:

$$Fault(x, close\_up, line01, no\_bypass) \leftarrow Alarm(instantaneous, line01)$$

where $x \in \{ground, phase\}$. If, in addition, the alarm that indicates an over-current in the ground line of Transmission Line 01 is activated then it is also known that the fault is due to a ground over-current. This relation is represented as:

$$Fault(ground, close\_up, line01, no\_bypass) \leftarrow$$
$$Alarm(instantaneous, line01), Alarm(ground, line01),$$

where $Fault(ground, close\_up, line01, no\_bypass)$ reads *there is a fault at transmission line 01, close to the power plant generator, due to an over-current in the ground line, which occurred when the system was not using the bypass circuit.*

Finally, certain alarms also eliminate the possibility of certain faults to occur in the main bus or in each of the transformers. For example, if Transformer 02 is connected to the auxiliary bus then it is not the case that Transformer 02 could be overloaded. Such an information requires the use of classical negation ($\neg$) in order to be properly coded.

The background knowledge used for power systems fault diagnosis includes:

- Twenty nine (instantiated) rules of the form:
  $Fault(occurrence, proximity, component, bypass\_use) \leftarrow$
  $\bigwedge_i (Alarm(type, local))_i$;
- Four rules of the form:
  $\neg Fault(main\_bus, proximity, component, bypass\_use) \leftarrow$
  $Alarm(auxiliary, x)$,
  where $x \in \{line01, line02, transformer01, transformer02\}$;
- Two rules of the form:
  $\neg Fault(overload, proximity, y, bypass\_use) \leftarrow$
  $Alarm(auxiliary, y)$,
  where $y \in \{transformer01, transformer02\}$.

The complete background knowledge for this task is listed in the Appendix at the end of this chapter. The 35 rules of the background knowledge associate 23 different alarms with 32 different faults. The list of input neurons (alarms) and output neurons (faults) is also given in the Appendix at the end of this chapter.

In addition, we have deliberately added noise to the background knowledge by changing the following rules (in which we use the character "*" to indicate "don't care"):

24 *Fault*(main_bus, *, transformer01, *) ←
(*Alarm*(auxiliary, transformer01), *Alarm*(breaker_open,bypass),
*Alarm*(breaker_open, line01), *Alarm* (breaker_open, line02),
*Alarm*(breaker_open, transformer02)).

25 *Fault*(main_bus, *, transformer02, *) ←
(*Alarm*(auxiliary, transformer02), *Alarm*(breaker_open,bypass),
*Alarm*(breaker_open, line01), *Alarm* (breaker_open, line02),
*Alarm*(breaker_open, transformer01)).

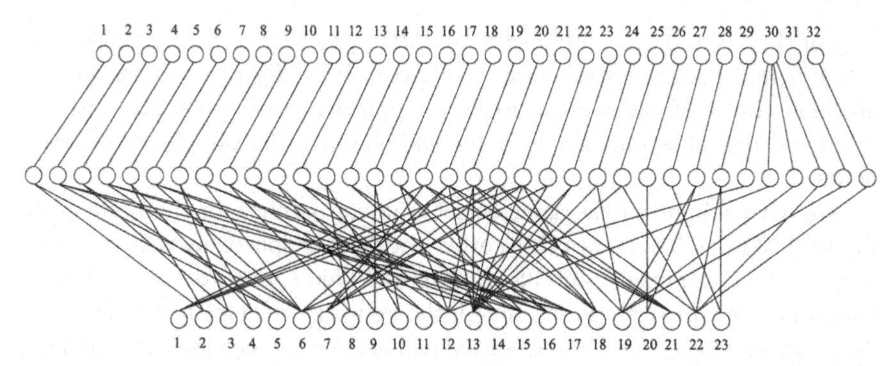


**Fig. 4.13.** Initial neural network for power system fault diagnosis. Input neurons are numbered from 1 to 23, according to the list of alarms presented in the Appendix. Output neurons are numbered from 1 to 32, according to the list of faults presented in the Appendix.

by the following rules:

24′ *Fault*(main_bus, ∗, transformer01, ∗)←
(*Alarm*(breaker_open, transformer01), *Alarm*(breaker_open,bypass),
*Alarm*(breaker_open, line01), *Alarm*(breaker_open, line02),
*Alarm*(breaker_open, transformer02)).

25′ *Fault*(main_bus, ∗, transformer02, ∗)←
(*Alarm*(breaker_open, transformer02), *Alarm*(breaker_open,bypass),
*Alarm*(breaker_open, line01), *Alarm*(breaker_open, line02),
*Alarm*(breaker_open, transformer01)).

in which *Alarm*(*auxiliary*, *transformer01*) and *Alarm*(*auxiliary*, *transformer02*) were replaced, respectively, by *Alarm*(*breaker_open*, *transformer01*) and *Alarm*(*breaker_ open*, *transformer02*).

Figure 4.13 contains the network in which the background knowledge (including rules 24′ and 25′, instead of 24 and 25) was set up. From the background knowledge and the *Translation Algorithm* of Chapter 3, we obtain $MAX_{\mathcal{P}} = 5$ and $A_{\min} > \frac{5-1}{5+1}$. Taking $A_{\min} = 0.7$, and assuming $\beta = 1$, we obtain $W \geq 2 \cdot (\frac{ln(1.7)-ln(0.3)}{5(-0.3)+0.7+1}) = 17.34$. In the network of Figure 4.13, we use $W = 18$. The thresholds of hidden and output neurons are calculated using Equations 3.2 and 3.3, respectively. Complementary literals, such as *Fault*(*main_bus*, ∗, ∗, ∗) and ¬*Fault(main_bus*, ∗, ∗, ∗*)*, are encoded in the network by two different output neurons.

We have trained the network of Figure 4.13 with 278 examples, which included single and multiple faults; that is, examples where each set of alarms is associated with a unique possible fault, and examples where each set of alarms is associated with many possible faults. In addition, approximately

10% of the training set contained noise. More precisely, in 28 examples, one of the faults' characteristic alarms was absent. In order to verify the generalisation ability of the system, we have applied two test sets: one with 92 examples containing only single faults, and one with 70 examples of multiple faults.

We have submitted the same set of training patterns to the network of Figure 4.13 (*CIL2P*), and to three other networks, respectively with 5 (*Backprop5*), 15 (*Backprop15*) and 35 (*Backprop35*) hidden neurons, in which no background knowledge was set up. All the networks were trained using standard backpropagation with momentum and exactly the same training parameters, until either the RMS error rate had been reduced to less than 0.01, or the network had been trained for 50,000 epochs. Among the networks above mentioned, only *Backprop5* did not present RMS error smaller than 0.01 after 50,000 epochs have elapsed.

In what follows, we present the results obtained for both test sets, according to:

- Test set performance (percentage of test set examples correctly classified);
- The average size of the ambiguity set (a measure of the frequency in which faults are wrongly accused); and
- Training set performance (as before, we use the root-mean-square (RMS) error rate decay).

Differently from the experiments on DNA sequence analysis, instead of analysing the output vector as a whole, here we will present the accuracy obtained by each output neuron individually. In this way, we can check how frequently the network accuses a fault correctly, regardless of the remaining faults. In order to investigate the network's overall performance w.r.t. the whole set of faults, we calculate the *Average Size of the Ambiguity Set*, commonly used to test fault diagnosis systems when they isolate failures from several possible fault modes, but fail to correctly identify the set of faults. We calculate the average size of the ambiguity set as follows: Let $x_i$ be the number of faults identified in a given example, i.e. the number of activated outputs in the network. Let $y_i$ be the number of faults wrongly identified in such an example, i.e. the number of wrongly activated outputs. Let $z$ be the number of testing examples. For each example $e_i$, $y_i/x_i$ is called the *Size of the Ambiguity Set* of $e_i$. The *Average Size of the Ambiguity Set* is then given by $((\sum_{i=1}^{z} y_i/x_i)/z) \times 100$. If $x_i = 0$ then example $e_i$ is left out of the calculation.

Firstly, let us present the test set performance of the networks. Figures 4.14, 4.15 and 4.16 compare the accuracy obtained by *CIL2P*, *Backprop35* and *Backprop15* w.r.t. the 92-examples test set with single faults. The results obtained by *Backprop5* were left out because it achieved a very poor

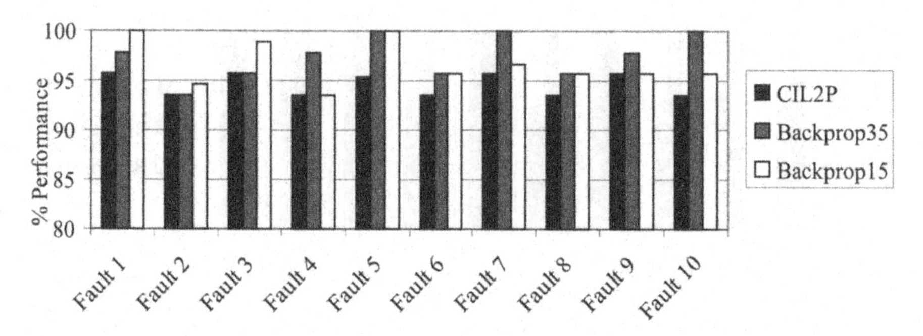


**Fig. 4.14.** Test set performance for the single fault test set (Faults 1 to 10).

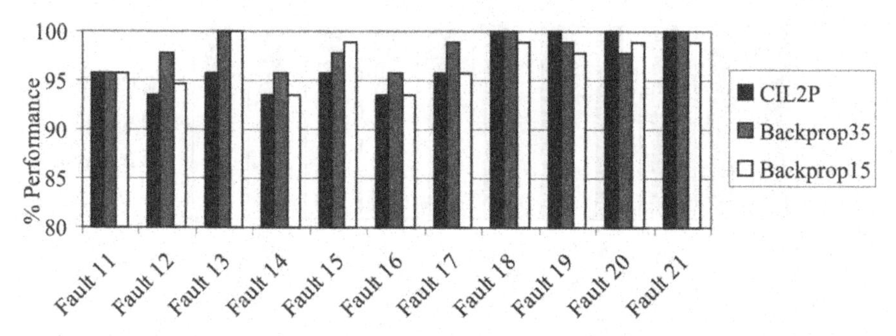


**Fig. 4.15.** Test set performance for the single fault test set (Faults 11 to 21).

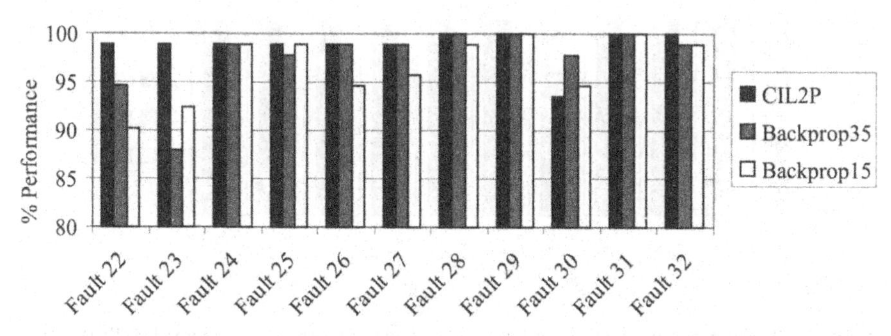


**Fig. 4.16.** Test set performance for the single faults test set (Faults 22 to 32).

generalisation. Similarly, Figures 4.17, 4.18 and 4.19 compare the accuracy obtained by *CIL2P*, *Backprop35* and *Backprop15* w.r.t. the 70-examples test set with multiple faults. Again, the results obtained by *Backprop5* were left out because it achieved a very poor generalisation.

In general, differences in accuracy between *CIL2P*, *Backprop35* and *Backprop15*, for each output neuron, are smaller than 5%. The exception is *Backprop15* in the multiple faults test set, which presents a much poorer accuracy

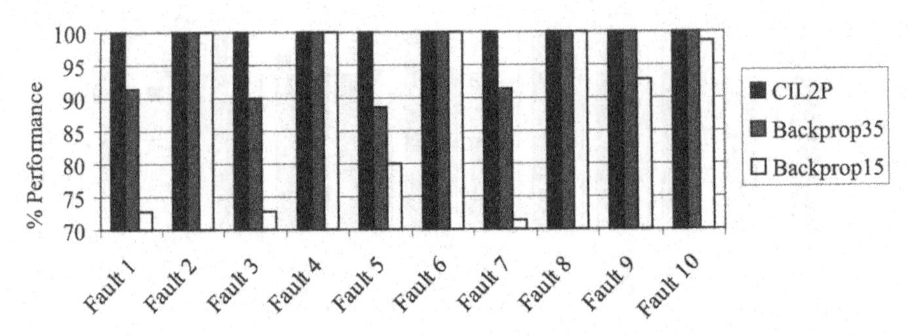


**Fig. 4.17.** Test set performance for the multiple faults test set (Faults 1 to 10).

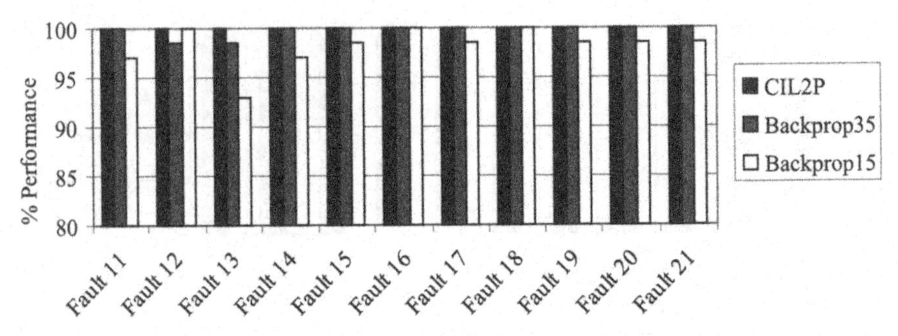


**Fig. 4.18.** Test set performance for the multiple faults test set (Faults 11 to 21).

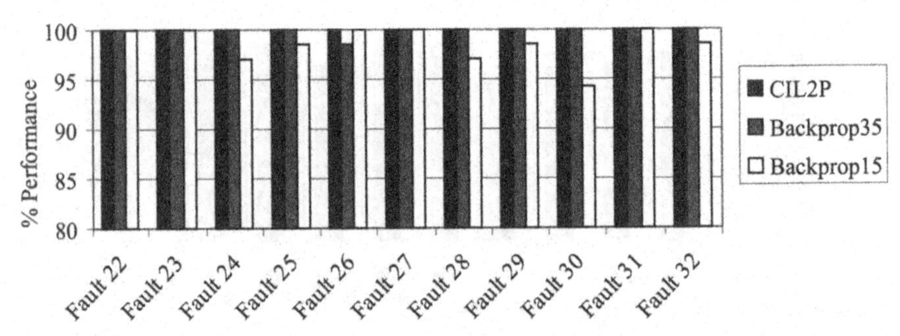


**Fig. 4.19.** Test set performance for the multiple faults test set (Faults 22 to 32).

for some outputs (see Figure 4.17). The *CIL2P* network is marginally worse than *Backprop35* in the single fault test set, and marginally better than *Backprop35* in the multiple faults test set. The graph of Figure 4.20 summarises the above results by showing the mean of the accuracy obtained by *CIL2P*,

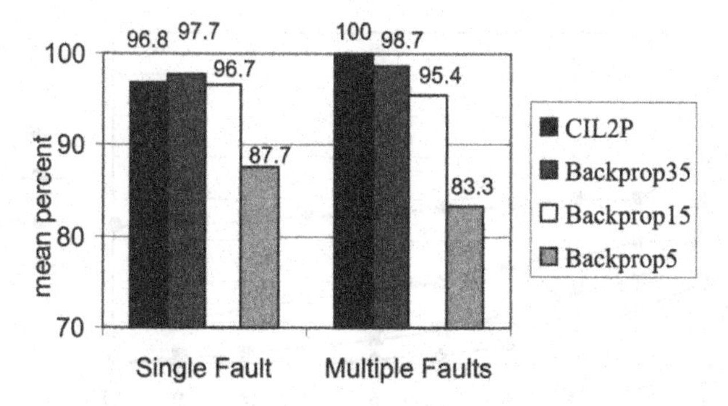


**Fig. 4.20.** Mean of the accuracy of Faults 1 to 32 obtained by *CIL2P*, *Backprop35*, *Backprop15* and *Backprop5*, in both the single fault and multiple faults test sets.

*Backprop35*, *Backprop15* and *Backprop5*. The results are not conclusive as to whether *CIL2P* is better or worse than *Backprop35*.[11]

Let us now compare the Ambiguity Sets of the networks. Table 4.4 shows the average sizes of the ambiguity sets of *CIL2P*, *Backprop35*, *Backprop15* and *Backprop5*, for each of the test sets (single fault and multiple faults). In the single fault test set, the ambiguity of *CIL2P* (0.5%) is smaller than that of any *Backprop* network, even though *CIL2P*'s mean accuracy is marginally poorer than *Backprop35*'s. This indicates that *CIL2P* is less subjected to wrongly indicating faults. Of course, it also indicates that *CIL2P*'s misclassifications generally occur because it might fail to identify faults. The ambiguity of *CIL2P* in the multiple faults test set is *nil*, since its mean accuracy was 100%.

**Table 4.4.** Average Size of Ambiguity Sets (For example, 0.5% of mean ambiguity indicates that, for each 200 faults accused by the network, 1 is wrong.)

| *Ambiguity Set* | *CIL2P* | *Backprop35* | *Backprop15* | *Backprop5* |
|---|---|---|---|---|
| *Single Fault* | 0.5% | 9.4% | 20.3% | 72.3% |
| *Multiple Faults* | 0.0% | 1.0% | 9.1% | 65.9% |

Finally, let us evaluate the training set performance of the networks. Figure 4.21 compares the RMS error decay curves during the learning process of *CIL2P*, *Backprop35*, *Backprop15* and*Backprop5*, up until 10,000 training epochs have elapsed. As expected, among *Backprop* networks the convergence is slower for networks with fewer hidden neurons. However, the *CIL2P* net-

[11] The results suggest though that the choice of 35 hidden neurons is a good one. Recall that such a number is given by the length of the background knowledge for *CIL2P* networks. Albeit having considerably large hidden layers, *Backprop35* and *CIL2P* did not suffer from overfitting.

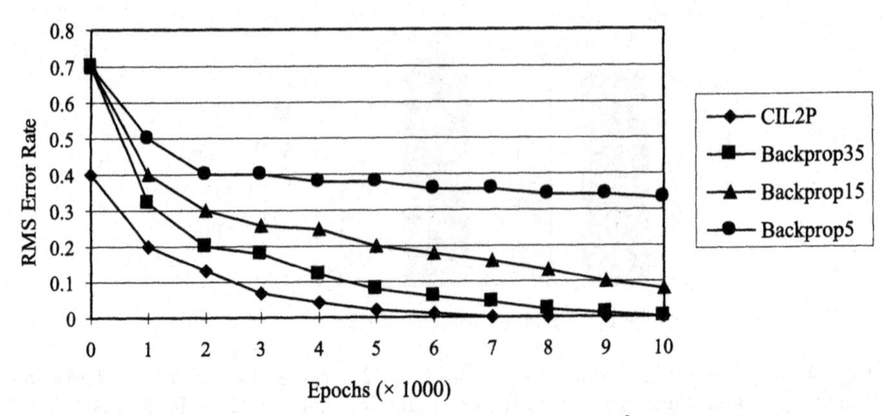


**Fig. 4.21.** RMS error in the learning process. For *C-IL²P*, zero training epochs represents the accuracy of the background knowledge w.r.t. the training set. For *Backpropagation*, zero training epochs represents random guess.

work converges considerably faster than *Backprop35*. This is clearly due to the presence of the background knowledge, and indicates that the learning performance of *CIL2P* is advantageous to that of *Backprop*.

It is important to note that the usefulness of background knowledge in a neural learning process is twofold: (1) the prior knowledge can improve training set performance, as seen in Figure 4.21, by facilitating the learning of examples, if certain training examples are already encoded in the prior knowledge, and (2) the prior knowledge can improve test set performance by correctly classifying certain test set examples if these examples are encoded in the prior knowledge but not catered for by the training examples. In the experiments of this section, case (1) was much more clearly verified than case (2). Of course, prior knowledge can also complicate the learning process if it is incorrect.

The above results are considered to be very promising as they indicate that *C-IL²P* also performs well in the presence of some incorrect background knowledge and training examples. The hypothesis that the background knowledge could reduce the network's performance when noisy training examples are used was not verified here (see [TS94a] for a discussion on the reduction of the performance of *KBANN* in the presence of noise).

## 4.3 Discussion

The 1990s saw much work in the field of hybrid learning systems. In particular, many fuzzy-neural systems were developed. It would have been impossible to compare the performance of *C-IL²P* with such a variety of systems.

The main difficulties of this task regard the employment by each system of different testing methodologies and very specific learning techniques. For example, in [Fu94], Fu described the Knowledge Based Conceptual Neural Network (*KBCNN*), a hybrid system similar to *KBANN*. The results obtained by *KBCNN* in the Promoter domain (98.1% accuracy) are marginally better than *C-IL²P's* (97.2% accuracy).[12] However, it seems that Fu has not used the method of cross-validation in his experiments with *KBCNN*. To further complicate any comparison, *KBCNN* is closely tied to its particular learning algorithm.

Nevertheless, the above analysis provides evidence for the usefulness of *C-IL²P*, by showing its efficiency in comparison with well-known machine learning systems. Furthermore, the overall performance of *C-IL²P* in absolute values is very satisfactory. The experiments described above suggest that *C-IL²P*'s effectiveness is a result of three of the system's features:

(1) *C-IL²P* is based on *Backpropagation*. As said before, backpropagation has been shown to outperform some of the main machine learning systems, specially in the presence of noisy data sets (see [TBB+91] for a comparison between many symbolic and non-symbolic learning algorithms[13]). Backpropagation has also been the most successfully and extensively used neural learning algorithm.

(2) *C-IL²P* uses *background knowledge*. The importance of using background knowledge in the process of inductive learning has been stressed by the symbolic machine learning community [HN94, LD92, PK92, TLI91]. The work on Inductive Logic Programming [LD94, MR94] has given the subject a new dimension. On the other hand, most of the work on neural networks has neglected this matter. The above experiments with *C-IL²P* corroborate the claims for using background knowledge in neural networks when it is available.

(3) *C-IL²P* provides an *accurate and compact translation* from symbolic knowledge to neural networks. The translation is accurate because the network computes the fixpoint operator $T_{\mathcal{P}}$ of the program used as background knowledge. It is compact because, whatever the background knowledge, the network derived contains a single hidden layer. Furthermore, no literal of the background knowledge is ever encoded in a hidden neuron. If this were the case, we would have no control over such a literal during learning.

---

[12] *KBCNN* has not been applied in the splice junction determination problem.

[13] [TBB+91] contains a collection of results obtained by a large group of researchers, each of whom is an advocate, and often the creator, of the technique tested.

## 4.4 Appendix

### Power System Fault Diagnosis[14]

– BACKGROUND KNOWLEDGE:

1. Fault (phase, close-up, line01, no_bypass) ← (Alarm (phase, line01), Alarm (instantaneous, line01)).
2. Fault (phase, close-up, line01, bypass) ← (Alarm (auxiliary, line01), Alarm (phase, bypass), Alarm (instantaneous, bypass)).
3. Fault (ground, close-up, line01, no_bypass) ← (Alarm (ground, line01), Alarm (instantaneous, line01)).
4. Fault (ground, close-up, line01, bypass) ← (Alarm (auxiliary, line01), Alarm (ground, bypass), Alarm (instantaneous, bypass)).
5. Fault (phase, distant, line01, no_bypass) ← (Alarm (phase, line01), Alarm (timer, line01)).
6. Fault (phase, distant, line01, bypass) ← (Alarm (auxiliary, line01), Alarm (phase, bypass), Alarm (timer, bypass)).
7. Fault (ground, distant, line01, no_bypass) ← (Alarm (ground, line01), Alarm (timer, line01)).
8. Fault (ground, distant, line01, bypass) ← (Alarm (auxiliary, line01), Alarm (ground, bypass), Alarm (timer, bypass)).
9. Fault (phase, close-up, line02, no_bypass) ← (Alarm (phase, line02), Alarm (instantaneous, line02)).
10. Fault (phase, close-up, line02, bypass) ← (Alarm (auxiliary, line02), Alarm (phase, bypass), Alarm (instantaneous, bypass)).
11. Fault (ground, close-up, line02, no_bypass) ← (Alarm (ground, line02), Alarm (instantaneous, line02)).
12. Fault (ground, close-up, line02, bypass) ← (Alarm (auxiliary, line02), Alarm (ground, bypass), Alarm (instantaneous, bypass)).
13. Fault (phase, distant, line02, no_bypass) ← (Alarm (phase, line02), Alarm (timer, line02)).
14. Fault (phase, distant, line02, bypass) ← (Alarm (auxiliary, line02), Alarm (phase, bypass), Alarm (timer, bypass)).
15. Fault (ground, distant, line02, no_bypass) ← (Alarm (ground, line02), Alarm (timer, line02)).
16. Fault (ground, distant, line02, bypass) ← (Alarm (auxiliary, line02), Alarm (ground, bypass), Alarm (timer, bypass)).
17. ¬Fault (main_bus, *, *, *) ← Alarm (auxiliary, line01).
18. ¬Fault (main_bus, *, *, *) ← Alarm (auxiliary, line02).
19. ¬Fault (main_bus, *, *, *) ← Alarm (auxiliary, transformer01).
20. ¬Fault (main_bus, *, *, *) ← Alarm (auxiliary, transformer02).
21. Fault (main_bus, *, *, *) ← (Alarm (breaker_open, line01), Alarm (breaker_open, line02), Alarm (breaker_open, transformer01), Alarm (breaker_open, transformer02)).
22. Fault (main_bus, *, line01, *) ← (Alarm (auxiliary, line01), Alarm (breaker_open, bypass), Alarm (breaker_open, line02), Alarm (breaker_open, transformer 01), Alarm (breaker_open, transformer02)).

[14] We use the character "*" to denote "*don't care*".

23. Fault (main_bus, ∗, line02, ∗) ← (Alarm (auxiliary, line02), Alarm (breaker_open, bypass), Alarm (breaker_open, line01), Alarm (breaker_open, transformer 01), Alarm (breaker_open, transformer02)).
24. Fault (main_bus, ∗, transformer01, ∗) ← (Alarm (auxiliary, transformer01), Alarm (breaker_open, bypass), Alarm (breaker_open, line01), Alarm (breaker_open, line02), Alarm (breaker_open, transformer02)).
25. Fault (main_bus, ∗, transformer02, ∗) ← (Alarm (auxiliary, transformer02), Alarm (breaker_ open, bypass), Alarm (breaker_open, line01), Alarm (breaker_open, line02), Alarm (breaker_open, transformer01)).
26. Fault (auxiliary_bus, ∗, line01, bypass) ← (Alarm (auxiliary, line01), Alarm (breaker_open, bypass)).
27. Fault (auxiliary_bus, ∗, line02, bypass) ← (Alarm (auxiliary, line02), Alarm (breaker_open, bypass)).
28. Fault (auxiliary_bus, ∗, transformer01, bypass) ← (Alarm (auxiliary, transformer01), Alarm (breaker_open, bypass)).
29. Fault (auxiliary_bus, ∗, transformer02, bypass) ← (Alarm (auxiliary, transformer02), Alarm (breaker_open, bypass)).
30. ¬Fault (overload, ∗, transformer01, no_bypass) ← Alarm (auxiliary, transformer01).
31. Fault (overload, ∗, transformer01, no_bypass) ← Alarm (overload, transformer 01).
32. ¬Fault (overload, ∗, transformer02, no_bypass) ← Alarm (auxiliary, transformer02).
33. Fault (overload, ∗, transformer02, no_bypass) ← Alarm (overload, transformer 02).
34. Fault (overload, ∗, transformer01, bypass) ← (Alarm (overload, transformer01), Alarm (auxiliary, transformer01)).
35. Fault (overload, ∗, transformer02, bypass) ← (Alarm (overload, transformer02), Alarm (auxiliary, transformer02)).

– List of Input Neurons (Alarms):

1. *Alarm(breaker_open, line01)*
2. *Alarm(phase, line01)*
3. *Alarm(ground, line01)*
4. *Alarm(timer, line01)*
5. *Alarm(instantaneous, line01)*
6. *Alarm(auxiliary, line01)*
7. *Alarm(breaker_open, line02)*
8. *Alarm(phase, line02)*
9. *Alarm(ground, line02)*
10. *Alarm(timer, line02)*
11. *Alarm(instantaneous, line02)*
12. *Alarm(auxiliary, line02)*
13. *Alarm(breaker_open, bypass)*
14. *Alarm(phase, bypass)*
15. *Alarm(ground, bypass)*
16. *Alarm(timer, bypass)*
17. *Alarm(instantaneous, bypass)*
18. *Alarm(breaker_open, transformer01)*

19. *Alarm(auxiliary, transformer01)*
20. *Alarm(overload, transformer01)*
21. *Alarm(breaker_open, transformer02)*
22. *Alarm(auxiliary, transformer02)*
23. *Alarm(overload, transformer02)*

– LIST OF OUTPUT NEURONS (FAULTS):

1. *Fault(phase, close-up, line01, no_bypass)*
2. *Fault(phase, close-up, line01, bypass)*
3. *Fault(ground, close-up, line01, no_bypass)*
4. *Fault(ground, close-up, line01, bypass)*
5. *Fault(phase, distant, line01, no_bypass)*
6. *Fault(phase, distant, line01, bypass)*
7. *Fault(ground, distant, line01, no_bypass)*
8. *Fault(ground, distant, line01, bypass)*
9. *Fault(phase, close-up, line02, no_bypass)*
10. *Fault(phase, close-up, line02, bypass)*
11. *Fault(ground, close-up, line02, no_bypass)*
12. *Fault(ground, close-up, line02, bypass)*
13. *Fault(phase, distant, line02, no_bypass)*
14. *Fault(phase, distant, line02, bypass)*
15. *Fault(ground, distant, line02, no_bypass)*
16. *Fault(ground, distant, line02, bypass)*
17. *Fault(main_bus, *, *, *)*
18. *Fault(main_bus, *, line01, *)*
19. *Fault(main_bus, *, line02, *)*
20. *Fault(main_bus, *, transformer01, *)*
21. *Fault(main_bus, *, transformer02, *)*
22. *Fault(auxiliary_bus, *, line01, bypass)*
23. *Fault(auxiliary_bus, *, line02, bypass)*
24. *Fault(auxiliary_bus, *, transformer01, bypass)*
25. *Fault(auxiliary_bus, *, transformer02, bypass)*
26. *Fault(overload, *, transformer01, no_bypass)*
27. *Fault(overload, *, transformer02, no_bypass)*
28. *Fault(overload, *, transformer01, bypass)*
29. *Fault(overload, *, transformer02, bypass)*
30. ¬*Fault (main_bus, *, *, *)*
31. ¬*Fault (overload, *, transformer01, no_bypass)*
32. ¬*Fault (overload, *, transformer02, no_bypass)*

# Part II

# Knowledge Extraction from Neural Networks

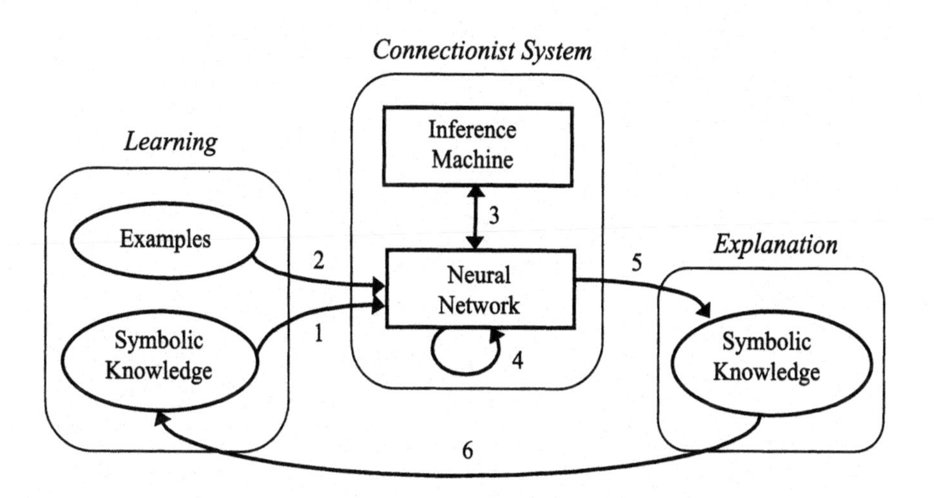


Part II of this book is about *Knowledge Extraction* from trained neural networks. It presents the theoretical and practical aspects of process (5) above. It contains a new Extraction Algorithm, responsible for deriving comprehensible logical rules from trained neural networks. It shows that the process of rule extraction is correct. The extraction algorithm does not depend on the learning strategy adopted, nor on the application domain. It is, therefore, of interest with respect to the application of neural networks to *Data Mining* tasks.[DL01] Part II concludes with an analysis of the results of applying process (5) to traditional examples and real-world problems of DNA sequence analysis and Power Systems Fault Diagnosis.

# 5. Knowledge Extraction from Trained Networks

In this chapter, we tackle the problem of extracting symbolic knowledge from trained neural networks; that is, the problem of finding "logical representations" for such networks. We present a new method of extraction that captures nonmonotonic rules encoded in the network and show that such a method is sound. The ideas presented here comply with the concept that, in machine learning, *Feature Construction* should be decoupled from *Model Construction* [KLF01]. Part I of this book has dealt with feature construction. Part II will deal with model construction.

Although neural networks have shown very good performance in many application domains, one of their main drawbacks lies in their incapacity to provide an explanation for the underlying reasoning mechanisms. It is now commonly accepted that such an *Explanation Capability* can be achieved by the extraction of symbolic knowledge from trained networks, using the so-called *Rule Extraction* methods. By allowing the explanation of the decision making process, the extraction of symbolic knowledge contributes to solving the *Knowledge Acquisition Bottleneck* problem. The domain theory extracted, which is obtained by learning with examples, can be added to an existing knowledge base or used in the solution of analogous domains problems.

We start by discussing some of the main problems of knowledge extraction methods. We then discuss how these problems may be ameliorated. To this end, a partial ordering on the set of input vectors is defined, as well as a number of pruning and simplification rules. The pruning rules are then used to reduce the search space of the extraction algorithm during a pedagogical extraction, whereas the simplification rules are used to reduce the size of the extracted set of rules. We show that, in the case of regular networks, the extraction algorithm is sound and complete.[1]

We proceed to extend the extraction algorithm to the class of non-regular networks, the general case. We show that non-regular networks always contain regularities in their subnetworks. As a result, the underlying extraction method for regular networks can be applied, but now in a decompositional fashion. In order to combine the sets of rules extracted from each subnetwork

[1] Following [Fu94], we say that an extraction algorithm is sound and complete if the set of rules is equivalent to the network.

into the final set of rules, we use a method whereby we are able to preserve the soundness of the extraction algorithm, although we have to forego completeness.

In Section 5.1 we define the extraction problem precisely and discuss the main problems of the task of extracting knowledge from trained networks. In Section 5.2 we present our solution to the extraction problem, culminating with the outline of the extraction algorithm for the class of regular networks, and the proofs of soundness and completeness of the method, here referred to as the ABG extraction method. In Section 5.3 we extend the extraction algorithm to the class of non-regular networks – the general case – and show that the extended ABG method of extraction is sound. Section 5.5 summarises the chapter and presents some references to further reading.

## 5.1 The Extraction Problem

We define the extraction problem as follows:

> *Given a particular set of weights $W_{ij}$ and thresholds $\theta_i$, resulting from a training process on a neural network, find for each input vector* $\mathbf{i}$, *all the outputs $o_j$ in the corresponding output vector* $\mathbf{o}$ *such that the activation of $o_j$ is greater than $A_{\min}$, where $A_{\min} \in (0,1)$ is a predefined value* (in this case, we say that output neuron $o_j$ is "*active*" for input vector $\mathbf{i}$).

Intuitively, the extraction task is to find the relations between input and output concepts in a trained network, in the sense that certain inputs *cause* a particular output. We argue that neural networks are non-monotonic systems, i.e. they jump to conclusions that might be withdrawn when new information is available [MT93a]. Thus, the set of rules extracted may contain default negation ($\sim$). Each neuron can represent a concept or its classical negation ($\neg$). Consequently, we expect to extract rules of the form: $L_1, ..., L_n, \sim L_{n+1}, ..., \sim L_m \rightarrow L_{m+1}$, where each $L_i$ is a literal, $L_j$ $(1 \leq j \leq m)$ represents a neuron in the network's input layer, $L_{m+1}$ represents a neuron in the network's output layer, $\sim$ stands for *default negation*, and $\rightarrow$ means *causal implication*[2].

We assume that for each input $i_j$ in the input vector $\mathbf{i}$, either $i_j = 1$ or $i_j = -1$. This is done because we consider bipolar activation functions. We then associate each input neuron with a concept, say $a$, and if $i_j = 1$ then $a$ is *true*, while $i_j = -1$ means that $a$ is $false$. For example, consider a network with input neurons $a$ and $b$. If $a = 1$ and $b = -1$ activate output neuron $c$ then we derive the rule $a \sim b \rightarrow c$.

[2] Notice that this is the language of Extended Logic Programming [GL91].

Briefly, the problem of extraction lies on the complexity of the extraction algorithm. In Chapter 3, we have seen that each logic program is equivalent to a single hidden layer neural network. In one direction of this equivalence relation, a translation algorithm derives a neat neural network structure when a logic program is given. The problem arises in the converse direction, i.e. given a trained neural network, how could we find out the equivalent logic program? Unfortunately, it is very unlikely that a neat network will result from the learning process. Furthermore, a typical real-world application network may contain hundreds of input neurons and thousands of connections.

The knowledge acquired by a neural network during its training phase is encoded in: (*i*) the network's architecture; (*ii*) the activation function associated with it; and (*iii*) the value of its weights. As pointed out in [ADT95], "the task of extracting explanations from trained neural networks is the one of interpreting in a comprehensible form the collective effect of (*i*), (*ii*), and (*iii*)". Also in [ADT95], a *classification* scheme for extraction algorithms is given, and it is based on: (*a*) the expressive power of the extracted rules; (*b*) the "translucency" of the network; (*c*) the quality of the extracted rules; and (*d*) the algorithmic complexity. Item (*a*) is about the symbolic knowledge presented to the user from the extraction process. It is, in general, in the form of "if then else" rules. Item (*b*) contains two basic categories: *decompositional* and *pedagogical*. In the *decompositional*, rule extraction is done for each hidden and output neuron within the trained network. In the *pedagogical*, the network is seen as a "black box", and the extraction is done by mapping input neurons directly into output neurons. Classification item (*c*) attempts to measure how well the task of extracting the rules has been performed, considering the *accuracy*, *consistency* and *comprehensibility* of the extracted set of rules. Finally, classification item (*d*) is about the efficiency of the extraction algorithm. In this sense, a crucial issue in the development of an extraction algorithm is how to constrain its *search space*.

Complementary to the above classification scheme, in [Thr94], Thrun defines the following desirable *properties* of an extraction algorithm: (*i*) No architectural requirements: a general extraction mechanism should be able to operate with all types of neural networks; (*ii*) No training requirements: the algorithm should not make assumptions about the way the network has been built, and how its weights and thresholds have been learned; (*iii*) Correctness: the extracted rules should describe the underlying network as correctly as possible; (*iv*) High expressive power: more powerful languages and more compact rule sets are highly desirable.

Figure 5.1 gives a general idea about the pieces of knowledge represented in a neural network and their relations. The training set (2) corroborates part of the background knowledge (1) and revises another part, while the network's generalisation set (3) normally embodies the training set. We say

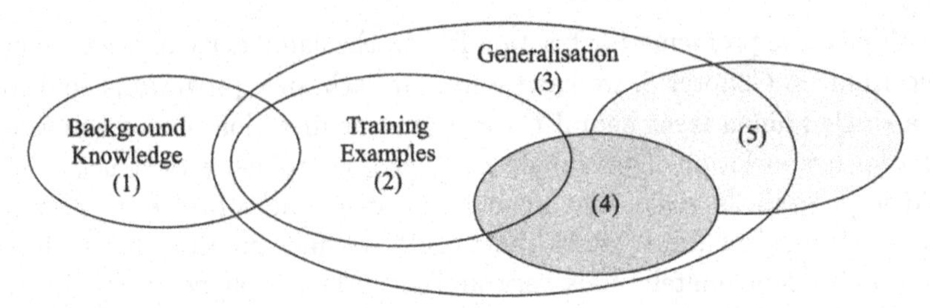


**Fig. 5.1.** Knowledge encoded in a neural network and extraction.

that an extraction algorithm is sound and complete if the rule set extracted is equivalent to the network's generalisation set. If, however, the rule set (see (4), below) is equivalent to a subset of (3), then the extraction is sound but incomplete. Rule set (5) is an example of unsound and incomplete extraction.

In the *C-IL²P* system, after learning takes place, the network $N$ encodes a knowledge $P'$ that contains the background knowledge $P$ complemented or even revised by the knowledge learned with training examples. We want to derive $P'$ from $N$. At the moment, only pedagogical approaches can guarantee that the knowledge extracted is equivalent to the network, i.e. that the extraction process is sound and complete. In [HK94], for instance, all possible combinations of the input vector $\mathbf{i}$ of $N$ are taken into account in the process of rule generation. In this way, the method must consider $2^n$ different input vectors, where $n$ is the number of neurons in the input layer of $N$.

Obviously, pedagogical approaches are not efficient when the size of the neural network increases, as in real-world applications. In order to overcome this limitation, decompositional methods, in general, apply heuristically guided searches to the process of extraction. The "Subset" method [Fu94], for instance, attempts to search for subsets of weights of each neuron in the hidden and output layers of $N$, such that the neuron's input potential exceeds its threshold. Each subset that satisfies the above condition is written as a rule. Based on the Subset method, the "MofN" technique [TS93] uses weight clustering and pruning in order to facilitate the extraction of rules. It also generates a smaller number of rules by taking advantage of the *M of N* representation, in which $m(A_1, ..., A_n) \rightarrow A$ indicates that if $m$ *of* $(A_1, ..., A_n)$ are *true* then $A$ is *true*, where $m \leq n$.

Decompositional methods, such as [TS93], in general use weight pruning mechanisms prior to extraction. However, there is no guarantee that a pruned network will be equivalent to the original one. That is why these methods usually require retraining the network. During retraining, some restrictions must be imposed on the learning process – for instance, allowing only the thresholds, but not the weights, to change – for the network to keep its "well behaved" pruned structure. At this point, there is no guarantee that retrain-

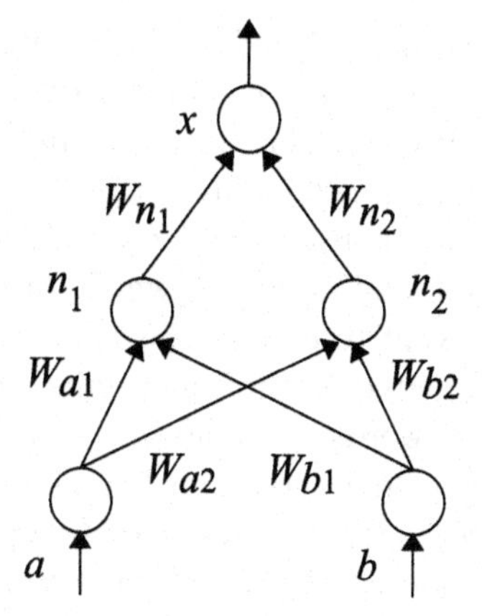


**Fig. 5.2.** A fully connected network with two input neurons $(a, b)$, two hidden neurons $(n_1, n_2)$ and a single output neuron $(x)$.

ing will be successful under such restrictions. In addition, some extraction methods, e.g. [Set97b, Set97a], apply *penalty functions* during training in order to try to maintain, after learning new examples, the initial "well behaved" structure of the network and, therefore, facilitate rule extraction. Such methods are bound to restrict the network's learning capability, as they would not be applicable to networks trained with *off the shelf* learning algorithms. Even if we avoid the use of penalty functions and weight clustering and pruning, the simple task of decomposing the network into smaller subnetworks from which rules are extracted and then assembled has to be carried out carefully. This is so because, in general, the collective effect of the network is different from the effect of the superposition of its parts [ADT95]. As a result, most decompositional methods are unsound, as the following example illustrates.

*Example 5.1.1. (unsoundness and incompleteness of decompositional extraction algorithms)* Consider the network $N$ of Figure 5.2. Let us assume that the weights are such that $a = 1$ and $b = 1$ activate neither $n_1$ nor $n_2$, but that the composition of the activation values of $n_1$ and $n_2$ activates $x$. As a result, we would expect to extract the rule $ab \rightarrow x$ from $N$. For example, suppose that $a = 1$ and $b = 1$ gives $n_1 = 0.3$ and $n_2 = 0.4$, and that these activation values result in $x = 0.99$.[3] A decompositional method would most probably derive a unique rule from such a network, namely, $n_1 n_2 \rightarrow x$, not being able to establish the correct relation between $a$, $b$ and $x$; that is, $ab \rightarrow x$.

[3] For example, let $i_j \in \{0, 1\}$. If $f(x) = \frac{1}{1+e^{-x}}$ is the activation function of the neurons in $N$ and the thresholds of $n_1$, $n_2$ and $x$ are all *zero* then $W_{a1} = -0.5$, $W_{b1} = -0.35$, $W_{a2} = -0.2$, $W_{b2} = -0.2$, $W_{n_1} = 3$ and $W_{n_2} = 9.25$ is a set of weights that makes $N$ behave as described above for inputs $a = 1$ and $b = 1$.

Now, assume that inputs $a = 1$ and $b = 1$ activate $n_1$ and $n_2$, but $n_1$ and $n_2$ together do not activate $x$ (say, $x < 0.5$). For example, assume $W_{a1} = 0.2$, $W_{b1} = 0.2$, $W_{a2} = 0.4$, $W_{b2} = 0.45$, $W_{n_1} = 9$ and $W_{n_2} = -8.1$, and take $f(x) = \frac{1}{1+e^{-x}}$ as the activation function of the neurons of $N$. Also, assume that the thresholds of neurons $n_1$, $n_2$ and $x$ are all *zero*. In this case, $a = 1$ and $b = 1$ makes $n_1 \simeq 0.6$ and $n_2 \simeq 0.7$, which, in turn, make $x \simeq 0.4$. As a result, now we do not want to extract the rule $ab \rightarrow x$ from $N$. However, if $n_1$ and $n_2$ are approximated as threshold units then $n_1 = 1$ and $n_2 = 1$ produce $x \simeq 0.7$. In other words, although $a = 1$ and $b = 1$ do not activate $x$, approximating the sigmoidal activation function of $n_1$ and $n_2$ by a step function results in $x$ being activated. Hence decompositional methods that do so, such as [TS93], would conclude that $ab \rightarrow x$ when, in fact, $ab \nrightarrow x$.

The first of the above cases is an example of incompleteness. The second one shows how decompositional methods may turn out to be unsound. Even the Subset method [Fu94], which is sound w.r.t. each hidden and output neuron, may become unsound w.r.t. the whole network due to the assumption that the activation function of the hidden neurons can be approximated by a step function.

Clearly, the classification of rule extraction methods as *pedagogical* or *decompositional* reflects a trade-off between the *complexity* of the extraction method and the *quality* of the knowledge extracted. In general, highly accurate pedagogical methods of extraction present exponential complexity, while more efficient decompositional methods of extraction are unsound, and hence could present poor accuracy, depending on the application domain. In our view, an alternative is to prune the set of input vectors, rather than the set of weights, of the network from which we want to extract rules. Our goal is to reduce complexity in the average case by applying the extraction algorithm on a smaller search space, yet maintaining the highest possible quality and, in particular, maintaining soundness.

Finally, assuming that neural networks are nonmonotonic systems, we also need to be able to capture nonmonotonic rules encoded in the network. In order to do so, we add negation by default ($\sim$) to the language. We argue that one cannot derive an adequate set of rules from a network without having ($\sim$) in the language, as the following example illustrates.

*Example 5.1.2. (nonmonotonicity of neural networks)* Consider the neural network $N$ of Figure 5.3. Let $W_{n_1a} = 5$, $W_{n_1b} = -5$ and $W_{xn_1} = 1$. Assume that the activation function of $a$ and $b$ is the identity function $g(x) = x$, the activation function of $n_1$ and $x$ is the standard sigmoidal function $f(x) = \frac{1}{1+e^{-x}}$, $i_j \in \{0, 1\}$, and let $\theta_{n_1} = \theta_x = 0.5$, where $\theta_{n_1}$ and $\theta_x$ are the thresholds of neurons $n_1$ and $x$, respectively. Hence, inputs $a = 1$ and $b = 0$ activate $x$ ($x > 0.5$). If one concludes, from that, that $a \rightarrow x$, one should be able to

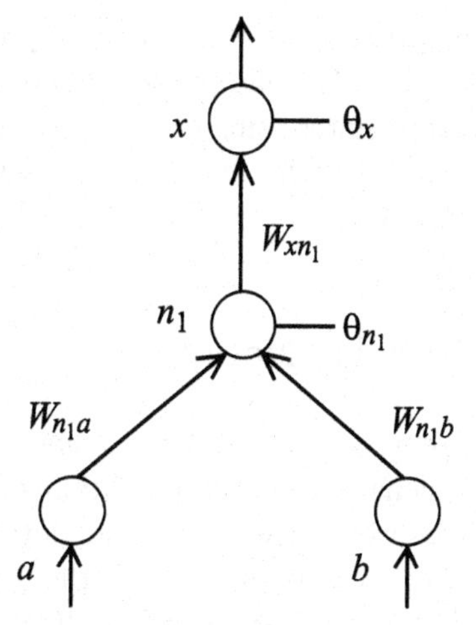


**Fig. 5.3.** A fully connected network with two input neurons $(a, b)$, a single hidden neuron $(n_1)$ and a single output neuron $(x)$.

conclude as well that $ab \rightarrow x$, since the latter rule is subsumed by the former. However, inputs $a = 1$ and $b = 1$ do not activate $x$ $(x < 0.5)$. In this case, one would conclude that $ab \nrightarrow x$, a contradiction!

Therefore, the correct rule to be extracted in the first place, when $a = 1$ and $b = 0$ activate $x$, is $a \sim b \rightarrow x$. The meaning of such a rule should be: $x$ fires in the presence of $a$, provided that $b$ is not present. In fact, if $b$ turns out to be *true* then the conclusion of $x$ is overruled, because $ab \nrightarrow x$. Such a nonmonotonic behaviour should be captured by the extraction of rules with default negation $(\sim)$, as opposed to classical negation $(\neg)$, which is logically stronger than $\sim$ in the sense that a literal should be *proved* instead of *assumed* by default. Classical negation should be explicitly represented in the network by a neuron labelled $\neg x$, as we will exemplify in Chapter 6 with the experiments on knowledge extraction from the network used in Chapter 4 for fault diagnosis.

An immediate result of the above observation is that, in order to conclude that a network $N'$ with two input neurons, say $a$ and $b$, encodes the rule $a \rightarrow x$, firstly we need to make sure that the following rules: $ab \rightarrow x$ and $a \sim b \rightarrow x$, are both encoded in $N'$. In other words, $a \rightarrow x$ should be seen as a *simplification* of the rules $ab \rightarrow x$ and $a \sim b \rightarrow x$ of $N'$, which indicate that $b$ is a *don't care*. In this scenario, the use of *zeros* as input values could be misleading, as for example, when $a = 1$ and $b = 0$ led us to conclude that $a \rightarrow x$ could be a rule of $N$. This is one of the reasons why we find the use of $\{-1, 1\}$ inputs more appropriate.

Summarising, this chapter presents an *eclectic* approach to rule extraction from trained neural networks that allows the *reduction of complexity* of the extraction algorithm in some interesting cases, yet executing a *sound extraction*, and *capturing nonmonotonicity* in the set of rules extracted, by adding *default negation* to the language.

## 5.2 The Case of Regular Networks

Having identified the above problems, let us now start working towards an outline of their solutions. Given the definition of the extraction problem (Section 5.1), firstly we realise that each output neuron $o_j$ has an associated constraint $\mathbf{C}o_j$. We want to find the activation value of $o_j$, $Act(o_j) = h(\sum_{i=1}^{r}(W_{ji}^2 n_i) - \theta_{o_j})$, such that $Act(o_j) > A_{\min}$, where $n_i = h(\sum_{k=1}^{p}(W_{ik}^1 . i_k) - \theta_{n_i})$.

Recall from Chapter 2 that $\mathbf{i} = (i_1, i_2, ..., i_p)$ is the network's input vector ($i_{j(1 \leq j \leq p)} \in \{-1, 1\}$), $\mathbf{o} = (o_1, o_2, ..., o_q)$ is the network's output vector ($o_{j(1 \leq j \leq q)} \in [-1, 1]$), $\mathbf{n} = (n_1, n_2, ..., n_r)$ is the hidden layer vector ($n_{j(1 \leq j \leq r)} \in [-1, 1]$), $\theta_{n_j(1 \leq j \leq r)}$ is the threshold of the $j$th hidden neuron ($\theta_{n_j} \in \Re$), $\theta_{o_j(1 \leq j \leq q)}$ is the threshold of the $j$th output neuron ($\theta_{o_j} \in \Re$), $-\theta_{n_j}$ (resp. $-\theta_{o_j}$) is called the bias of the $j$th hidden neuron (resp. output neuron), $W_{ij(1 \leq i \leq r, 1 \leq j \leq p)}^1$ is the weight of the connection from the $j$th neuron in the input layer to the $i$th neuron in the hidden layer ($W_{ij}^1 \in \Re$), $W_{ij(1 \leq i \leq q, 1 \leq j \leq r)}^2$ is the weight of the connection from the $j$th neuron in the hidden layer to the $i$th neuron in the output layer ($W_{ij}^2 \in \Re$), and finally $h(x) = \frac{2}{1+e^{-\beta x}} - 1$ is the standard bipolar (semi-linear) activation function.

Considering the monotonically crescent characteristic of the activation function $h(x)$ and given that $0 < A_{\min} < 1$ and $\beta > 0$, we can rewrite $h(x) > A_{\min}$ as $x > h^{-1}(A_{\min})$.[4] Hence, the above constraint, $Act(o_j) > A_{\min}$, on each output neuron $o_j$ can be rewritten as Equation 5.1 below, which is given in terms of the hidden neurons' activation values.[5]

$o_j$ *is active for* $\mathbf{i}$ *iff*

$$W_{j1}^2 n_1 + W_{j2}^2 n_2 + \cdots + W_{jr}^2 n_r > h^{-1}(A_{\min}) + \theta_{o_j} \quad (5.1)$$

We can equivalently define the extraction problem as follows: Let $\mathbf{I}$ be the set of input vectors and $\mathbf{O}$ be the set of output vectors of a network $N$.

[4] Note that in order to satisfy $o_j > A_{\min}$, it is required that $x = \sum_{i=1}^{r}(W_{ji}^2 n_i) - \theta_{o_j} > 0$.

[5] Given $h(x) = \frac{2}{1+e^{-\beta x}} - 1$, we obtain $h^{-1}(x) = -\frac{1}{\beta} \ln\left(\frac{1-x}{1+x}\right)$. We use the bipolar semi-linear activation function for convenience; any monotonically crescent activation function could have been used here.

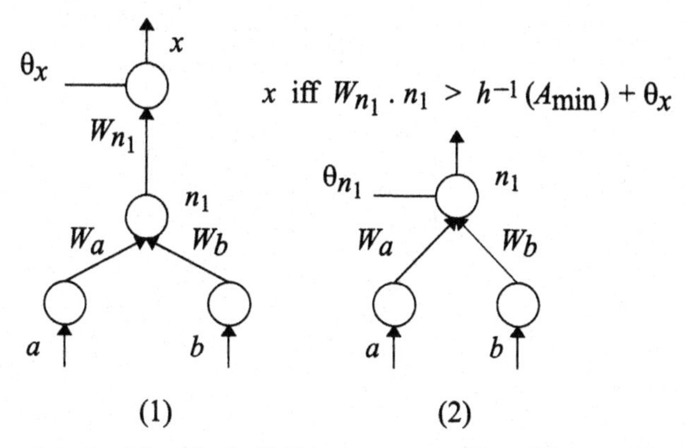


**Fig. 5.4.** A network with a single hidden neuron $n_1$ (1) and its associated constraint (2) w.r.t. output $x$. $W_a, W_b, W_{n_1} \in \Re^+$.

Define a binary relation $\xi$ on $\mathbf{I} \times \mathbf{O}$ such that $\mathbf{o}\xi\mathbf{i} \leftrightarrow \mathbf{o} = \delta(\mathbf{i})$, where $\delta$ is the function computed by $N$. The extraction problem can be stated as: for each $o_j$ in $\mathbf{o} \in \mathbf{O}$ such that $Act(o_j) > A_{\min}$, find the set $\mathbf{I}' \subseteq \mathbf{I}$, where $\mathbf{i} \in \mathbf{I}'$ if $\mathbf{o}\xi\mathbf{i}$.

### 5.2.1 Positive Networks

We start by considering a very simple network where all weights are positive real numbers. As a result, given two input vectors $\mathbf{i}_m$ and $\mathbf{i}_n$, if for all $i$, $1 \leq i \leq r$, $n_i(\mathbf{i}_m) > n_i(\mathbf{i}_n)$ then for all $j$, $1 \leq j \leq q$, $o_j(\mathbf{i}_m) > o_j(\mathbf{i}_n)$, where $n_i(\mathbf{i})$ and $o_j(\mathbf{i})$ denote, respectively, the activation values of hidden neuron $n_i$ and output neuron $o_j$, given input vector $\mathbf{i}$. Moreover, if $\mathbf{i}_m = (1, 1, ..., 1)$, the activation value of each neuron $n_i$ is maximum and, therefore, the activation value of each neuron $o_j$ is maximum as well. Similarly, if $\mathbf{i}_n = (-1, -1, ..., -1)$ then the activation of each $n_i$ is minimum, and thus so is the activation of each $o_j$. That results also from the monotonically crescent characteristic of the activation function $h(x)$, as we will see in detail later. Let us firstly present a simple example to help clarify the above ideas.

*Example 5.2.1.* Consider the network of Figure 5.4(1) and its associated constraint of Figure 5.4(2).

We know that $n_1 = h(W_a.a + W_b.b - \theta_{n_1})$. Since $W_a, W_b > 0$, it is easy to verify that the ordering of Figure 5.5 on the set of input vectors $\mathbf{I}$ holds w.r.t. the output $x$.

The ordering says, for instance, that the activation of $n_1$ is maximum if $\mathbf{i} = (1, 1)$, that $n_1((1, 1)) \geq n_1((1, -1))$, and that $n_1$ is minimum if $\mathbf{i} = (-1, -1)$. Since $W_{n_1} > 0$, the activation of $x$ is also maximum if $\mathbf{i} = (1, 1)$, and minimum if $\mathbf{i} = (-1, -1)$. In other words, the activation value of $x$ is governed by the ordering of Figure 5.5.

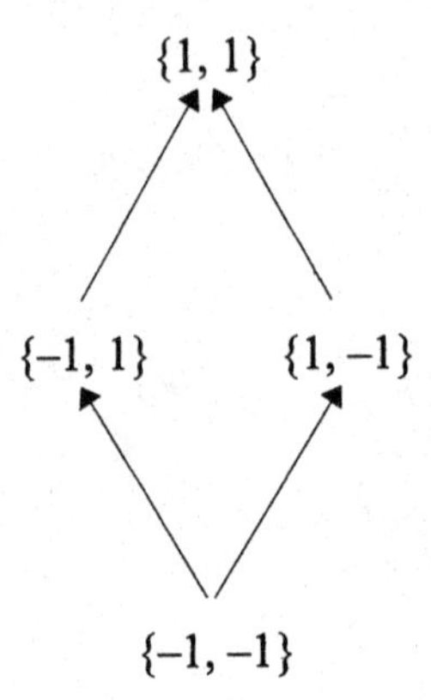


**Fig. 5.5.** Ordering on the set of input vectors (**I**) of $N$.

Given such an ordering, we can draw some conclusions. If the minimum element $(-1, -1)$ is given as the network's input (representing $\sim a \wedge \sim b$), and it activates $x$, satisfying the constraint $W_{n_1} \cdot n_1 > h^{-1}(A_{\min}) + \theta_x$, then any other element in the ordering will also activate $x$. In this case, since all possible input vectors are in the ordering, we can conclude that $x$ is a fact ($\rightarrow x$). If, on the other hand, the maximum element $(1, 1)$ (representing $a \wedge b$) does not activate $x$ then no other element in the ordering does. As a result, no rule with conclusion $x$ should be obtained from the network. Similarly, if it is the case that both $(1, 1)$ (representing $a \wedge b$) and $(1, -1)$ (representing $a \wedge \sim b$) activate $x$, that is, $a \wedge b \rightarrow x$ and $a \wedge \sim b \rightarrow x$, then we can conclude that $a \rightarrow x$, regardless of the activation value of $b$. In this case, the rule $a \rightarrow x$ has been derived as a simplification of the rules $a \wedge b \rightarrow x$ and $a \wedge \sim b \rightarrow x$, which, in turn, have been obtained from *querying* the network.[6]

We have identified, therefore, that if $\forall ij, W_{ij} \in \Re^+$ it is easy to find an ordering on the set of input vectors **I** w.r.t. the set of output vectors **O**. Such information can be very useful to guide a pedagogical extraction of symbolic knowledge from the network. The ordering can help reduce the search space, so that we can safely avoid checking irrelevant input vectors, in the sense that those vectors that are not checked would not generate new rules. Moreover, each rule obtained is sound because the extraction is done by querying the actual network.

Notice that in the worst case we still have to check $2^n$ input vectors, and in the best case we only need to check one input vector (either the minimum or the maximum element in the ordering). Note also that there is, in fact, a

[6] Throughout, we use the term "to *query* the network" as shorthand for "to present an input vector to a network and obtain its output vector".

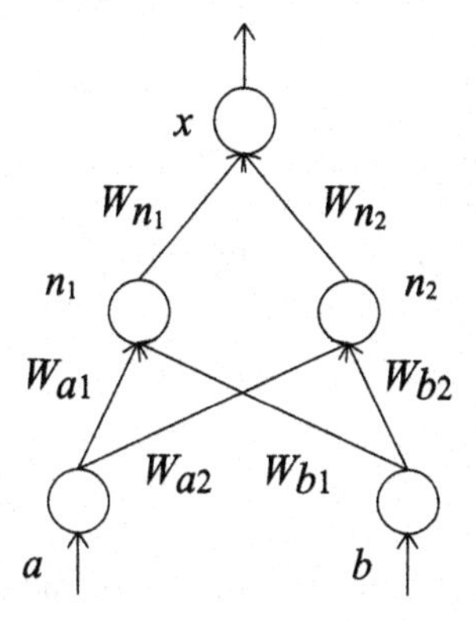


**Fig. 5.6.** A network with two hidden neurons $n_1$ and $n_2$. $W_{ij} \in \Re^+$.

pre-ordering[7] on the set of input vectors, which, however, may be impossible to find without querying each input vector for a particular set of weights. Thus we will focus initially on the analysis of a group of networks where an ordering can be easily found. The following example illustrates what we mean by "an ordering easily found".

*Example 5.2.2.* Consider the network of Figure 5.6. If we know, for instance, that $W_{a1} > W_{b1}$ then we can derive the linear ordering of Figure 5.7(1) on the set of input vectors w.r.t. the activation value of neuron $n_1$. In the same way, if we know that $W_{a2} = W_{b2}$ then we can derive the partial ordering of Figure 5.7(2) w.r.t. $n_2$.

If $W_{a1} > W_{b1}$ and $W_{a2} > W_{b2}$, one can derive a linear ordering on the set of input vectors w.r.t. the output ($x$). However, if $W_{a1} > W_{b1}$ (see Figure 5.8(1)) and $W_{a2} < W_{b2}$ (Figure 5.8(2)), one can only derive the partial ordering of Figure 5.8(3) w.r.t. $x$. Notice that the ordering of Figure 5.8(3) is equal to the ordering used in Example 5.2.1.

When a particular set of weights is given, for example if $W_{a1} = 10, W_{b1} = 5, W_{a2} = 2$ and $W_{b2} = 8$, then one can verify, by querying the network, that $x((-1,1)) > x((1,-1))$. For the time being, however, the partial ordering of Figure 5.8(3) will be used because it does not depend on the weights' values.

Examples 5.2.1 and 5.2.2 indicate that the partial ordering on the set of input vectors is the same for a network with two hidden neurons and for a network with only one hidden neuron. In fact, we will see later that if $W_{ij} \in \Re^+$ then the partial ordering on the set of input vectors w.r.t. the outputs' activations are not affected by the number of hidden neurons.

Let us now try to see if we can find an ordering easily in the case where there are three inputs $\{a, b, c\}$, but still with $W_{ij} \in \Re^+$. It seems reasonable to

[7] A pre-order is a reflexive and transitive relation.

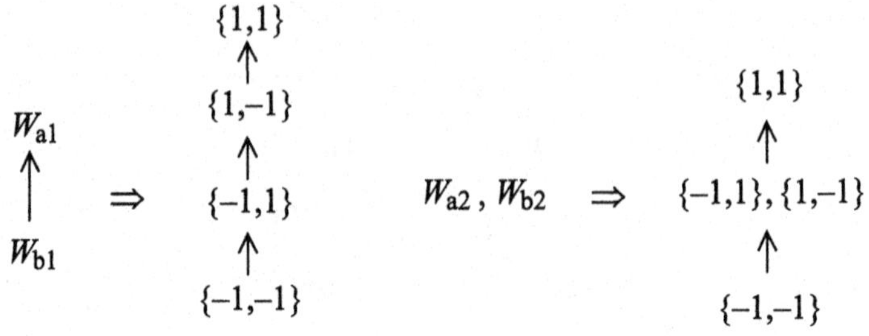

{1,1} ↑ {1,−1} ↑ {−1,1} ↑ {−1,−1}

$W_{a1}$ ↑ $W_{b1}$ ⇒

$W_{a2}$, $W_{b2}$ ⇒ {1,1} ↑ {−1,1}, {1,−1} ↑ {−1,−1}

(1) (2)

**Fig. 5.7.** Orders on the set of input vectors w.r.t. hidden neurons $n_1$ and $n_2$.

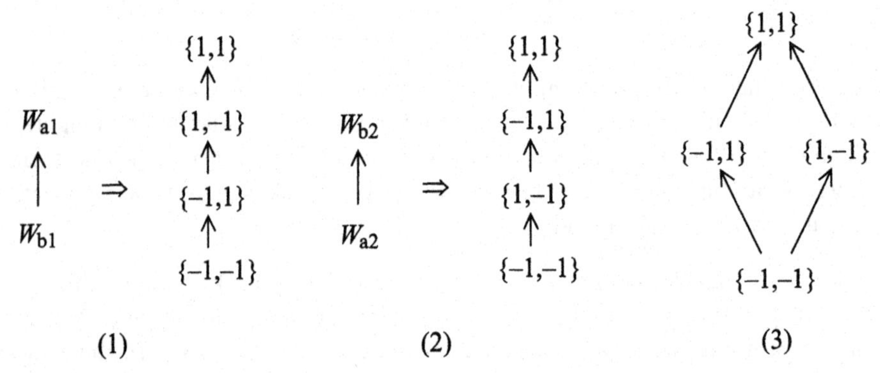


**Fig. 5.8.** Orders on the set of input vectors w.r.t. hidden neurons $n_1$ and $n_2$, and output neuron $x$.

consider the ordering of Figure 5.9, since we do not have any extra information regarding the number of hidden neurons and the network's weights. The ordering is built starting from element $(-1,-1,-1)$ and then flipping each input at a time from -1 to 1 until $(1,1,1)$ is obtained.

It seems that, for an arbitrary number of input and hidden neurons, if $W_{ij} \in \Re^+$ then there exists a unique minimal element $(-1,-1,...,-1)$ and a unique maximum element $(1,1,...,1)$ in the ordering on the set of input vectors w.r.t. the activation values of the output neurons. Let us see if we can confirm this.

We assume the following conventions. Let $\mathbf{P}$ be a finite set of literals. Recall from Chapter 2 that an interpretation is a function from $\mathbf{P}$ to $\{true, false\}$. Given a neural network, we associate each input and output neuron with a unique literal in $\mathbf{P}$. Let $\mathcal{I}$ be the set of input neurons and $\mathcal{O}$ the set of output neurons. Then, each input vector $\mathbf{i}$ can be seen as an interpretation. Suppose $\mathcal{I} = \{p, q, r\}$. We fix a linear ordering on the symbols of $\mathcal{I}$ and

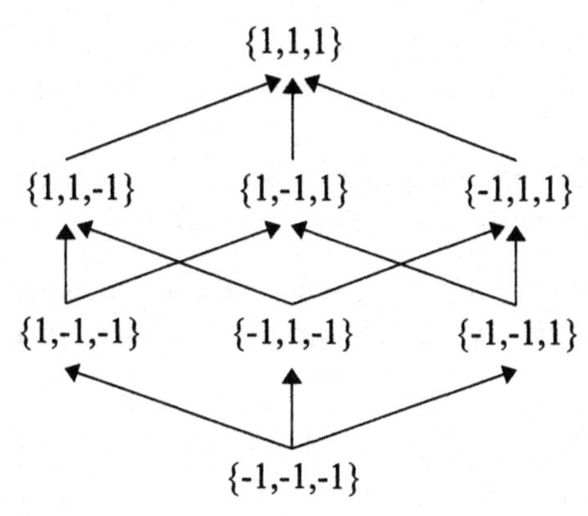


**Fig. 5.9.** Partial ordering w.r.t. set inclusion on the powerset of $\{a, b, c\}$.

represent it as a list, say $[p, q, r]$. This will allow us to refer to interpretations and input vectors interchangeably in the following way. We represent $\mathbf{i}$ as a string of 1's and -1's, where the value 1 in a particular position in the string means that the literal at the corresponding position in the list of symbols is assigned *true*, and the value -1 means that it is assigned *false*. For example, if $\mathbf{i} = (1, -1, 1)$ then $\mathbf{i}(p) = \mathbf{i}(r) = true$ and $\mathbf{i}(q) = false$.

Each input vector $\mathbf{i}$ can be seen as an abstract representation of a subset of the set of input neurons, with '1's denoting the presence and '-1's denoting the absence of a neuron in the set. For example, given the set of input neurons $\mathcal{I}$ as the list $[p, q, r]$, if $\mathbf{i} = (1, -1, 1)$ then it represents the set $\{p, r\}$, if $\mathbf{i} = (-1, -1, -1)$, it represents $\emptyset$, if $\mathbf{i} = (1, 1, 1)$, it represents $\{p, q, r\}$, and so on. We conclude that the set of input vectors $\mathbf{I}$ is an abstract representation of the power set of the set of input neurons $\mathcal{I}$. We write it as $\mathbf{I} = \wp(\mathcal{I})$.

We are now in a position to formalise the above concepts. We start by defining a distance function between input vectors. The distance between two input vectors is the number of neurons assigned different inputs by each vector. In terms of the above analogy between input vectors and interpretations, the same distance function can be defined as the number of propositional variables with different truth-values.

**Definition 5.2.1.** ($dist : \mathbf{I} \times \mathbf{I} \rightarrow \aleph$) *Let $\mathbf{i}_m$ and $\mathbf{i}_n$ be two input vectors in $\mathbf{I}$. The distance $dist(\mathbf{i}_m, \mathbf{i}_n)$ between $\mathbf{i}_m$ and $\mathbf{i}_n$ is the number of inputs $i_j$ for which $\mathbf{i}_m(i_j) \neq \mathbf{i}_n(i_j)$, where $\mathbf{i}(i_j)$ denotes the input value $i_j$ of vector $\mathbf{i}$.*

For example, the distance between $\mathbf{i}_1 = (-1, -1, 1)$ and $\mathbf{i}_2 = (1, 1, -1)$ is $dist(\mathbf{i}_1, \mathbf{i}_2) = 3$. The distance between $\mathbf{i}_3 = (-1, 1, -1)$ and $\mathbf{i}_4 = (1, -1, -1)$ is $dist(\mathbf{i}_3, \mathbf{i}_4) = 2$.

**Proposition 5.2.1.** *[Rod97] The function dist is a metric on* $\mathbf{I}$.

Clearly, the function *dist* is also a bounded metric on $\mathbf{I}$. That is, $dist(\mathbf{i}_m, \mathbf{i}_n) \leq p$ for all $\mathbf{i}_m, \mathbf{i}_n \in \mathbf{I}$, where $p$ is the length of the input vectors $\mathbf{i}_m$ and $\mathbf{i}_n$.

Another concept that will prove to be important is the sum of the input elements in a input vector. We define it as follows.

**Definition 5.2.2.** ($\langle\rangle : \mathbf{I} \rightarrow \mathcal{Z}$) *Let* $\mathbf{i}_m$ *be a p-ary input vector in* $\mathbf{I}$. *The sum* $\langle \mathbf{i}_m \rangle$ *of* $\mathbf{i}_m$ *is the sum of all input elements* $i_j$ *in* $\mathbf{i}_m$*, that is* $\langle \mathbf{i}_m \rangle = \sum_{j=1}^{p} \mathbf{i}_m(i_j)$.

For example, the sum of $\mathbf{i}_1 = (-1, -1, 1)$ is $\langle \mathbf{i}_1 \rangle = -1$. The sum of $\mathbf{i}_2 = (1, 1, -1)$ is $\langle \mathbf{i}_2 \rangle = 1$.

Now we define the ordering $\leq_{\mathbf{I}}$ on $\mathbf{I} = \wp(\mathcal{I})$ w.r.t. set inclusion. Recall that $\mathbf{i}_m \in \mathbf{I}$ is an abstract representation of a subset of $\mathcal{I}$. We say that $\mathbf{i}_m \subseteq \mathbf{i}_n$ if the set represented by $\mathbf{i}_m$ is a subset of the set represented by $\mathbf{i}_n$.

**Definition 5.2.3.** *Let* $\mathbf{i}_m$ *and* $\mathbf{i}_n$ *be input vectors in* $\mathbf{I}$. $\mathbf{i}_m \leq_{\mathbf{I}} \mathbf{i}_n$ *iff* $\mathbf{i}_m \subseteq \mathbf{i}_n$.

Clearly, for a finite set $\mathcal{I}$, $\mathbf{I}$ is a finite partially ordered set w.r.t. $\leq_{\mathbf{I}}$ having $\mathcal{I}$ as its maximum element and the empty set $\emptyset$ as its minimum element, that is, $sup(\mathbf{I}) = \{1, 1, ..., 1\}$ and $inf(\mathbf{I}) = \{-1, -1, ..., -1\}$. In fact, $[\mathbf{I}, \leq_{\mathbf{I}}]$ is a lattice.

**Proposition 5.2.2.** *[PY73] The partially ordered set* $[\mathbf{I}, \leq_{\mathbf{I}}]$ *is a distributive lattice.*

Note that $\mathbf{I}$ is the $n$-cube in the Cartesian $n$-dimensional space of coordinates $x_1, x_2, ..., x_n$ where the generic $x_j (1 \leq j \leq n)$ is either $-1$ or $1$.

Finally, Proposition 5.2.3 shows that $\leq_{\mathbf{I}}$ is the ordering of our interest w.r.t. the network's output.

**Proposition 5.2.3.** *If* $W_{ji} \in \Re^+$ *then* $\mathbf{i}_m \leq_{\mathbf{I}} \mathbf{i}_n$ *implies* $o_j(\mathbf{i}_m) \leq o_j(\mathbf{i}_n)$*, for all* $1 \leq j \leq q$.

*Proof. Let* $\mathbf{i}_m \leq_{\mathbf{I}} \mathbf{i}_n$ *and* $dist(\mathbf{i}_m, \mathbf{i}_n) = 1$*, then* $\mathbf{i}_m(i_i) = -1$ *and* $\mathbf{i}_n(i_i) = 1$ *for some input* $i_i$. *Let* $r$ *be the number of hidden neurons in the network. Firstly, we have to show that:*

$$h(\sum_{i=1}^{p}(W^1_{1i}\mathbf{i}_m(i_i) - \theta_{n_1})) + h(\sum_{i=1}^{p}(W^1_{2i}\mathbf{i}_m(i_i) - \theta_{n_2})) + \cdots +$$

$$h(\sum_{i=1}^{p}(W^1_{ri}\mathbf{i}_m(i_i) - \theta_{n_r})) \leq h(\sum_{i=1}^{p}(W^1_{1i}\mathbf{i}_n(i_i) - \theta_{n_1}))+$$

$$h(\sum_{i=1}^{p}(W^1_{2i}\mathbf{i}_n(i_i) - \theta_{n_2})) + \cdots + h(\sum_{i=1}^{p}(W^1_{ri}\mathbf{i}_n(i_i) - \theta_{n_r}))$$

*By the definition of* $\leq_{\mathbf{I}}$ *and since* $W_{ji} \in \Re^+$ *we derive immediately that for all* $j (1 \leq j \leq r)$ $\sum_{i=1}^{p}(W_{ji}^1 \mathbf{i}_m(i_i) - \theta_{n_j}) \leq \sum_{i=1}^{p}(W_{ji}^1 \mathbf{i}_n(i_i) - \theta_{n_j})$, *and by the monotonically crescent characteristic of* $h(x)$, *we obtain* $\forall j (1 \leq j \leq r)$ $h(\sum_{i=1}^{p}(W_{ji}^1 \mathbf{i}_m(i_i) - \theta_{n_j})) \leq h(\sum_{i=1}^{p}(W_{ji}^1 \mathbf{i}_n(i_i) - \theta_{n_j}))$. *This proves that if* $\mathbf{i}_m \leq_{\mathbf{I}} \mathbf{i}_n$ *and* $dist(\mathbf{i}_m, \mathbf{i}_n) = 1$ *then* $n_j(\mathbf{i}_m) \leq n_j(\mathbf{i}_n)$ *for all* $1 \leq j \leq r$. *In the same way, we obtain that* $h(\sum_{i=1}^{r}(W_{ji}^2 \mathbf{n}_m(n_i) - \theta_{o_j})) \leq h(\sum_{i=1}^{r}(W_{ji}^2 \mathbf{n}_n(n_i) - \theta_{o_j}))$, *and, therefore, that:*

$$\text{If } \mathbf{i}_m \leq_{\mathbf{I}} \mathbf{i}_n \text{ and } dist(\mathbf{i}_m, \mathbf{i}_n) = 1 \text{ then } o_j(\mathbf{i}_m) \leq o_j(\mathbf{i}_n), \text{ for } 1 \leq j \leq q \quad (5.2)$$

*Now, let* $\mathbf{i}_m \leq_{\mathbf{I}} \mathbf{i}_n$ *and* $dist(\mathbf{i}_m, \mathbf{i}_n) = k$ $(1 < k \leq p)$. *There are* $k-1$ *vectors* $\mathbf{i}_\xi, ..., \mathbf{i}_\zeta$ *such that* $\mathbf{i}_m \leq_{\mathbf{I}} \mathbf{i}_\xi \leq_{\mathbf{I}} ... \leq_{\mathbf{I}} \mathbf{i}_\zeta \leq_{\mathbf{I}} \mathbf{i}_n$. *From 5.2 above, and since* $\leq$ *is transitive, it follows that if* $\mathbf{i}_m \leq_{\mathbf{I}} \mathbf{i}_n$ *then* $o_j(\mathbf{i}_m) \leq o_j(\mathbf{i}_n)$, *for all* $1 \leq j \leq q$. □

### 5.2.2 Regular Networks

Let us see now if we can relax the condition $W_{ji} \in \Re^+$ and still find easily an ordering on the set of input vectors of a network. We start by giving an example.

*Example 5.2.3.* Consider the network given at Example 5.2.2 (Figure 5.6), but now assume $W_{b1}, W_{b2} < 0$. Although some weights are negative, we can find a "regularity" in the network. For example, input neuron $b$ contributes negatively to the activation of both $n_1$ and $n_2$, and there are no negative connections from the hidden to the output layer of the network. Following [Fu94], we can transform the network of Figure 5.6 into the network of Figure 5.10, where all weights are positive and input neuron $b$ is negated.

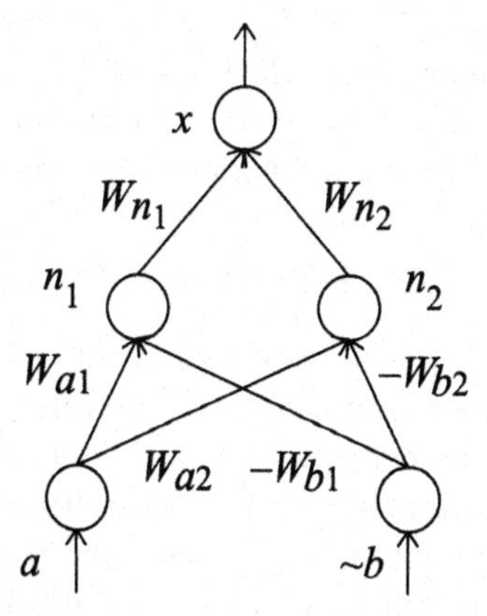


**Fig. 5.10.** The positive form of a (regular) network.

Given the network of Figure 5.10, we can find an ordering on the set of input vectors in the same way as before. The only difference is that now $\mathcal{I} = \{a, \sim b\}$. We will see later that, if we account for the fact that $\mathcal{I}$ may now have negated literals (default negation), then the networks of Figures 5.6 and 5.10 are equivalent.

Let us analyse what we have done in the above example. We continue to assume that the weights from the hidden layer to any one neuron in the output layer of the network are all positive real numbers. Then, for each input neuron $y$, we do the following:

1. If $y$ is linked to the hidden layer through connections with positive weights only:
   a) do nothing.
2. If $y$ is linked to the hidden layer through connections with negative weights $W_{jy}$ only:
   a) change each $W_{jy}$ to $-W_{jy}$ and rename $y$ by $\sim y$.
3. If $y$ is linked to the hidden layer through positive and negative connections:
   a) add a neuron named $\sim y$ to the input layer, and
   b) for each negative connection with weight $W_{jy}$ from $y$ to $n_j$:
      i. add a new connection with weight $-W_{jy}$ from $\sim y$ to $n_j$, and
      ii. delete the connection with weight $W_{jy}$ from $y$ to $n_j$.

We call the above procedure the *Transformation Algorithm.*

*Example 5.2.4.* Consider again the network given at Example 5.2.2 (Figure 5.6), but now assume that only $W_{a2} < 0$. Applying the Transformation Algorithm we obtain the network of Figure 5.11.

Although the network of Figure 5.11 has positive weights only, it is clearly not equivalent to the original network (Figure 5.6). In this case, the combination of $n_1$ and $n_2$ is not straightforward. Note that, $\mathbf{i}_1 = (1,1)$ in the original network provides the maximum activation of $n_1$, but not the maximum activation of $n_2$, which is given by $\mathbf{i}_2 = (-1,1)$. We cannot affirm any more that $x(\mathbf{i}_1) > x(\mathbf{i}_2)$ without querying the network with such input vectors.

Examples 5.2.3 and 5.2.4 indicate that if the Transformation Algorithm generates a network where complementary literals (say, $a$ and $\sim a$) appear in the input layer (see the network of Figure 5.11) then the ordering $\leq_{\mathbf{I}}$ on $\mathbf{I}$ is not applicable. However, if complementary literals do not appear in the input layer of the network obtained with the above transformation (see Figure 5.10), it seems that $\leq_{\mathbf{I}}$ is still valid. This motivates the following definition.

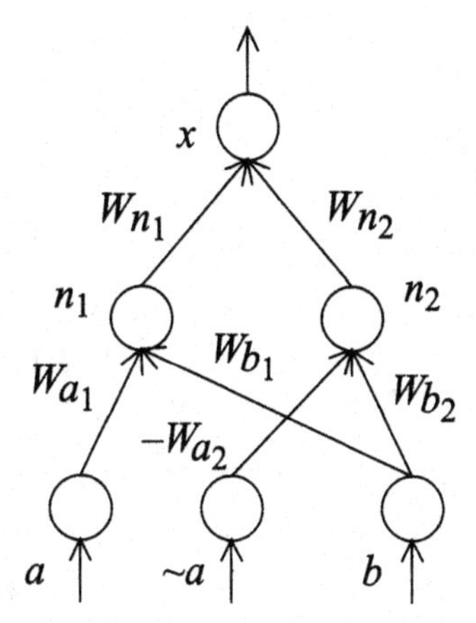


**Fig. 5.11.** The positive form of a (non-regular) network.

**Definition 5.2.4.** *A single hidden layer neural network is said to be* regular *if its connections from the hidden layer to each output neuron have either all positive or all negative weights, and if the above Transformation Algorithm generates on it a network without complementary literals in the input layer.*

Returning to Example 5.2.3, we have seen that the positive form $N_+$ of a regular network $N$ may have negated literals in the set of input neurons (e.g. $\mathcal{I}_+ = \{a, \sim b\}$). In this case, if we represent $\mathcal{I}_+$ as a list, say $[a, \sim b]$, and refer to an input vector $\mathbf{i} = (-1, 1)$ w.r.t. $\mathcal{I}_+$, then we consider $\mathbf{i}$ as the abstract representation of the set $\{\sim b\}$. In the same way, $\mathbf{i} = (1, -1)$ represents $\{a\}$, and so on. In this sense, the set of input vectors of $N_+$ can be ordered w.r.t. set inclusion exactly as before, i.e. using Definition 5.2.3, as the following example illustrates.

*Example 5.2.5.* Consider the network $N_+$ of Figure 5.10. Given $\mathcal{I}_+ = [a, \sim b]$, we obtain the ordering of Figure 5.12(1) w.r.t. set inclusion. The ordering of Figure 5.12(2) on the set of input vectors of the original network $N$ is obtained by mapping each element of (1) into (2) using $\sim b = 1$ implies $b = -1$, and $\sim b = -1$ implies $b = 1$. As a result, querying $N_+$ with $\mathbf{i} = (1, 1)$ is equivalent to querying $N$ with $\mathbf{i} = (1, -1)$, querying $N_+$ with $\mathbf{i} = (-1, 1)$ is equivalent to querying $N$ with $\mathbf{i} = (-1, -1)$, and so on.

More precisely, we define the function $\sigma$ mapping input vectors of the positive form into input vectors of the original network, as follows. Let $\mathbf{I}$ be the set of input vectors of $s$ tuples. Given the set of input neurons $\mathcal{I}_+$ and an abstract representation $\mathbf{I}_+$ of $\wp(\mathcal{I}_+)$, each element $x_i \in \mathcal{I}_+$, $1 \leq i \leq s$, is mapped to the set $\{-1, 1\}$ such that $\sigma_{[x_1,...,x_s]}(i_1, ..., i_s) = (i'_1, ..., i'_s)$, where $i'_i = i_i$ if $x_i$ is a positive literal and $i'_i = -i_i$ if $x_i$ is a negative literal. For example $\sigma_{[a,\sim b,c,\sim d]}(1, 1, -1, -1) = (1, -1, -1, 1)$.

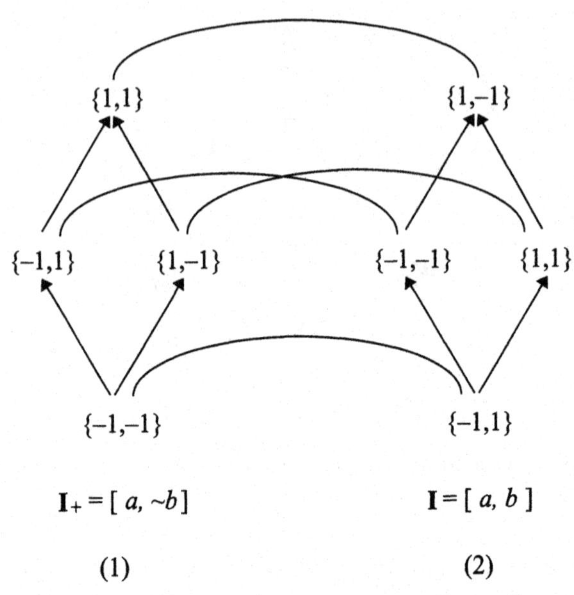


**Fig. 5.12.** The set inclusion ordering on the positive form of a network (1) and the ordering on the original network (2) w.r.t. output neuron $x$.

Note that the correspondence between input vectors and interpretations is still valid. We only need to define $\mathbf{i}(\sim p) = false$ iff $\mathbf{i}(p) = true$ and $\sim\sim p = p$. For example, for $\mathcal{I}_+ = [a, \sim b]$, if $\mathbf{i} = (-1, -1)$ then $\mathbf{i}(a) = false$ and $\mathbf{i}(b) = true$.

**Proposition 5.2.4.** *Let $\mathcal{I}_+$ be the set of input neurons of the positive form $N_+$ of a regular network $N$. Let $\mathbf{I}_+ = \wp(\mathcal{I}_+)$ be ordered under the set inclusion relation $\leq_{\mathbf{I}_+}$, and $\mathbf{i}_m, \mathbf{i}_n \in \mathbf{I}_+$. Thus, $\mathbf{i}_m \leq_{\mathbf{I}_+} \mathbf{i}_n$ implies $o_j(\sigma_{[\mathcal{I}_+]}(\mathbf{i}_m)) \leq o_j(\sigma_{[\mathcal{I}_+]}(\mathbf{i}_n))$, for all $1 \leq j \leq q$ in $N$.*

*Proof. Straightforward by Proposition 5.2.3 and the above definition of the mapping function $\sigma$.*□

Proposition 5.2.4 establishes the correlation between regular networks and their positive counterpart. For example, considering the mapping of Figure 5.12, if $\mathbf{i} = (-1, 1)$ activates $x$ in $N_+$ then $\mathbf{i} = (-1, -1)$ activates $x$ in $N$. The rule to be extracted for this query is $\sim a \sim b \rightarrow x$. As a result, the extraction procedure can either use the set inclusion ordering on $\mathcal{I}_+$ (as, e.g. in Figure 5.12(1)), and query directly the positive form of the network, or use the mapping function $\sigma$ to obtain the ordering on the regular, original network (Figure 5.12(2)), and query the original network. We will adopt the first policy. Note that if the network is already positive then $\sigma$ is the identity function.

We have seen briefly that if we can find an ordering on the set of input vectors of a network then there are some properties that can help in reducing the search space of input vectors during a pedagogical extraction of rules. Let us now define precisely these properties.

**Proposition 5.2.5.** *(Search Space Pruning Rule 1) Let* $\mathbf{i}_m$ *and* $\mathbf{i}_n$ *be input vectors of the positive form of a regular neural network* $N$ *such that* $dist(\mathbf{i}_m, \mathbf{i}_n) = 1$ *and* $\langle \mathbf{i}_m \rangle < \langle \mathbf{i}_n \rangle$. *If* $\mathbf{i}_n$ *does not satisfy the constraint* $\mathbf{Co}_j$ *(see Equation 5.1) on the* $j$*th output neuron of* $N$, *then* $\mathbf{i}_m$ *does not satisfy* $\mathbf{Co}_j$ *either.*

*Proof. Directly by Definitions 5.2.1, 5.2.2 and 5.2.3, if* $dist(\mathbf{i}_m, \mathbf{i}_n) = 1$ *and* $\langle \mathbf{i}_m \rangle < \langle \mathbf{i}_n \rangle$ *then* $\mathbf{i}_m \leq_{\mathbf{I}} \mathbf{i}_n$. *By Proposition 5.2.3,* $o_j(\mathbf{i}_m) \leq o_j(\mathbf{i}_n)$. *This completes the proof.* □

**Proposition 5.2.6.** *(Search Space Pruning Rule 2) Let* $\mathbf{i}_m$ *and* $\mathbf{i}_n$ *be input vectors of the positive form of a regular neural network* $N$, *such that* $dist(\mathbf{i}_m, \mathbf{i}_n) = 1$ *and* $\langle \mathbf{i}_m \rangle < \langle \mathbf{i}_n \rangle$. *If* $\mathbf{i}_m$ *satisfies the constraint* $\mathbf{Co}_j$ *on the* $j$*th output neuron of* $N$, *then* $\mathbf{i}_n$ *also satisfies* $\mathbf{Co}_j$.

*Proof. This is the contrapositive of Proposition 5.2.5.* □

Proposition 5.2.5 says that for any $\mathbf{i} \in \mathbf{I}$, starting from $sup(\mathbf{I})$, if $\mathbf{i}$ does not activate the $j$th output neuron $o_j$, then the immediate predecessors of $\mathbf{i}$ do not activate $o_j$ either. Similarly, Proposition 5.2.6 says that for any $\mathbf{i} \in \mathbf{I}$, starting from $inf(\mathbf{I})$, if $\mathbf{i}$ activates the $j$th output neuron $o_j$, then the immediate successors of $\mathbf{i}$ do as well.

In Example 5.2.1, we have seen briefly that the extracted rules $ab \rightarrow x$ and $a \sim b \rightarrow x$ could be simplified to obtain a single rule, namely $a \rightarrow x$. Let us now define a group of *simplification rules* that will help in the extraction of a smaller and clearer set of rules. They will also help to reduce the number of premises per rule, an important aspect of readability.

**Definition 5.2.5.** *(*Subsumption*) A rule* $r_1$ *subsumes a rule* $r_2$ *iff* $r_1$ *and* $r_2$ *have the same conclusion and the set of premises of* $r_1$ *is a subset of the set of premises of* $r_2$.

For example, $a \rightarrow x$ subsumes $ab \rightarrow x$ and $a \sim b \rightarrow x$.

**Definition 5.2.6.** *(*Complementary Literals*) Let* $r_1 = L_1, ..., L_i, ..., L_j \rightarrow L_{j+1}$ *and* $r_2 = L_1, ..., \sim L_i, ..., L_j \rightarrow L_{j+1}$ *be extracted rules, where* $j \leq |\mathcal{I}|$. *Then,* $r_3 = L_1, ..., L_{i-1}, L_{i+1}, ..., L_j \rightarrow L_{j+1}$ *is also an extracted rule. Note that* $r_3$ *subsumes* $r_1$ *and* $r_2$.

For example, if $\mathcal{I} = \{a, b, c\}$ and we write $a \sim b \rightarrow x$, then it simplifies $a \sim bc \rightarrow x$ and $a \sim b \sim c \rightarrow x$. Note that, considering the ordering on $\mathbf{I}$, the above property requires that two adjacent input vectors, $\mathbf{i}_m = (1, -1, 1)$ and $\mathbf{i}_n = (1, -1, -1)$, activate $x$.

**Definition 5.2.7.** (Fact) *If a literal $L_{j+1}$ holds in the presence of any combination of the truth values of literals $L_1, ..., L_j$ in $\mathcal{I}$ then we derive a rule of the form $\rightarrow L_{j+1}$ ($L_{j+1}$ is a fact).*

Definition 5.2.7 is an important special case of Definition 5.2.6. Considering the ordering on **I**, an output neuron $x$ is a fact iff $inf(\mathbf{I})$ activates $x$. Note that, by Proposition 5.2.6, if $inf(\mathbf{I})$ activates $x$ then any other input vector in **I** also does.

Another interesting special case occurs when $sup(\mathbf{I})$ does not activate $x$. In this case, by Proposition 5.2.5, no other input vector in **I** activates $x$ and, thus, there are no rules with conclusion $x$ to be derived from the network.

**Definition 5.2.8.** (M of N) *Let $m, n \in \aleph, \mathcal{I}' \subseteq \mathcal{I}, |\mathcal{I}'| = n, m \leq n$. If any combination of $m$ elements chosen from $\mathcal{I}'$ implies $L_{j+1}$, we derive a rule of the form $m(\mathcal{I}') \rightarrow L_{j+1}$.*

The above Definition 5.2.8 may be very useful in helping to reduce the number of rules extracted. It states that, for example, $2(abc) \rightarrow x$ represents $ab \rightarrow x, ac \rightarrow x$, and $bc \rightarrow x$. In this way, if for example we write $3(abcdef) \rightarrow x$ then this rule is a short representation of at least $C_3^6 = 20$ rules [8].

There is a rather intricate relation between each rule of the form *M of N* and the ordering on the set of input vectors **I**, in the sense that each valid *M of N* rule represents a subset of **I**. Here is a flavour of that relation in an example where it is easy to identify it. Suppose $\mathcal{I} = \{a, b, c\}$ and assume that $\mathcal{I}' = \mathcal{I}$. Let us say that the output neuron in question is $x$ and that constraint $\mathbf{C}_{o_x}$ is satisfied by at least one input vector in **I**. If only $sup(\mathbf{I})$ satisfies $\mathbf{C}_{o_x}$, we derive the rule $abc \rightarrow x$. Clearly, this rule is equivalent to $3(abc) \rightarrow x$. If all the immediate predecessors of $sup(\mathbf{I})$ also satisfy $\mathbf{C}_{o_x}$, it is not difficult to verify that the four rules obtained ($r_1 : abc \rightarrow x$, $r_2 : ab \sim c \rightarrow x$, $r_3 : a \sim bc \rightarrow x$, $r_4 :\sim abc \rightarrow x$) can be represented by $2(abc) \rightarrow x$. This is because, by Definition 5.2.6, each rule $r_2$, $r_3$ and $r_4$ can be simplified together with $r_1$, deriving $abc \rightarrow x$, $ab \rightarrow x$, $ac \rightarrow x$ and $bc \rightarrow x$. Since, by Definition 5.2.5, $abc \rightarrow x$ is subsumed by any of the other three rules, we obtain $2(abc) \rightarrow x$. Moreover, $2(abc) \rightarrow x$ subsumes $3(abc) \rightarrow x$. This motivates the definition of yet another simplification rule, as follows.

**Definition 5.2.9.** (M of N Subsumption) *Let $m, p \in \aleph, \mathcal{I}' \subseteq \mathcal{I}$. $m(\mathcal{I}') \rightarrow L_{j+1}$ subsumes $p(\mathcal{I}') \rightarrow L_{j+1}$ iff $m < p$.*

Returning to the illustration about the relation between *M of N* rules and subsets of **I**, let us see what happens if the elements at distance 2 from

[8] Note that if $\mathcal{I} = \{a, b, c\}$ and we write $1(ab) \rightarrow x$ then such an *M of N* rule is a simplification of $C_1^2 = 2$ rules: $a \rightarrow x$ and $b \rightarrow x$. However, by Definition 5.2.6, $a \rightarrow x$ and $b \rightarrow x$ are already simplifications of $abc \rightarrow x$, $ab \sim c \rightarrow x$, $a \sim bc \rightarrow x$, $a \sim b \sim c \rightarrow x$, $\sim abc \rightarrow x$, *and* $\sim ab \sim c \rightarrow x$.

$sup(\mathbf{I})$ all satisfy $\mathbf{C}_{o_x}$. We expect that the set of rules obtained from **I** could be represented by $1(abc) \rightarrow x$, and in fact it is. From the elements at distance 2 from $sup(\mathbf{I})$, we obtain the following rules: $r_1 : a \sim b \sim c \rightarrow x$, $r_2 : \sim ab \sim c \rightarrow x$, and $r_3 : \sim a \sim bc \rightarrow x$. By Proposition 5.2.6, we know that the elements at distance 1 from $sup(\mathbf{I})$ also satisfy $\mathbf{C}_{o_x}$, and we derive the rules: $r_4 : ab \sim c \rightarrow x$, $r_5 : a \sim bc \rightarrow x$, and $r_6 : \sim abc \rightarrow x$. Again by Proposition 5.2.6, $sup(\mathbf{I})$ itself also satisfies $\mathbf{C}_{o_x}$, and we derive $r_7 : abc \rightarrow x$. Now, applying Definition 5.2.6 over $r_1$ and $r_4$, we obtain the simplified rule $r_8 : a \sim c \rightarrow x$; taking $r_5$ and $r_7$, we obtain $r_9 : ac \rightarrow x$; and from $r_8$ and $r_9$, we derive $r_a : a \rightarrow x$. Similarly, from $r_2, r_4, r_6$ and $r_7$, we derive $r_b : b \rightarrow x$, and from $r_3, r_5, r_6$ and $r_7$, we derive $r_c : c \rightarrow x$. Finally, since $r_a$, $r_b$ and $r_c$ together subsume any rule previously obtained, by Definition 5.2.8 we may derive the single *M of N* rule $1(abc) \rightarrow x$.

We have identified a pattern in the ordering on **I** w.r.t. a group of *M of N* rules, the ones where $\mathcal{I}' = \mathcal{I}$. More generally, given $|\mathcal{I}| = k$, if all the elements in **I** that are at distance $d$ from $sup(\mathbf{I})$ satisfy constraint $\mathbf{C}_{o_x}$ then derive the *M of N* rule $(k - d)(\mathcal{I}) \rightarrow x$. Note that there are $C^k_{k-d}$ elements at distance $d$ from $sup(\mathbf{I})$, and that, as a result of Proposition 5.2.6, if all the elements in **I** at distance $d$ from $sup(\mathbf{I})$ satisfy $\mathbf{C}_{o_x}$ then any other element at distance $d'$ from $sup(\mathbf{I})$ such that $0 \leq d' < d$ also satisfies $\mathbf{C}_{o_x}$.

*Remark 5.2.1.* We have defined regular networks (see Definition 5.2.4) either with all the weights from the hidden layer to each output neuron positive, or with all of them negative. We have, although, considered in the above examples and definitions only the ones where all such weights are positive. However, it is not difficult to verify that the constraint $\mathbf{C}_{o_j}$ on the $j$th output of a regular network with negative weights from hidden to output layer is $W^2_{j1}n_1 + W^2_{j2}n_2 + \cdots + W^2_{jr}n_r < h^{-1}(A_{\min}) + \theta_{o_j}$. As a result, the only difference now is on the sign ($<$) of the constraint. In other words, in this case we only need to invert the signs at Propositions 5.2.5 and 5.2.6. All remaining definitions and propositions are still valid.

We have so far referred to soundness and completeness of the extraction algorithm in a somewhat vague manner. Let us define these concepts precisely.

**Definition 5.2.10.** (Extraction Algorithm Soundness) *A rule extraction algorithm from a neural network $N$ is* sound *iff for each rule $r_i$ extracted, whenever the premise of $r_i$ is presented to $N$ as input vector, in the presence of any combination of the input values of literals not referenced by rule $r_i$, the conclusion of $r_i$ presents activation greater than $A_{\min}$ in the output vector of $N$.*

**Definition 5.2.11.** (Extraction Algorithm Completeness) *A rule extraction algorithm from a neural network $N$ is* complete *iff each rule extracted by* *exhaustively verifying all the combinations of the input vector of $N$ either belongs to, or is subsumed by, a rule in the set of rules generated by the extraction algorithm.*

We are finally in a position to present the extraction algorithm for regular networks, which will be refined in Section 5.3 for the general case extraction.

– *Knowledge Extraction Algorithm for Regular Networks*[9]

1. Apply the *Transformation Algorithm* over $N$, obtaining its positive form $N_+$;
2. Let $inf(\mathbf{I}) = \{-1, -1, ..., -1\}$ and $sup(\mathbf{I}) = \{1, 1, ..., 1\}$;
3. For each neuron $o_j$ in the output layer of $N_+$ do:
   (a) Query $N_+$ with input vector $inf(\mathbf{I})$. If $o_j > A_{\min}$, apply the Simplification Rule *Fact* and stop.
   (b) Query $N_+$ with input vector $sup(\mathbf{I})$. If $o_j \leq A_{\min}$, stop.
   /* *Search the input vectors' space* **I**.
   (c) $\mathbf{i}_\perp := inf(\mathbf{I})$; $\mathbf{i}_\top := sup(\mathbf{I})$;
   (d) While $dist(\mathbf{i}_\perp, inf(\mathbf{I})) \leq n\text{DIV}2$ or $dist(\mathbf{i}_\top, sup(\mathbf{I})) \leq n\text{DIV}2 + n\text{MOD}2$, where $n$ is the number of input neurons of $N_+$, and still generating new $\mathbf{i}_\perp$ or $\mathbf{i}_\top$, do:
   /* *Generate the successors of* $\mathbf{i}_\perp$ *and query the network*

      i. set new $\mathbf{i}_\perp :=$ old $\mathbf{i}_\perp$ flipped according to the ordering $\leq_{\mathbf{I}}$ on $\mathbf{I}$;[10]
      ii. Query $N_+$ with input vector $\mathbf{i}_\perp$;
      iii. If Search Space *Pruning Rule 2* is applicable, stop generating new $\mathbf{i}_\perp$;
      iv. Apply the Simplification Rule *Complementary Literals*, and Add the rules derived according to $\sigma$ to the rule set.

   /* *Generate the predecessors of* $\mathbf{i}_\top$ *and query the network*
      v. set new $\mathbf{i}_\top :=$ old $\mathbf{i}_\top$ flipped according to the ordering $\leq_{\mathbf{I}}$ on $\mathbf{I}$;[11]
      vi. Query $N_+$ with input vector $\mathbf{i}_\top$;
      vii. If Search Space *Pruning Rule 1* is applicable, stop generating new $\mathbf{i}_\top$;
      viii. Apply the Simplification Rule *M of N*, and Add the rules derived according to $\sigma$ to the rule set.

   (e) Apply the Simplification Rules *Subsumption* and *M of N Subsumption* on the rule set regarding $o_j$.

---

[9] The algorithm is kept simple for clarity, and is not necessarily the most efficient.
[10] From $inf(\mathbf{I})$, we generate new $\mathbf{i}_\perp$ by flipping the elements at old $\mathbf{i}_\perp$ from right to left. For example, if old $\mathbf{i}_\perp = \{-1, -1, 1, 1\}$, we obtain new $\mathbf{i}_\perp = \{-1, 1, -1, 1\}$.
[11] From $sup(\mathbf{I})$, we generate new $\mathbf{i}_\top$ by flipping the elements at old $\mathbf{i}_\top$ from left to right. For example, if old $\mathbf{i}_\top = \{-1, -1, 1, 1\}$, we obtain new $\mathbf{i}_\top = \{-1, 1, -1, 1\}$.

Note that if the weights from the hidden to the output layer of $N$ are negative, we simply substitute $inf(\mathbf{I})$ by $sup(\mathbf{I})$ and vice versa. In a given application, the above extraction algorithm can be halted if a desired degree of accuracy is achieved in the set of rules. The algorithm is such that the exact symbolic representation of the network is being approximated at each cycle.

*Example 5.2.6.* Suppose $\mathcal{I} = \{a, b, c\}$ and let $\mathbf{I} = \wp(\mathcal{I})$ be ordered w.r.t. set inclusion. Assume $\sigma_{[a,b,c]}(1,1,1) = (1,1,1)$. We start by checking $inf(\mathbf{I})$ w.r.t. an output neuron $x$. If $inf(\mathbf{I})$ activates $x$, i.e. $inf(\mathbf{I})$ satisfies constraint $\mathbf{C}_{o_x}$, then by Proposition 5.2.6 any other input vector activates $x$ and by Definition 5.2.7 we can extract $\rightarrow x$ and stop. If, on the other hand, $inf(\mathbf{I})$ does not activate $x$, then we may need to query the network with the immediate successors of $inf(\mathbf{I})$. Let us call these input vectors $\mathbf{I}^*$, where $dist(inf(\mathbf{I}), \mathbf{I}^*) = 1$.

We proceed to check the element $sup(\mathbf{I})$. If $sup(\mathbf{I})$ does not satisfy $\mathbf{C}_{o_x}$, by Proposition 5.2.5 we can stop, extracting no rules with conclusion $x$. If $sup(\mathbf{I})$ activates $x$, we conclude that $abc \rightarrow x$, but we still have to check the input vectors $\mathbf{I}^{**}$ at distance 1 from $sup(\mathbf{I})$. We may also later apply some simplification on $abc \rightarrow x$, if at least one of the input vectors in $\mathbf{I}^{**}$ activates $x$. Hence, we keep $abc \rightarrow x$ in stand by and proceed.

Let us say that we choose to start by checking $\mathbf{i}_1 = (-1,-1,1)$ in $\mathbf{I}^*$. If $\mathbf{i}_1$ does not satisfy $\mathbf{C}_{o_x}$, we have to check the remaining inputs in $\mathbf{I}^*$. However, if $\mathbf{i}_1$ activates $x$ then, again by Proposition 5.2.6, we know that $(-1,1,1)$ and $(1,-1,1)$ also do. This tells us that not all the inputs in $\mathbf{I}^{**}$ need to be checked. Moreover, if all the elements in $\mathbf{I}^*$ activate $x$ then we can use Definition 5.2.8 to derive $1(abc) \rightarrow x$ and stop the search.

Analogously, when checking $\mathbf{I}^{**}$ we can obtain information about $\mathbf{I}^*$. If, for instance, $\mathbf{i}_2 = (1,1,-1)$ does not activate $x$ then $(-1,1,-1)$ and $(1,-1,-1)$ in $\mathbf{I}^*$ do not either, now by Proposition 5.2.5. If, on the contrary, $\mathbf{i}_2$ activates $x$, we can derive $ab \rightarrow x$, using Proposition 5.2.6 and Definition 5.2.6. If not only $\mathbf{i}_2$ but also the other inputs in $\mathbf{I}^{**}$ activate $x$ then we obtain $2(abc) \rightarrow x$, which subsumes $abc \rightarrow x$ by Definitions 5.2.8 and 5.2.5. In this case, we still need to query the network with inputs $\mathbf{i}$ at distance 1 from $\mathbf{i}_2$ such that $\langle \mathbf{i} \rangle < \langle \mathbf{i}_2 \rangle$, but those inputs are already the ones in $\mathbf{I}^{**}$ and therefore we can stop. Note that the stopping criteria are the following: either all elements in the ordering are visited or, if not, for each element not visited, Propositions 5.2.5 and 5.2.6 guarantee that it is safe not to consider it, in the sense that it is either already represented in the set of rules or irrelevant and will not give rise to any new rule.

**Theorem 5.2.1.** (Soundness) *The extraction algorithm for regular networks is sound (satisfies Definition 5.2.10).*

*Proof.* *We have to show that, whether a rule $r$ is extracted by querying the network (*Case 1*) or by a simplification of rules (*Case 2*), any rule $r'$ that is subsumed by $r$, including $r$ itself, can be obtained by querying the network. We prove this by contradiction. Consider a set $\mathbf{I}$ of $p$-ary input vectors. Assume that there exist rules $r$ and $r'$ such that $r'$ is subsumed by $r$, and $r'$ is not obtainable by querying the network. Assume also that $r$ contains the largest number of premises of such a rule. Let $X_i$ denote $L_i$ or $\sim L_i (1 \leq i \leq p)$.*

Case 1: *If $r$ is itself obtained by querying the network, then the only possible subsumed rule is $r$, and obviously this yields a contradiction.*

Case 2: *$r$ is either a simplification by Complementary Literals, or a Fact, or a M of N rule. It is shown that each assumption yields a contradiction.*

*Let $r = L_1, ..., L_q \to L_j$ $(1 \leq q < p)$ be a simplification by Complementary Literals. Then $r$ is derived from two rules $r'_1 = L_1, ..., L_s, ..., L_q \to L_j$ and $r'_2 = L_1, ..., \sim L_s, ..., L_q \to L_j$, $(1 \leq s \leq q)$. Each of these has more premises than $r$. So, by assumption, all rules subsumed by $r'_1$ and $r'_2$ are obtainable by querying the network. By Proposition 5.2.6, $r$ is also obtained by querying the network. Since, by Definition 5.2.5, any other rule subsumed by $r$ is also subsumed by either $r'_1$ or by $r'_2$, this leads to a contradiction.*

*Let $r = \to L_j$ be a simplification by Fact. Then, $r$ must have been obtained by querying the network with $inf(\mathbf{I})$. By Proposition 5.2.6, any rule of the form $X_1, ..., X_p \to L_j$ is also obtainable by querying the network, contradicting the assumption about $r'$.*

*Finally, if a further simplification is made, to obtain $r = m(L_1, ..., L_n) \to L_j$ $(1 \leq m < n \leq p)$ by M of N simplification, then $r$ is obtained from a set of rules of the form $L_1, ..., L_m \to L_j$, where $L_1, ..., L_m$ are $m$ elements chosen from $\{L_1, ..., L_n\}$. By the previous cases, all subsumed rules are obtainable by querying the network.* □

**Theorem 5.2.2.** *(*Completeness*) The extraction algorithm for regular networks is complete (satisfies Definition 5.2.11).*

*Proof.* *We have to show that the extraction algorithm terminates either when all possible combinations of the input vector have been queried in the network (*Case 1*) or the set of rules extracted subsumes any rule that would be derived from an element not queried (*Case 2*).* Case 1 *is trivial. In* Case 2, *we have to show that any element not queried either would not generate a rule (*Case 2(*i*)*) or would generate a rule that is subsumed by some rule extracted (*Case 2(*ii*)*).*

*Consider a set $\mathbf{I}$ of $p$-ary input vectors.*

Case 2(i): *Let $\mathbf{i}_m, \mathbf{i}_n \in \mathbf{I}$, $dist(\mathbf{i}_m, \mathbf{i}_n) = q$ $(1 \leq q \leq p)$ and $\langle \mathbf{i}_m \rangle < \langle \mathbf{i}_n \rangle$. Assume that $\mathbf{i}_n$ is queried in the network and that $\mathbf{i}_n$ does not generate a rule. By Proposition 5.2.5 $q$ times, $\mathbf{i}_m$ would not generate a rule either.*

Case 2(ii): *Let $\mathbf{i}_k, \mathbf{i}_o \in \mathbf{I}$, $dist(\mathbf{i}_k, \mathbf{i}_o) = t$ $(1 \leq t \leq p)$ and $\langle \mathbf{i}_k \rangle < \langle \mathbf{i}_o \rangle$. Assume that $\mathbf{i}_k$ is queried in the network and that $\mathbf{i}_k$ derives a rule $r_k$. Let*

*$S = \{L_1, ..., L_s\}$ be the set of positive literals in the body of $r_k$, where $s \in [1, p]$. By Definition 5.2.6, the rule $r = L_1, ..., L_s \rightarrow L_j$ can be obtained from $r_k$. Clearly, $r$ subsumes $r_k$. Now, by Proposition 5.2.6 $t$ times, $\mathbf{i}_o$ would also derive a rule $r_o$. Let $U = \{L_1, ..., L_u\}$ be the set of positive literals in the body of $r_o$, where $u \in [1, p]$. Since $\langle \mathbf{i}_k \rangle < \langle \mathbf{i}_o \rangle$ then $S \subset U$ and, by Definition 5.2.5, $r$ also subsumes $r_o$.*

*That completes the proof since all the stopping criteria of the extraction algorithm have been covered.* □

## 5.3 The General Case Extraction

So far, we have seen that for the case of regular networks it is possible to apply an ordering on the set of input vectors, and use a sound and complete pedagogical extraction algorithm that searches for relevant input vectors in this ordering. Furthermore, the neural network and its set of rules can be shown equivalent (that results directly from the proofs of soundness and completeness of the extraction algorithm).

Despite the above results being highly desirable, it is much more likely that a non-regular network will result from an unbiased training process. In order to overcome this limitation, in the sequel we present the extension of our extraction algorithm to the general case, the case of non-regular networks. The idea is to investigate fragments of a non-regular network in order to find regularities over which the above described extraction algorithm could be applied. We would then split a non-regular network into regular subnetworks, extract the symbolic knowledge from each subnetwork, and finally assemble the rule set of the original non-regular network. That, however, is a decompositional approach, and we need to bear in mind that the collective behaviour of a network is not equivalent to the behaviour of its parts grouped together. We will need, therefore, to be especially careful when assembling the network's final set of rules.

The problem with non-regular networks is that it is difficult to find the ordering on the set of input vectors without having to actually check each input. In this case, the gain obtained in terms of complexity could be lost. By considering its regular subnetworks, the main problem we have to tackle is how to combine the information obtained into the network's rule set. This problem is due mainly to the non-discrete nature of the network's hidden neurons. As we have seen in Example 5.1.1, this is the reason why a decompositional approach may be unsound (see Section 5.1). In order to solve this problem, we will assume that hidden neurons present four possible activation values $(-1, A_{\text{max}}, A_{\text{min}}, 1)$. Performing a kind of worst case analysis, we will be able to show that the general case extraction is sound, although we will have to trade completeness for efficiency.

### 5.3.1 Regular Subnetworks

We start by defining precisely the above intuitive concept of a subnetwork.

**Definition 5.3.1.** (Subnetworks) *Let $N$ be a neural network with $p$ input neurons, $r$ hidden neurons and $q$ output neurons, respectively $\{i_1, ..., i_p\}$, $\{n_1, ..., n_r\}$ and $\{o_1, ..., o_q\}$. Let $N'$ be a neural network with $p'$ input neurons $\{i'_1, ..., i'_{p'}\}$, $r'$ hidden neurons $\{n'_1, ..., n'_{r'}\}$ and $q'$ output neurons $\{o'_1, ..., o'_{q'}\}$. $N'$ is a* subnetwork *of $N$ iff $0 \leq p' \leq p$, $0 \leq r' \leq r$, $0 \leq q' \leq q$, and for all $i'_i$, $n'_j$, $o'_k$ in $N'$, $W_{n'_j i'_i} = W_{n_j i_i}$, $W_{o'_k n'_j} = W_{o_k n_j}$, $\theta_{n'_j} = \theta_{n_j}$ and $\theta_{o'_k} = \theta_{o_k}$.*

Our first task is to find the regular subnetworks of a non-regular network. It is not difficult to verify that any network containing a single hidden neuron is regular. As a result, we could be tempted to split a non-regular network with $r$ hidden neurons into $r$ subnetworks, each containing the same input and output neurons as the original network plus only one of its hidden neurons.

However, let us briefly analyse what could happen if we were to extract rules from each of the above subnetworks. Suppose that, for a given output neuron $x$, from the subnetwork containing hidden neuron $n_1$, the extraction algorithm obtains the rules $a, b \rightarrow_{n_1} x$ and $c, d \rightarrow_{n_1} x$, while from the subnetwork containing hidden neuron $n_2$, it obtains the rule $c, d \rightarrow_{n_2} x$. The problem is that the information that $[a, b]$ activates $x$ through $n_1$ is not very useful. It may be the case that the same input $[a, b]$ has no effect on the activation of $x$ through $n_2$, or that it actually blocks the activation of $x$ through $n_2$. It may also be the case that, for example, $a, d \rightarrow x$ as a result of the combination of the activation values of $n_1$ and $n_2$, but not through each one of them individually. If, therefore, we take the intersection of the rules derived from each subnetwork, we would be extracting only the rules that are encoded in every hidden neuron individually, but not the rules derived from each hidden neuron or from the collective effect of the hidden neurons. If, on the other hand, we take the union of the rules derived from each subnetwork, then the extraction could clearly be unsound.

It seems that we need to analyse a non-regular network first from the input layer to each of the hidden neurons, and then from the hidden layer to each of the output neurons. That motivates the following definition of "*Basic Neural Structures*".

**Definition 5.3.2.** (Basic Neural Structures) *Let $N$ be a neural network with $p$ input neurons, $r$ hidden neurons and $q$ output neurons, respectively $\{i_1, ..., i_p\}$, $\{n_1, ..., n_r\}$ and $\{o_1, ..., o_q\}$. A subnetwork $N'$ of $N$ is a* Basic Neural Structure *(*BNS*) iff either $N'$ contains exactly $p$ input neurons,* 1 *hidden neuron and* 0 *output neurons of $N$, or $N'$ contains exactly* 0 *input neurons, $r$ hidden neurons and* 1 *output neuron of $N$.*

Note that a *BNS* is a neural network with no hidden neurons and a single neuron in its output layer. Note also that a network $N$ with $r$ hidden neurons and $q$ output neurons contains $r+q$ *BNSs*. We call a *BNS* containing no output neurons of $N$, an *Input to Hidden BNS*, and a *BNS* containing no input neurons of $N$, a *Hidden to Output BNS*.

Interestingly, if we apply the *Transformation Algorithm* on a network $N$, we obtain a network $N'$ whose *BNSs* will all be positive (sub)networks of $N'$. This is precisely what makes *BNSs* interesting subnetworks for rule extraction. For the sake of simplicity, we may as well apply a restricted version of the *Transformation Algorithm* over *BNSs*. For each input neuron $y$ of a *BNS* we do the following:

1. If $y$ is linked to the output neuron of the *BNS* through a positive weight:
   a) do nothing.
2. If $y$ is linked to the output neuron of the *BNS* through a negative weight $W_{jy}$:
   a) change $W_{jy}$ to $-W_{jy}$ and rename $y$ by $\sim y$.

We say that the *BNS* obtained is the positive form of the original *BNS*. Furthermore, we say that the original *BNS* is regular if its positive form does not contain pairs of neurons labelled as complementary literals in its input layer.

**Proposition 5.3.1.** *Any* BNS *is regular.*[12]

*Proof. By Definition 5.3.2, any* BNS *contains a single output neuron and no complementary literals in its input layer. Since no input neuron is added by the* Transformation Algorithm *on a* BNS, *the positive form of any* BNS *does not contain pairs of neurons labelled as complementary literals in its input layer, and thus the original* BNS *is regular.* □

The above result indicates that *BNSs*, which can be easily obtained from a network $N$, are suitable subnetworks for applying the extraction algorithm, when $N$ is a non-regular network.[13]

### 5.3.2 Knowledge Extraction from Subnetworks

We have seen that, if we split a non-regular network into *BNSs*, there is always an ordering easily found in each subnetwork. The problem, now, is

[12] Sometimes, we refer to single hidden layer regular networks and to Basic Neural Structures simply as *regular networks*.

[13] Note that applying the Transformation Algorithm to a network $N$ and then splitting the result into $BNSs$ gives the same result as splitting $N$ into $BNSs$ and then applying the simplified Transformation Algorithm to the $BNSs$.

that *Hidden to Output BNSs* do not present discrete activation $\{-1, 1\}$ in their input layers. Instead, each input neuron may present activation values in the ranges $[-1, A_{\max})$ or $(A_{\min}, 1]$, where $A_{\max} \in (-1, 0)$ and $A_{\min} \in (0, 1)$ are predefined values. We will, hence, need to consider this during the extraction from *Hidden to Output BNSs*. For the time being, let us simply assume that each neuron in the input layer of a *Hidden to Output BNS* is labelled $n_i$, and if $n_i$ is connected to the neuron in the output layer of the *BNS* through a negative weight, then we rename it $\sim n_i$ when applying the Transformation Algorithm, as done for *Input to Hidden BNSs*. Moreover, let us assume that neurons in the input layer of the positive form of *Hidden to Output BNSs* present activation values $-1$ or $A_{\min}$ only. This results from the above mentioned worst case analysis, as we will see later in this section.

We need to rewrite Search Space Pruning Rules 1 and 2 for *BNSs*. Now, given a *BNS* with $s$ input neurons $\{i_1, ..., i_s\}$ and the output neuron $o_j$, the constraint $\mathbf{C}_{o_j}$ on the activation of $o_j$ for an input vector $\mathbf{i}$ is simply given by:

$o_j$ *is active for* $\mathbf{i}$ *iff*

$$W_{o_j i_1} i_1 + W_{o_j i_2} i_2 + ... + W_{o_j i_s} i_s > h^{-1}(A_{\min}) + \theta_{o_j} \tag{5.3}$$

**Proposition 5.3.2.** *Let $\mathcal{I}_+$ be the set of input neurons of the positive form $B_+$ of a* BNS *$B$ with output $o_j$. Let $\mathbf{I}_+ = \wp(\mathcal{I}_+)$ be ordered under the set inclusion relation $\leq_{\mathbf{I}_+}$, and $\mathbf{i}_m, \mathbf{i}_n \in \mathbf{I}_+$. If $\mathbf{i}_m \leq_{\mathbf{I}_+} \mathbf{i}_n$ then $o_j(\sigma_{[\mathcal{I}_+]}(\mathbf{i}_m)) \leq o_j(\sigma_{[\mathcal{I}_+]}(\mathbf{i}_n))$ in $B$.*

*Proof. If $B$ is an* Input to Hidden BNS *then the proof is trivial, by Proposition 5.3.1 and Proposition 5.2.3. If $B$ is a* Hidden to Output BNS, *assume $\mathbf{i}_m(i_k) = -1$ and $\mathbf{i}_n(i_k) = A_{\min}$. Since all the weights in $B_+$ are positive real numbers and $A_{\min} > 0$, we obtain $(W_{o_j i_k}(-1) - \theta_{o_j}) \leq (W_{o_j i_k}(A_{\min}) - \theta_{o_j})$. Since $\mathbf{i}_m \leq_{\mathbf{I}_+} \mathbf{i}_n$, we also have $(\sum_{i=1}^{p}(W_{o_j i_i}\mathbf{i}_m(i_i) - \theta_{o_j})) \leq (\sum_{i=1}^{p}(W_{o_j i_i}\mathbf{i}_n(i_i) - \theta_{o_j}))$, and by the monotonically crescent characteristic of $h(x)$, $h(\sum_{i=1}^{p}(W_{o_j i_i}\mathbf{i}_m(i_i) - \theta_{o_j})) \leq h(\sum_{i=1}^{p}(W_{o_j i_i}\mathbf{i}_n(i_i) - \theta_{o_j}))$, i.e. $o_j(\mathbf{i}_m) \leq o_j(\mathbf{i}_n)$ in $B_+$. Finally, from the definition of $\sigma$, mapping input vectors of $B_+$ into input vectors of $B$, it follows directly that $o_j(\sigma_{[\mathcal{I}_+]}(\mathbf{i}_m)) \leq o_j(\sigma_{[\mathcal{I}_+]}(\mathbf{i}_n))$ in $B$.* □

**Corollary 5.3.1.** *(*BNS *Pruning Rule 1) Let $\mathbf{i}_m \leq_{\mathbf{I}} \mathbf{i}_n$. If $\mathbf{i}_n$ does not satisfy the constraint $\mathbf{C}o_j$ on a* BNS*'s output neuron, then $\mathbf{i}_m$ does not satisfy $\mathbf{C}o_j$ either.*

*Proof. Directly from Proposition 5.3.2.*

**Corollary 5.3.2.** *(*BNS *Pruning Rule 2) Let $\mathbf{i}_m \leq_{\mathbf{I}} \mathbf{i}_n$. If $\mathbf{i}_m$ satisfies the constraint $\mathbf{C}o_j$ on a* BNS*'s output neuron, then $\mathbf{i}_n$ also satisfies $\mathbf{C}o_j$.*

*Proof. Directly from Proposition 5.3.2.*

The particular characteristic of *BNSs*, specifically because they have no hidden neurons, allows us to define a new ordering that can be very useful in helping to reduce the search space of the extraction algorithm. If now, in addition, we take into account the values of the weights of the *BNS*, we may be able to assess, given two input vectors $\mathbf{i}_m$ and $\mathbf{i}_n$ such that $\langle \mathbf{i}_m \rangle = \langle \mathbf{i}_n \rangle$, whether $o_j(\mathbf{i}_m) \leq o_j(\mathbf{i}_n)$ or vice versa.[14] Assume, for instance, that $\mathbf{i}_m$ and $\mathbf{i}_n$ differ only on inputs $i_i$ and $i_k$, where $i_i = 1$ in $\mathbf{i}_n$ and $i_k = 1$ in $\mathbf{i}_m$. Thus, if $|W_{o_j i_i}| \leq |W_{o_j i_k}|$, it is not difficult to see that $o_j(\mathbf{i}_n) \leq o_j(\mathbf{i}_m)$. Let us formalise this idea.

In the following propositions, we consider the positive form of a *BNS* and the partial ordering w.r.t. set inclusion.

**Proposition 5.3.3.** *(*BNS *Pruning Rule 3) Let $\mathbf{i}_m, \mathbf{i}_n$ and $\mathbf{i}_o$ be three different input vectors in $\mathbf{I}$ such that $dist(\mathbf{i}_m, \mathbf{i}_o) = 1$, $dist(\mathbf{i}_n, \mathbf{i}_o) = 1$ and $\langle \mathbf{i}_m \rangle, \langle \mathbf{i}_n \rangle < \langle \mathbf{i}_o \rangle$, that is, both $\mathbf{i}_m$ and $\mathbf{i}_n$ are immediate predecessors of $\mathbf{i}_o$. Let $\mathbf{i}_m$ be obtained from $\mathbf{i}_o$ by flipping the $i$th input from 1 (resp. $A_{\min}$ for* Hidden to Output BNSs*) to $-1$, while $\mathbf{i}_n$ is obtained from $\mathbf{i}_o$ by flipping the $k$th input from 1 (resp. $A_{\min}$ for* Hidden to Output BNSs*) to $-1$. If $W_{o_j i_k} \leq W_{o_j i_i}$ then $o_j(\mathbf{i}_m) \leq o_j(\mathbf{i}_n)$. In this case, we write $\mathbf{i}_m \leq_{\langle\rangle} \mathbf{i}_n$.*

*Proof. We know that both $\mathbf{i}_m$ and $\mathbf{i}_n$ are obtained from $\mathbf{i}_o$ by flipping, respectively, inputs $\mathbf{i}_o(i)$ and $\mathbf{i}_o(k)$ from 1 (resp. $A_{\min}$) to $-1$. We also know that $o_j(\mathbf{i}_o) = h(W_{o_j i_i}\mathbf{i}_o(i) + W_{o_j i_k}\mathbf{i}_o(k) + \Delta + \theta_{o_j})$, and that $A_{\min} > 0$. For* Input to Hidden BNSs, *$o_j(\mathbf{i}_m) = h(-W_{o_j i_i} + W_{o_j i_k} + \Delta + \theta_{o_j})$ and $o_j(\mathbf{i}_n) = h(W_{o_j i_i} - W_{o_j i_k} + \Delta + \theta_{o_j})$. For* Hidden to Output BNSs, *$o_j(\mathbf{i}_m) = h(-W_{o_j i_i} + A_{\min} W_{o_j i_k} + \Delta + \theta_{o_j})$ and $o_j(\mathbf{i}_n) = h(A_{\min} W_{o_j i_i} - W_{o_j i_k} + \Delta + \theta_{o_j})$. Since $W_{o_j i_k} \leq W_{o_j i_i}$, and from the monotonically crescent characteristic of $h(x)$, we obtain $o_j(\mathbf{i}_m) \leq o_j(\mathbf{i}_n)$ in both cases.* □

As before, a direct result of Proposition 5.3.3 is that: if $\mathbf{i}_m$ satisfies the constraint $\mathbf{C}_{o_j}$ on the output neuron of the *BNS*, then $\mathbf{i}_n$ also satisfies $\mathbf{C}_{o_j}$. By contraposition, if $\mathbf{i}_n$ does not satisfy $\mathbf{C}_{o_j}$ then $\mathbf{i}_m$ does not satisfy $\mathbf{C}_{o_j}$ either.

**Proposition 5.3.4.** *(*BNS *Pruning Rule 4) Let $\mathbf{i}_m, \mathbf{i}_n$ and $\mathbf{i}_o$ be three different input vectors in $\mathbf{I}$ such that $dist(\mathbf{i}_m, \mathbf{i}_o) = 1$, $dist(\mathbf{i}_n, \mathbf{i}_o) = 1$ and $\langle \mathbf{i}_o \rangle < \langle \mathbf{i}_m \rangle, \langle \mathbf{i}_n \rangle$, that is, both $\mathbf{i}_m$ and $\mathbf{i}_n$ are immediate successors of $\mathbf{i}_o$. Let $\mathbf{i}_m$ be obtained from $\mathbf{i}_o$ by flipping the $i$th input from $-1$ to 1 (resp. $A_{\min}$ for* Hidden to Output BNSs*), while $\mathbf{i}_n$ is obtained from $\mathbf{i}_o$ by flipping the $k$-th input from $-1$ to 1 (resp. $A_{\min}$ for* Hidden to Output BNSs*). If $W_{o_j i_k} \leq W_{o_j i_i}$, then $o_j(\mathbf{i}_n) \leq o_j(\mathbf{i}_m)$. In this case, we write $\mathbf{i}_n \leq_{\langle\rangle} \mathbf{i}_m$.*

[14] Recall that, previously, two input vectors $\mathbf{i}_m$ and $\mathbf{i}_n$ such that $\langle \mathbf{i}_m \rangle = \langle \mathbf{i}_n \rangle$ were incomparable.

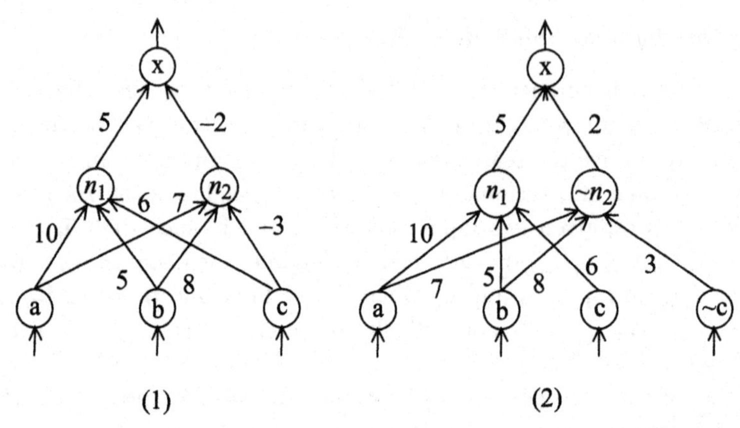


**Fig. 5.13.** A non-regular network (1) and its positive form (2) obtained by applying the Transformation Algorithm on its *BNSs*.

*Proof. This is the contrapositive of Proposition 5.3.3.* □

*Example 5.3.1.* Consider the network $N$ of Figure 5.13(1) and its positive form $N_+$ at Figure 5.13(2), obtained by applying the Transformation Algorithm over each *BNS* of $N$. $N_+$ contains three *BNSs* – two *Input to Hidden BNSs*, one with inputs $[a, b, c]$ and output $n_1$, and the other with inputs $[a, b, \sim c]$ and output $n_2$, and one *Hidden to Output BNS*, having inputs $[n_1, \sim n_2]$ and output $x$.

Considering the ordering on set inclusion and the mapping $\sigma$, we verify that $[a, b, c] = (1, 1, 1)$ is the maximum element of the *BNS* with output $n_1$, $[a, b, \sim c] = (1, 1, 1)$ is the maximum element of the *BNS* with output $n_2$, and $[n_1 \sim n_2] = (A_{\min}, A_{\min})$ is the maximum element of the *BNS* with output $x$.

If now we add information about the weights, we can apply Pruning Rules 3 and 4 as well. Take, for example, the *BNS* with output $n_1$, where $W_{n_1 b} \leq W_{n_1 c} \leq W_{n_1 a}$. Using Pruning Rules 3 and 4, we can obtain a new ordering on input vectors $\mathbf{i}_m$ and $\mathbf{i}_n$ such that $\langle \mathbf{i}_m \rangle = \langle \mathbf{i}_n \rangle$.[15] We obtain $(-1, 1, 1) \leq_{\langle\rangle} (1, 1, -1) \leq_{\langle\rangle} (1, -1, 1)$ and $(-1, 1, -1) \leq_{\langle\rangle} (-1, -1, 1) \leq_{\langle\rangle} (1, -1, -1)$. Similarly, given $W_{xn_2} \leq W_{xn_1}$, we obtain $\{\sim n_1, \sim n_2\} \leq_{\langle\rangle} \{n_1, n_2\}$ for the *Hidden to Output BNS*[16]. Figure 5.14 contains two diagrams in which this new ordering is superimposed on the previous set inclusion ordering for the *BNSs* with outputs $n_1$ and $x$.

---

[15] Recall that such input vectors are incomparable under the set inclusion ordering.

[16] Here, we have deliberately used $\{n_i, \sim n_i\}$, instead of $\{1, -1\}$, to stress the fact that hidden neurons do not present discrete activation values. We will come back to this later in this chapter.

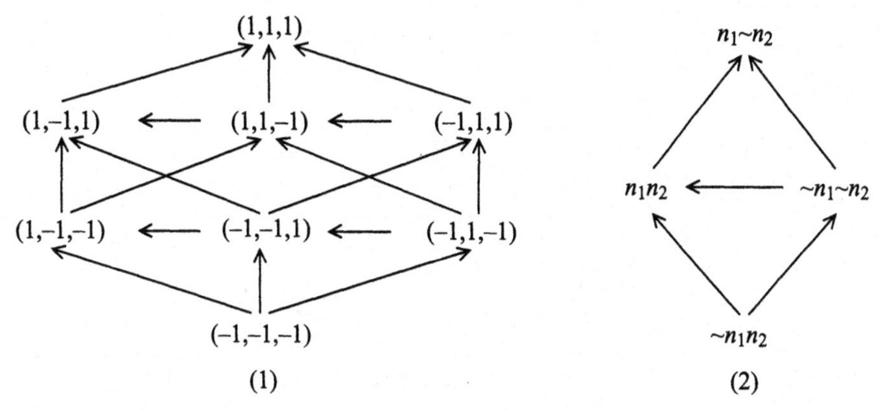


**Fig. 5.14.** Adding information about the weights of the *BNSs* with output $n_1$ (1) and $x(2)$.

The above example illustrates the ordering $\preceq$ on the set of input vectors $\mathbf{I}$ of *BNSs*. The ordering results from the superimposition of the ordering $\leq_{\langle\rangle}$, obtained from Pruning Rules 3 and 4, on the set inclusion ordering $\leq_{\mathbf{I}}$, obtained from Pruning Rules 1 and 2. Let us define $\preceq$ more precisely.

**Definition 5.3.3.** *For all* $\mathbf{i}_m, \mathbf{i}_n \in \mathbf{I}$, *we say that* $\mathbf{i}_m \preceq \mathbf{i}_n$ *iff* $\mathbf{i}_m \leq_{\mathbf{I}} \mathbf{i}_n$ *or* $\mathbf{i}_m \leq_{\langle\rangle} \mathbf{i}_n$.[17]

**Proposition 5.3.5.** $\preceq$ *is a partial ordering on the set of input vectors* $\mathbf{I}$ *of a* BNS.

*Proof. We need to show that* $\preceq$ *is a reflexive, transitive and anti-symmetric relation on* $\mathbf{I}$. *Let* $x, y, z \in \mathbf{I}$. *Reflexivity is trivial. Transitivity: if* $x \preceq y$ *and*

[17] An alternative way to define $\preceq$ is as follows. For all $x, y \in \mathcal{I}$ we say that $x \preceq y$ iff $W_x \leq W_y$, where $W_x$ and $W_y$ are the weights associated with $x$ and $y$, respectively. We represent each element $\mathbf{i}$ of the power set of $\mathcal{I}$ as an ordered list such that $\mathbf{i} = [x_1, x_2, ..., x_n]$ iff $x_n \preceq ... \preceq x_2 \preceq x_1$, where $x_i \in \mathcal{I}$ for $1 \leq i \leq n$. The power set of $\mathcal{I}$ is ordered according to the following rules:

(i) Let $x_i, y_i (1 \leq i \leq n) \in \mathcal{I}$. $[x_1, ..., x_n] \preceq [y_1, ..., y_n]$ in $\wp(\mathcal{I})$ $iff$ $\forall i, x_i \preceq y_i$ in $\mathcal{I}$.

(ii) If $\|[x_1, ..., x_n]\| \neq \|[y_1, ..., y_m]\|$ then complete the smaller list with $\otimes$ and define $\otimes \preceq z$ for any $z \in \mathcal{I}$.

As an example, if $c \preceq b \preceq a$ then $c \preceq ab$ because $c \preceq a$ and $\otimes \preceq b$. Also, $bc \npreceq a$ because $c \npreceq \otimes$, even though $b \preceq a$. In addition, $a \npreceq bc$ because $a \npreceq b$. In this case, $a$ and $bc$ are incomparable. Similarly, given $b \preceq a$, we have $a \preceq ab$. In this case, we say that $\{a\} \preceq \{a, b\}$, even though $a \npreceq ba$. This is the reason why we have to represent the elements of $\wp(\mathcal{I})$ as an ordered list, as described above.

Note that the above definition is given in terms of the set of input neurons ($\mathcal{I}$) and not in terms of the set of input vectors ($\mathbf{I}$), even though there is a one to one correspondence.

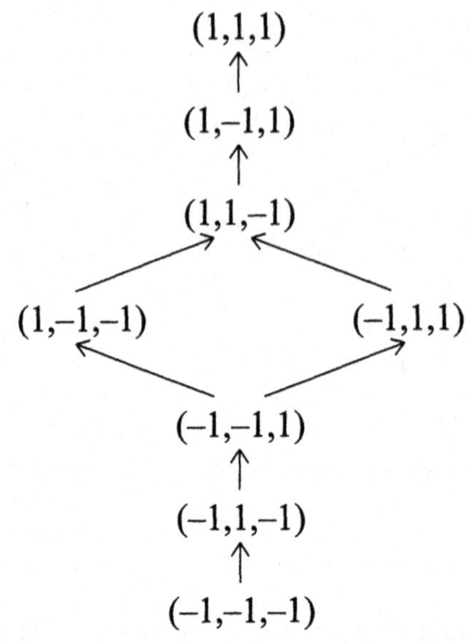


**Fig. 5.15.** The ordering $\preceq$ on the set of input vectors of the *BNS* with output $n_1$.

*$y \preceq z$ then the interesting cases are when $x \leq_{\mathbf{I}} y$ and $y \leq_{\langle\rangle} z$, or when $x \leq_{\langle\rangle} y$ and $y \leq_{\mathbf{I}} z$. In the first case, since $y \leq_{\langle\rangle} z$ then there exists $w \in \mathbf{I}$ such that $dist(y, w) = 1$ and $dist(z, w) = 1$. Hence, either $x = y$ and $x \leq_{\langle\rangle} z$, or $x \subset y$ (i.e. $x$ is a proper subset of $y$), and $x \leq_{\mathbf{I}} w$ and so $x \leq_{\mathbf{I}} z$ and $x \preceq z$. The proof is analogous for the second case. Anti-symmetry: the interesting case is when $x \leq_{\mathbf{I}} y$ and $y \leq_{\langle\rangle} x$. Assume $x \neq y$. As a result, $x \subset y$ and, for all $w \in \mathbf{I}$, $dist(x, w) = dist(y, w)$, which contradicts the assumption.* □

Returning to Example 5.3.1, it is not difficult to see that the ordering $\preceq$ on the *BNS* with output $n_1$ is given by the diagram of Figure 5.15 (see also Figure 5.14(1)). Incomparable elements in $\preceq$, as $\mathbf{i}_1 = (1, -1, -1)$ and $\mathbf{i}_2 = (-1, 1, 1)$ at Figure 5.15, indicate that it is not easy to establish whether $\mathbf{i}_1 \preceq \mathbf{i}_2$ without actually querying the *BNS* with both inputs. Note also that $\preceq$ is a chain for the *BNS* with output $x$, i.e. $\{\sim n_1, n_2\} \preceq \{\sim n_1, \sim n_2\} \preceq \{n_1, n_2\} \preceq \{n_1, \sim n_2\}$.

Figure 5.16 displays $\preceq$ on $\mathbf{I} = \wp(\mathcal{I})$ for $\mathcal{I} = \{a, b, c, d\}$, given $(1, 1, 1, 1) = [a, b, c, d]$ and $W_d \leq W_c \leq W_b \leq W_a$. Note that $\preceq$ follows the ordering on $W_a + W_b + W_c + W_d$.

$\preceq$ provides a systematic way of searching the set of input vectors $\mathbf{I}$. Let us illustrate this with the following example, which also gives a glance about the implementation of the extraction algorithm in the general case.

*Example 5.3.2.* Consider the *Input to Hidden BNS* of Figure 5.17(1), and its positive form 5.17(2). The ordering's maximum element is input vector

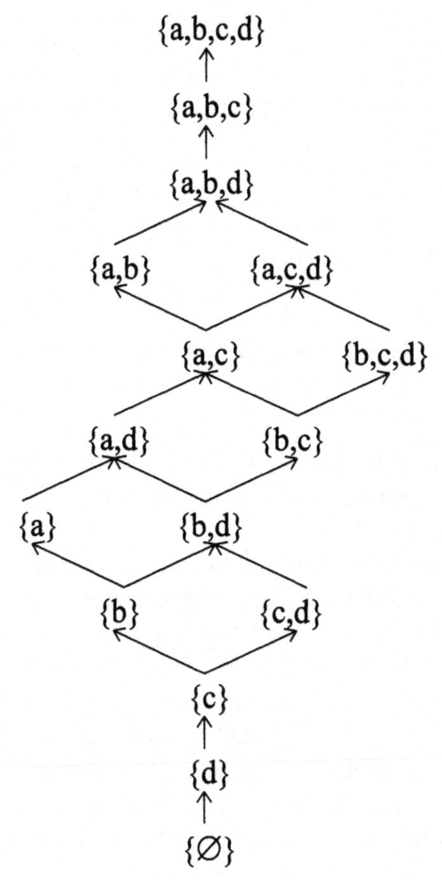


**Fig. 5.16.** $\preceq$ on $\wp(\mathcal{I})$, given $\mathcal{I} = \{a, b, c, d\}$ and $(1, 1, 1, 1) = [a, b, c, d]$.

$\mathbf{i}_\top = (1, 1, 1, 1) = (a, b, \sim c, \sim d)$. Taking the *BNS* of Figure 5.17(2), if $\mathbf{i}_\top$ does not activate $n_i$ then we proceed to generate the elements $\mathbf{i}_m$ such that $dist(\mathbf{i}_m, \mathbf{i}_\top) = 1$. However, Pruning Rule 3 says that there is an ordering among elements $\mathbf{i}_m$. For example, it says that $(1, 1, 1, -1) = (a, b, \sim c, d)$ provides a smaller activation value to $n_i$ than $(1, 1, -1, 1) = (a, b, c, \sim d)$.

Therefore, given $W_{n_i \sim c} \leq W_{n_i a} \leq W_{n_i \sim d} \leq W_{n_i b}$,[18] we start from $\mathbf{i}_\top$ by flipping from 1 to -1 the input $\sim c$ with the smallest weight $W_{n_i \sim c}$, and obtain the input vector $\mathbf{i}_1 = (1, 1, -1, 1)$. By *Pruning Rule 3*, the activation of $n_i$ given $\mathbf{i}_1$ is greater than the activation of $n_i$ given any other element $\mathbf{i}_m$ such that $\langle \mathbf{i}_m \rangle = \langle \mathbf{i}_1 \rangle$. Thus, if $n_i(\mathbf{i}_1) \leq A_{\min}$ then $n_i(\mathbf{i}_m) \leq A_{\min}$. In this case we could stop the search. Otherwise, we would have to apply $\sigma$ and derive the rule $a, b, c, \sim d \rightarrow n_i$, and carry on generating and querying the remaining elements $\mathbf{i}_m$ such that $\langle \mathbf{i}_m \rangle = \langle \mathbf{i}_1 \rangle$. Again, due to *Pruning Rule 3*, we could

[18] In the original *BNS*, we have: $|W_{n_i c}| \leq |W_{n_i a}| \leq |W_{n_i d}| \leq |W_{n_i b}|$.

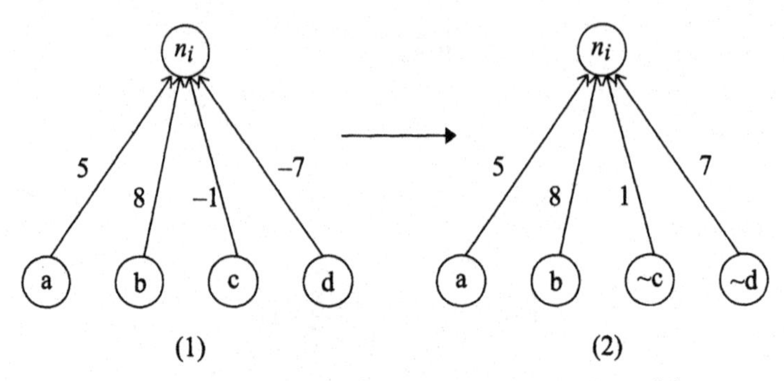


**Fig. 5.17.** An *Input to Hidden BNS* (1), and its positive form (2).

do so by flipping, from $\mathbf{i}_\top$, the input $a$ with the next smallest weight, $W_{n_i a}$, and repeat the above process until either we can stop or we flip the input $b$ with the largest weight, $W_{n_i b}$.

Similarly, starting from the ordering's minimum element $\mathbf{i}_\bot = (-1, -1, -1, -1) = (\sim a, \sim b, c, d)$, if $\mathbf{i}_\bot$ does not activate $n_i$ then we flip from $-1$ to 1 the input $\sim c$ with the smallest weight $W_{n_i \sim c}$, to obtain input vector $\mathbf{i}_2 = (-1, -1, 1, -1)$. By *Pruning Rule 4*, if $n_i(\mathbf{i}_2) > A_{\min}$ then $n_i(\mathbf{i}_n) > A_{\min}$ for all $\mathbf{i}_n$ such that $\langle \mathbf{i}_n \rangle = \langle \mathbf{i}_2 \rangle$. In this case, we could apply $\sigma$ to derive the rule $1(a, b, \sim c, \sim d) \to n_i$, using simplification *M of N*, and stop the search. Otherwise, we would need to generate another element from $\mathbf{i}_\bot$, this time by flipping the input $a$ with the next smallest weight, and repeat the above process until either we can stop or we flip the input $b$ with the largest weight, $W_{n_i b}$.

Let us now focus on the problem of knowledge extraction from *Hidden to Output BNSs*. The problem lies on the fact that hidden neurons do not present discrete activation. As a result, we need to provide a special treatment for the procedure of knowledge extraction from *Hidden to Output BNSs*. We have seen already that, if we simply assume that hidden neurons are either fully active or non-active, then the extraction algorithm loses soundness.

We say that a hidden neuron is *active* if its activation value lies in the interval $(A_{\min}, 1]$, or *non-active* if its activation value lies in the interval $[-1, A_{\max})$.[19] Trying to find an ordering on such intervals of activation is not easy. For example, taking the *Hidden to Output BNS* of the network of Figure 5.13(1), one can not say that having $n_1 < A_{\max}$ and $n_2 < A_{\max}$ results in a smaller activation value for $x$ than having $n_1 < A_{\max}$ and $n_2 > A_{\min}$. If $A_{\max} = -A_{\min} = -0.2$ then $n_1 = -0.3$ and $n_2 = -0.3$ may provide a greater activation in $x$ than $n_1 = -0.95$ and $n_2 = 0.25$.

[19] Recall that $A_{\min} \in (0, 1)$ and $A_{\max} \in (-1, 0)$.

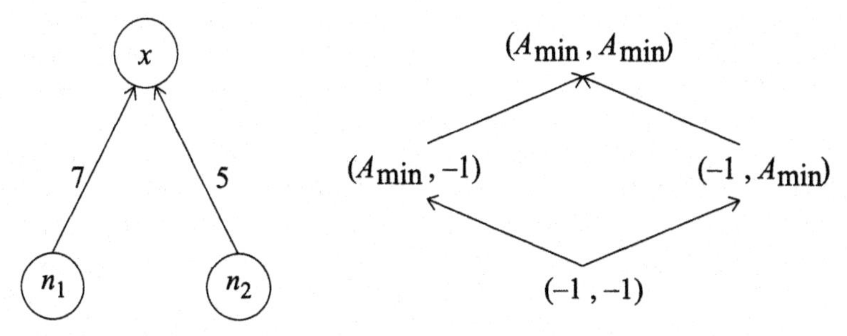


**Fig. 5.18.** A *Hidden to Output BNS* and the corresponding set inclusion ordering on the activation values of its input neurons in the worst cases.

At this stage, we need to compromise in order to keep soundness. Roughly, we have to analyse the activation values of the hidden neurons in the "worst cases". Such values are given by $-1$ and $A_{\min}$ in the case of a hidden neuron connected through a positive weight to the output, and by $A_{\max}$ and 1 in the case of a hidden neuron connected through a negative weight to the output.

*Example 5.3.3.* Consider the *Hidden to Output BNS* of Figure 5.18. The intuition behind its corresponding ordering is as follows: either both $n_1$ and $n_2$ present activation greater than $A_{\min}$, or one of them presents activation greater than $A_{\min}$ while the other presents activation smaller than $A_{\max}$, or both of them present activation smaller than $A_{\max}$.

Considering the activation values in the worst cases, since the weights from $n_1$ and $n_2$ to $x$ are both positive, if the activation of $n_i$, $i \in \{1, 2\}$, is smaller than $A_{\max}$ then we assume that it is $-1$. On the other hand, if the activation of $n_i$ is greater than $A_{\min}$ then we consider that it is equal to $A_{\min}$. In this way, we can derive the ordering of Figure 5.18 safely, as we show in the sequel. In addition, given that $W_{xn_2} \leq W_{xn_1}$, we also obtain $(-1, A_{\min}) \preceq (A_{\min}, -1)$. As before, in this case $\preceq$ is a chain.

The recipe for performing a sound extraction from non-regular networks, concerning *Hidden to Output BNSs*, is: *If the weight from $n_i$ to $o_j$ is positive then assume $n_i = A_{\min}$ and $\sim n_i = -1$. If the weight from $n_i$ to $o_j$ is negative then assume $n_i = 1$ and $\sim n_i = A_{\max}$.* These are the worst cases analyses, which means that we consider the minimum contribution of each hidden neuron to the activation of an output neuron.

*Remark 5.3.1.* Note that when we consider that the activation values of hidden neurons are either positive in the interval $(A_{\min}, 1]$ or negative in the interval $[-1, A_{\max})$, we assume, without loss of generality, that the network's learning algorithm is such that no hidden neuron presents activation in the range $[A_{\max}, A_{\min}]$. An alternative would be to make both $A_{\max}$ and $A_{\min}$

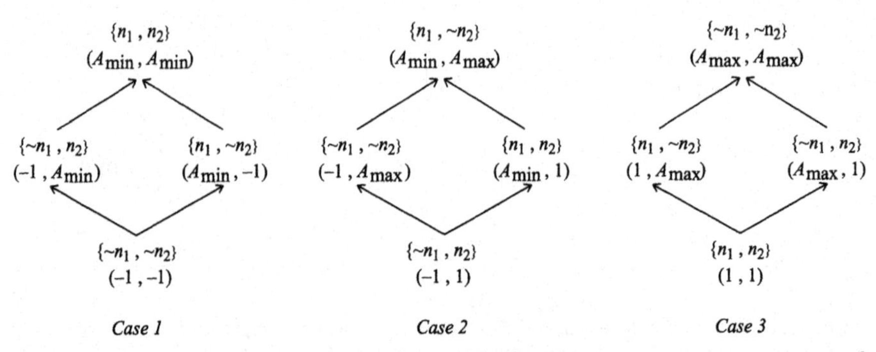


**Fig. 5.19.** Orderings on *Hidden to Output BNSs* with two input neurons $n_1$ and $n_2$, using worst case analyses on $[-1, A_{\max})$ and $(A_{\min}, 1]$.

approach *zero*.[20] We will discuss other alternatives, which include the addition of a penalty function to the learning algorithm, in Chapter 9, Section 9.7. We will discuss the practical aspects of fine-tuning the value of $A_{\min}$ as a parameter of the extraction algorithm in Chapter 6.

In the sequel, we exemplify how to obtain the ordering on a *Hidden to Output BNS* with two input neurons $n_1$ and $n_2$, connected to an output neuron $x$ with positive and negative weights.

We start by applying the Transformation Algorithm on the *Hidden to Output BNS.* We obtain the *BNS*'s positive form and check the labels of its input neurons (the network's hidden neurons). If they are labelled $n_1$ and $n_2$, i.e. $sup(\mathbf{I}) = (n_1, n_2)$, then the weights from both of them to $x$ are positive in the original *BNS.* Thus, we assume that $\sim n_i = -1$ and $n_i = A_{\min}$ for $i = \{1, 2\}$. As a result, we derive the ordering of Figure 5.19(*Case* 1). If, however, the Transformation Algorithm tells us that $sup(\mathbf{I}) = (n_1, \sim n_2)$ then we consider $\sim n_1 = -1$ and $n_1 = A_{\min}$ for the activation values of $n_1$, and $\sim n_2 = A_{\max}$ and $n_2 = 1$ for the activation values of $n_2$. Figure 5.19(*Case* 2) shows the ordering obtained in the original *BNS* when $sup(\mathbf{I}) = (n_1, \sim n_2)$. Finally, if $sup(\mathbf{I}) = (\sim n_1, \sim n_2)$, we assume that $\sim n_i = A_{\max}$ and $n_i = 1$ for $i = \{1, 2\}$, as shown in Figure 5.19(*Case* 3). Note that if, in addition, we have $\left|W_{o_j n_2}\right| \leq \left|W_{o_j n_1}\right|$, we also obtain $(A_{\min}, -1) \leq_{\langle\rangle} (-1, A_{\min})$ in Figure 5.19(Case 1), $(A_{\min}, 1) \leq_{\langle\rangle} (-1, A_{\max})$ in 5.19(Case 2), and $(A_{\max}, 1) \leq_{\langle\rangle} (1, A_{\max})$ in 5.19(Case 3). In this case, the resulting orders $\preceq$ are chains, as expected.

Let us now see if we can define a mapping for *Hidden to Output BNSs,* analogous to the mapping $\sigma$ for Regular Networks and *Input to Hidden BNSs.* In fact, if we assume, without loss of generality, that $A_{\max} = -A_{\min}$

[20] Recall that, in general, we allow training examples to change background knowledge. In this case, the value of $A_{\min}$ could also have been changed.

then the same function $\sigma$ mapping input vectors of the positive form into input vectors of the *BNS* can be used here. Let $i_i \in \{-1, A_{\min}\}$, $i'_i \in \{-1, -A_{\min}, A_{\min}, 1\}$, $x_i \in \mathcal{I}_+, 1 \leq i \leq p$. Recall that $\sigma_{[x_1,...,x_p]}(i_1, ..., i_p) = (i'_1, ..., i'_p)$, where $i'_i = i_i$ if $x_i$ is a positive literal and $i'_i = -i_i$ otherwise. Thus, $\sigma_{[a,\sim b,c,\sim d]}(A_{\min}, A_{\min}, -1, -1) = (A_{\min}, -A_{\min}, -1, 1)$. The following example illustrates the use of $\sigma$ for *Hidden to Output BNSs.*

*Example 5.3.4.* Consider a *Hidden to Output BNS* ($B$) with three input neurons ($n_1, n_2, n_3$) and output $o$. Let $W_{on_1} > 0$, $W_{on_2} < 0$ and $W_{on_3} > 0$. Thus, the positive form ($B^+$) of $B$ contains $n_1, \sim n_2$ and $n_3$ as input neurons. Using the mapping $\sigma$ above, we obtain $\sigma_{[n_1,\sim n_2,n_3]}(A_{\min}, A_{\min}, A_{\min}) = (A_{\min}, -A_{\min}, A_{\min})$. In other words, querying the original *BNS* ($B$) with $[n_1, n_2, n_3] = (A_{\min}, -A_{\min}, A_{\min})$ is equivalent to querying its positive form ($B^+$) with $[n_1, \sim n_2, n_3] = (A_{\min}, A_{\min}, A_{\min})$. Similarly, $\sigma_{[n_1,\sim n_2,n_3]}(-1, -1, -1) = (-1, 1, -1)$, $\sigma_{[n_1,\sim n_2,n_3]}(-1, -1, A_{\min}) = (-1, 1, A_{\min})$, and so on (see Figure 5.20).

Since we have taken the activation values in the worst cases and assumed $A_{\max} = -A_{\min}$, the extraction process can be carried out by querying the positive form of the *BNS* with values in $\{-1, A_{\min}\}$ only, i.e. following the ordering of Figure 5.20(b). In this way, the only difference between $B^+$ and the positive form of an *Input to Hidden BNS* is that input values 1 should be replaced by $A_{\min}$ (compare, for instance, Figures 5.15 and 5.20(b)).

We are finally in a position to present the extraction algorithm extended for non-regular networks.

– *Knowledge Extraction Algorithm – General Case*

1. Split the neural network $N$ into *BNSs*;
2. For each *BNS* $\mathcal{B}_i$ ($1 \leq i \leq r + q$) do:[21]
   a) Apply the *Transformation Algorithm* and find its positive form $\mathcal{B}_i^+$;
   b) Order $\mathcal{I}_+$ according to the weights associated with each input of $\mathcal{B}_i^+$;[22]
   c) If $\mathcal{B}_i^+$ is an *Input to Hidden BNS*, take $i_i \in \{-1, 1\}$;
   d) If $\mathcal{B}_i^+$ is a *Hidden to Output BNS*, take $i_i \in \{-1, A_{\min}\}$;
   e) Let $Inf(\mathbf{I}) = \{-1, -1, ..., -1\}$ and $Sup(\mathbf{I}) = \{1, 1, ..., 1\}$;
   f) Call the *Knowledge Extraction Algorithm for Regular Networks*, step 3, where $N_+ := \mathcal{B}_i^+$;
   /* *Recall that, now, we have to replace Search Space Pruning Rules 1 and 2, respectively, by BNS Pruning Rules 1 and 2.*

[21] Recall that $r$ is the number of hidden neurons and $q$ is the number of output neurons of $N$. As a result, $r + q$ is the number of *BNSs* in $N$.

[22] This is done because of $\leq_{\langle\rangle}$ (see Example 5.3.2).

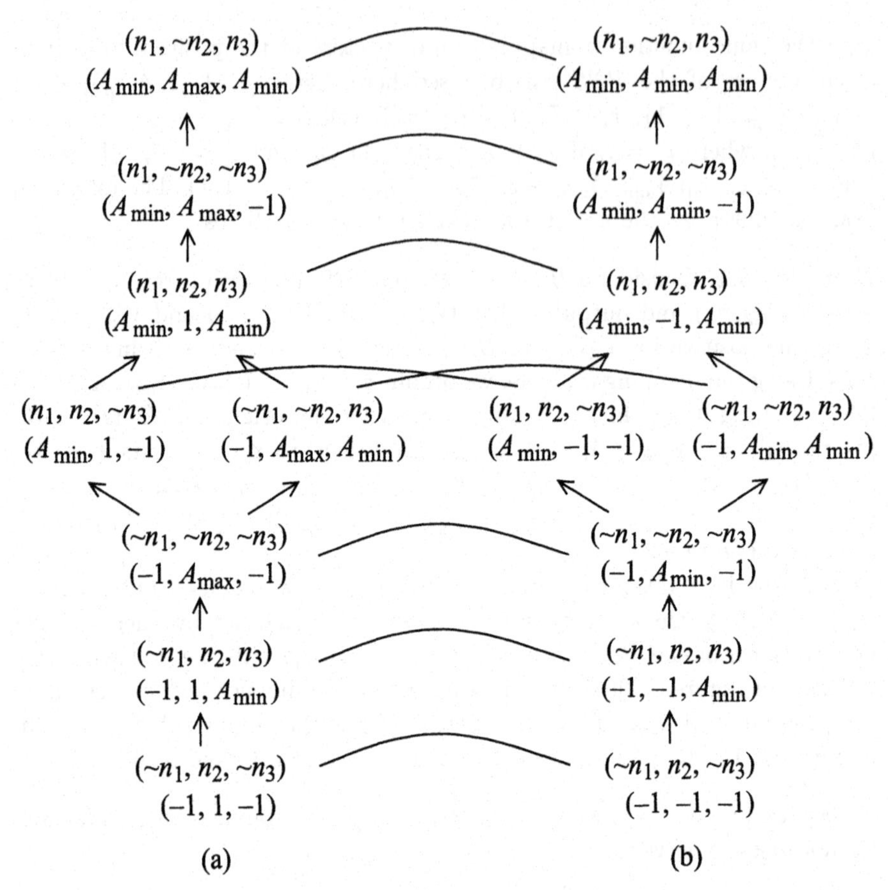


**Fig. 5.20.** (a) the ordering on a hidden to output BNS with three input neurons and the associated activation values in the worst case; (b) the ordering on the positive form of the BNS and the mapping between the orderings.

/* *We also need to add the following lines to the extraction algorithm for regular networks (step 3d):*

- If BNS *Pruning Rule 4* is applicable, stop generating the successors of $\mathbf{i}_\perp$;
- If BNS *Pruning Rule 3* is applicable, stop generating the predecessors of $\mathbf{i}_\top$;

3. Assemble the final Rule Set of $N$.

In what follows, we describe in detail step 3 of the above algorithm, and discuss the problems resulting from the worst case analysis of *Hidden to Output BNSs*.

### 5.3.3 Assembling the Final Rule Set

Steps 1 and 2 of the general case extraction algorithm generate local information about each hidden and output neuron. In step 3, such information needs to be carefully combined, in order to derive the final set of rules of $N$. So far, we have used $n_i$ and $\sim n_i$ to indicate, respectively, that the activation of hidden neuron $n_i$ is greater than $A_{\min}$ or smaller than $A_{\max}$. Bear in mind, however, that hidden neurons $n_i$ do not have concepts directly associated to them. Thus, the task of assembling the final set of rules is that of relating the concepts in the network's input layer directly to the ones in its output layer, by removing literals $n_i$. The following Lemma 5.3.1 will serve as basis for this task.

**Lemma 5.3.1.** *The extraction of rules from* Input to Hidden BNSs *is sound and complete.*

*Proof. From Proposition 5.3.1 and Theorem 5.2.1, we obtain soundness of the set of rules. From Proposition 5.3.1 and Theorem 5.2.2, we obtain completeness of the set of rules.* □

**Lemma 5.3.2.** *The extraction of rules from* Hidden to Output BNSs *is sound.*

*Proof. From Proposition 5.3.1 and Theorem 5.2.1, if we are able to derive a rule $r$ taking $n_i \in \{-1, A_{\min}\}$ then, from the monotonically crescent characteristic of $h(x)$, $r$ will still be valid if $n_i \in \{(-1, -A_{\min}), (A_{\min}, 1)\}$, where $A_{\min} > 0$.* □

Lemma 5.3.1 allows us to use the *completion* of the rules extracted from *Input to Hidden BNSs* to assemble the set of rules of the network, i.e. it allows an extracted rule of the form $X_1, ..., X_p \to L_j$ to be substituted by the stronger $X_1, ..., X_p \leftrightarrow L_j$. For example, assume that the extraction algorithm derives $a \to n_1$ from a *BNS* $\mathcal{B}_1$ and $b \sim c \to n_2$ from a *BNS* $\mathcal{B}_2$. By Lemma 5.3.1, we have $a \leftrightarrow n_1$ and $b \sim c \leftrightarrow n_2$. By contraposition, we have $\sim a \leftrightarrow \sim n_1$ from $\mathcal{B}_1$, and $\sim b \vee c \leftrightarrow \sim n_2$ from $\mathcal{B}_2$. Now that we have the necessary information regarding the activation values of $n_1$ and $n_2$, assume that we have derived the rule $n_1 \sim n_2 \to x$ from a *Hidden to Output BNS* $\mathcal{B}_3$. We know that $a \to n_1$ and $\sim b \vee c \to \sim n_2$. As a result, we may assemble the final set of rules regarding output $x$, namely, $\{a \sim b \to x,\ ac \to x\}$.

The following example illustrates how to assemble the final set of rules of a network in a sound mode. It also illustrates the incompleteness of the general case extraction, which we prove in the sequel.

*Example 5.3.5.* Consider a neural network $N$ with two input neurons $a$ and $b$, two hidden neurons $n_1$ and $n_2$ and one output neuron $x$. Assume that the

**Table 5.1.** Possible activation values of $n_1$, $n_2$ and $x$, given input vectors $[a, b]$.

| $a$ | $b$ | $n_1$ | $n_2$ | $x$ |
|---|---|---|---|---|
| -1 | -1 | $< A_{\max}$ | $< A_{\max}$ | $< A_{\max}$ |
| -1 | 1 | $> A_{\min}$ | $> A_{\min}$ | $< A_{\max}$ |
| 1 | -1 | $< A_{\max}$ | $< A_{\max}$ | $< A_{\max}$ |
| 1 | 1 | $> A_{\min}$ | $< A_{\max}$ | $> A_{\min}$ |

set of weights is such that the activation values in Table 5.1 are obtained for each input vector.

An exhaustive pedagogical extraction algorithm, although inefficiently, would derive the unique rule $ab \rightarrow x$ from $N$. That is because $[a, b] = (1, 1)$ is the only input vector that activates $x$. A decompositional approach, on the other hand, would split the network into its *BNSs*. Since $[a, b] = (-1, 1)$ and $[a, b] = (1, 1)$ activate $n_1$, the rules $\sim ab \rightarrow n_1$ and $ab \rightarrow n_1$ would be derived, and hence $b \rightarrow n_1$. Similarly, the rule $\sim ab \rightarrow n_2$ would be derived, since $[a, b] = (-1, 1)$ also activates $n_2$.

Taking $A_{\min} = 0.5$, suppose that, given $[a, b] = (-1, 1)$, the activation values of $n_1$ and $n_2$ are, respectively, 0.6 and 0.95. As we have seen in Example 5.1.1, if we had assumed that the activation values of $n_1$ and $n_2$ were both 1, we could have wrongly derived the rule $n_1 n_2 \rightarrow x$ (unsoundness). To solve this problem, we have taken the activation values of the hidden neurons in the worst case, namely, $n_1 = A_{\min}$ and $n_2 = A_{\min}$.

Now, given $[a, b] = (1, 1)$, suppose that the activation values of $n_1$ and $n_2$ are, respectively, 0.9 and $-0.6$. If we take the activation values in the worst case, that is, $n_1 = A_{\min}$ and $n_2 = -1$, we might not be able to derive the rule $n_1 \sim n_2 \rightarrow x$, as expected (incompleteness).

Assuming that we do manage to derive $n_1 \sim n_2 \rightarrow x$, the final set of rules of $N$ can be assembled as follows: by Lemma 5.3.1, we derive $b \leftrightarrow n_1$ and $\sim a \wedge b \leftrightarrow n_2$. From $\sim a \wedge b \leftrightarrow n_2$, we obtain $a \vee \sim b \leftrightarrow \sim n_2$. From $b \leftrightarrow n_1$, $a \vee \sim b \leftrightarrow \sim n_2$ and $n_1 \sim n_2 \rightarrow x$, we have $b \wedge (a \vee \sim b) \rightarrow x$, which is equivalent to $ab \rightarrow x$, in accordance with the result of the exhaustive pedagogical extraction.

**Theorem 5.3.1.** *The extraction algorithm for non-regular networks is sound.*

*Proof. Directly from Lemmas 5.3.1 and 5.3.2.* □

**Theorem 5.3.2.** *The extraction algorithm for non-regular networks is incomplete.*

*Proof. We give a counter-example. Let $\mathcal{B}$ be a* Hidden to Output BNS *with input $n_1$ and output $x$. Let $\beta = 1$, $W_{xn_1} = 1$, $\theta_x = 0.1$. Assume $A_{\min} = 0.4$. Given $Act(n_1) = 1$, we obtain $Act(x) = 0.42$, i.e. $n_1 \rightarrow x$. Taking $Act(n_1) = A_{\min}$, we obtain $Act(x) = 0.15$ and, thus, we have lost $n_1 \rightarrow x$.* □

As far as efficiency is concerned, one can apply the extraction algorithm until a predefined number of input vectors is queried, and then test the accuracy of the set of rules derived against the accuracy of the network. If, for instance, in a particular application, the set of rules obtained classifies correctly, say, 95% of the training and testing examples of the network, then one could stop the extraction process. Theorem 5.3.1 will ensure that it is sound, that is, that any rule extracted is represented in the network.

## 5.4 Knowledge Representation Issues

In this section, we show that one hidden layer is a necessary condition for neural networks to encode logic programs. We also outline some considerations about the representation of *M of N* rules in neural networks.

In Chapter 3, we have presented a *Translation Algorithm* that converts each logic program $\mathcal{P}$ into a neat neural network structure $\mathcal{N}$ by associating each rule of $\mathcal{P}$ with a hidden neuron of $\mathcal{N}$. For this reason, we call $\mathcal{N}$ the canonical (neural) form of $\mathcal{P}$. In Chapter 4, however, we have seen that each hidden neuron can encode more than a single rule. Take, for example, the program $P = \{ab \rightarrow x, c \rightarrow x\}$. $P$ can be encoded (or trained) in a network with a single hidden neuron. The network of Figure 5.21 is a possible implementation of $P$. Similarly, $a \rightarrow x$ and $a \rightarrow y$ can both be encoded through a single hidden neuron. In the present chapter, the extraction task was to find out how the knowledge trained with examples was encoded in the network. This was necessary because, during training, the network's canonical form may, and probably will, be lost.[23]

We have already seen that a single hidden layer network is sufficient for representing general and extended logic programs. Let us see now why a hidden layer is also a necessary condition. Although rules of the form $i_1, ..., i_n \rightarrow o$ could be implemented in a network without hidden neurons, that is, *perceptrons*, this is not the case if, for example, $P = \{ab \rightarrow x, cd \rightarrow x\}$. We use $P$ as a counter-example in the following proof.

**Proposition 5.4.1.** *Let $\mathcal{N}$ be a* Perceptron. *There exists a logic program $\mathcal{P}$ such that $\mathcal{N}$ does not compute $T_{\mathcal{P}}$.*

*Proof. by Contradiction.*

*We give a counter-example. Let $P = \{ab \rightarrow x, cd \rightarrow x\}$. Assume that the network of Figure 5.22 computes $T_P$. Let $h(x)$ be a monotonically crescent*

[23] Alternatively, some extraction methods try to control the learning process in order to keep the initial neat structure of the network, thus facilitating the extraction process. On the other hand, such methods restrict the learning capabilities of the network and could not be used with off-the-shelf learning algorithms.

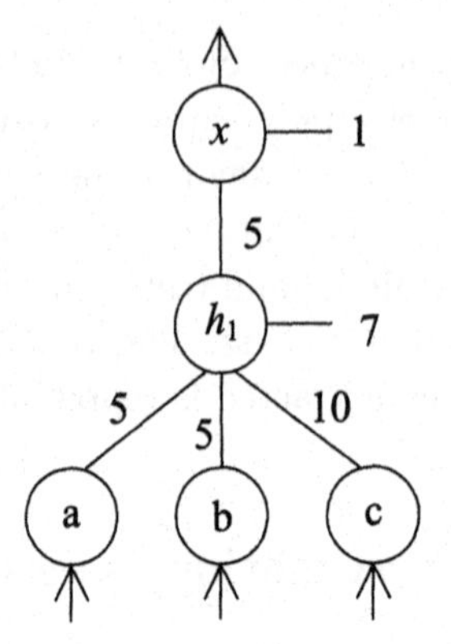


**Fig. 5.21.** A network that encodes $P = \{ab \to x, c \to x\}$ using a single hidden neuron.

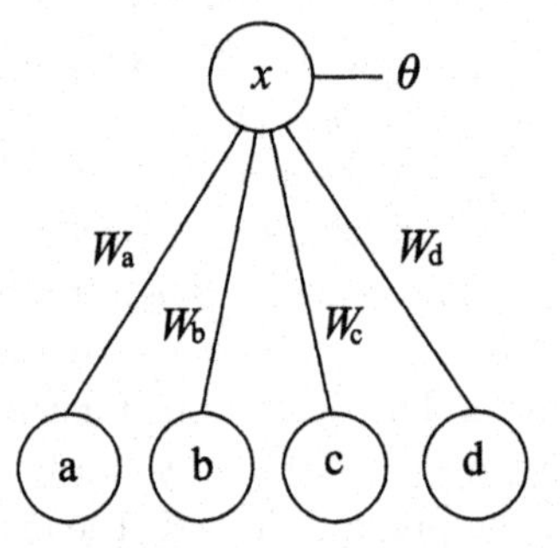


**Fig. 5.22.** A perceptron with four input neurons and a single output neuron.

*function, and* $(a, b, c, d) \in \{-1, 1\}^4$. *Hence,* $h(\alpha_1 - \theta_x) \geq A_{\min}$ *and* $h(\alpha_2 - \theta_x) \geq A_{\min}$, *where* $\alpha_1 = W_a + W_b - W_c - W_d$ *and* $\alpha_2 = -W_a - W_b + W_c + W_d$. *However, we do not want to activate* $x$ *when, for example,* $a$ *and* $c$ *are true. Thus,* $h(\alpha_3 - \theta_x) < A_{\min}$, *where* $\alpha_3 = W_a - W_b + W_c - W_d$, *should be verified. Similarly,* $h(\alpha_4 - \theta_x) < A_{\min}$, $h(\alpha_5 - \theta_x) < A_{\min}$ *and* $h(\alpha_6 - \theta_x) < A_{\min}$, *where* $\alpha_4 = W_a - W_b - W_c + W_d$, $\alpha_5 = -W_a + W_b + W_c - W_d$ *and* $\alpha_6 = -W_a + W_b - W_c + W_d$, *should also be verified. Putting the above inequalities together yields* $\alpha_3 < \alpha_1$ *and* $\alpha_6 < \alpha_2$. *As a result,* $W_c < W_b$ *and* $W_b < W_c$, *i.e. a contradiction. This shows that* $P$ *requires a network with two hidden neurons. The proof is similar if* $(a, b, c, d) \in \{0, 1\}^4$. □

The above proposition will be necessary in Chapter 7. Let us now concentrate on the representation of *M of N* rules. In fact, neural networks are very good at learning and encoding such rules. This is one of the main reasons of neural networks superiority over symbolic machine learning systems in application domains where *M of N* rules are useful. The following proposition shows why neural networks and *M of N* rules are made for each other.

**Proposition 5.4.2.** *Let $\mathcal{P} = \{m(i_1, ..., i_n) \rightarrow x\}$. There exists a* Perceptron *$\mathcal{N}$ such that $\mathcal{N}$ computes $T_{\mathcal{P}}$.*

*Proof.* *By definition, $0 \leq m \leq n$. Let $p \geq m$. We need to show that $x$ is activated iff $p$ of $i_1, ..., i_n$ are satisfied. Let $Wi_1, ..., Wi_i = W$, $Wi_{i+1}, ..., Wi_n = -W$, ($W \in \Re^+$). When $p$ of $i_1, ..., i_n$ are satisfied then the input potential of $x$ is $I_x = pW - (n-p)W - \theta_x$. In the worst case, $p = m$ and $h(mW - (n-m)W - \theta_x) > A_{\min}$ has to be satisfied. Similarly, when less than $m$ of $(i_1, ..., i_n)$ are satisfied then $x$ should not be activated. Again in the worst case, $h((m-1)W - (n-(m-1))W - \theta_x) < A_{\min}$ has to be satisfied. Hence, $(m-1)W - (n-(m-1))W < mW - (n-m)W$, and thus $-2W < 0$ should be true. Since $W > 0$, this condition can be easily satisfied.* □

For example, in order to represent $3(ab \sim c \sim d) \rightarrow x$ in the network of Figure 5.22, it suffices to define $W_a = W_b = 1$, $W_c = W_d = -1$, and $\theta = 0.5$, taking $h(x) = \frac{2}{1+e^{-x}} - 1$ and $(a, b, c, d) = \{-1, 1\}^4$.

Further development of the above results could guide us towards methods and heuristics to build more efficient learning and extraction algorithms. A flavour of it is given when we train a network to encode *Exactly* $2(abc) \rightarrow x$. Such a rule is equivalent to $2(abc) \wedge \sim 3(abc) \rightarrow x$.[24] $2(abc) \wedge \sim 3(abc)$ cannot be encoded through a single hidden neuron, since $2(abc)$ subsumes $3(abc)$. We then know that at least two hidden neurons are necessary for learning *Exactly* $2(abc) \rightarrow x$. In fact, two hidden neurons are sufficient, because $2(abc)$ and $3(abc)$ can be easily learned by a hidden neuron each, say $h_1$ and $h_2$, respectively, while the output neuron $x$ can as easily negate $h_2$ and perform an *and* between $h_1$ and $\sim h_2$. We will come back to this example in the next chapter.

## 5.5 Summary and Further Reading

We have seen that decompositional extraction can be unsound. On the other hand, sound and complete pedagogical extraction may present exponential complexity. We call this problem the *complexity* × *quality* trade-off. In order to ameliorate it, we started by analysing the cases where regularities can be found in the set of weights of a neural network. If such regularities are present, a number of *pruning rules* can be used to safely reduce the search space of the extraction algorithm. These pruning rules reduce the extraction algorithm's complexity in some interesting cases. Notwithstanding, we have been able to show that the extraction method is sound and complete w.r.t. an exhaustive pedagogical extraction. A number of *simplification rules* also help in reducing the size of the extracted set of rules.

---

[24] Note that $\sim 3(abc) \leftrightarrow 1(\sim a \sim b \sim c)$.

We then extended the extraction algorithm to the case where regularities are not present in the network as a whole. This is the general case, since we do not fix any constraints on the network's learning algorithm. However, we have identified subnetworks of non-regular networks that always contain regularities, by showing that the network's building block, here called Basic Neural Structure (*BNS*), is regular. As a result, using the same underlying ideas, we were able to derive rules from each *BNS*. In this case, however, we were applying a decompositional approach, and our problem was how to assemble the final set of rules of the network. We needed to provide a special treatment for *Hidden to Output BNSs*, since the activation values of hidden neurons are not discrete, but real numbers in the interval $[-1, 1]$. In order to deal with that, we assumed, without loss of generality, two possible intervals of activation, $[-1, A_{\max})$ and $(A_{\min}, 1]$, and performed a worst case analysis. Finally, we used the completeness of the extraction from *Input to Hidden BNSs* to assemble the final set of rules of the network, and show that the general case extraction method is still sound.

In this chapter, we have described the ABG method for extracting symbolic knowledge from trained neural networks. Although neural networks have shown very good performance in many application domains, one of their main drawbacks lies on the incapacity to explain the reasoning mechanisms that justify a given answer. As a result, their use has become limited. This motivated the first attempts towards finding the justification for neural networks' reasoning, dating back to the end of the 1980s. Nowadays, it seems to be a consensus that the way to try to solve this problem is to extract the symbolic knowledge from the trained network. The problem of knowledge extraction turned out to be one of the most interesting open problems in the field. So far, some extraction algorithms were proposed [AG95, CS94, Fu94, PHD94, Set97a, TS93] and had their effectiveness empirically confirmed using certain applications as benchmark. Some important theoretical results have also been obtained [dGZ99, Fu94, HK94, Thr94, SH97, Mai98]. However, we are not aware of any extraction method that fulfils the following list of desirable properties suggested by Thrun in [Thr94]: (1) no architectural requirements; (2) no training requirements; (3) correctness; and (4) high expressive power. The extraction algorithm presented here satisfies the above requirements 2 and 3. It does impose, however, some restriction on the network's architecture. For instance, it assumes that the network contains a single hidden layer. This, according to the results of Hornik *et al.*[HSW89], is not a drawback though. In what concerns the expressive power of the extracted set of rules, our extraction algorithm enriches the language commonly used by adding default negation. This is done because neural networks encode nonmonotonicity. In spite of that, we believe that item (4) is the subject, among the above, that needs most attention and further development.

The literature on neural-symbolic integration increasingly deals with the subject of knowledge extraction, one of the area's most attractive and difficult problems. The reader can find references to the early work on rule extraction in [ADT95]. More recently, [BCC+99] and [CZ00] describe a number of interesting approaches to neural-symbolic integration, some containing developments on rule extraction. In addition, [Kur00] is a special issue on neural networks and structured knowledge, and has been devoted to the problem of rule extraction and applications. Among the existing extraction methods, the ones presented in [HK94], [SN88] and [Gal93], the "Ruleneg" [PHD94], the "Validity Interval Analysis" algorithm [Thr94], and the "Rule Extraction as Learning" method [CS94] (see also [Cra96]) all use pedagogical approaches. In [HK94], Holldobler and Kalinke present an exhaustive pedagogical extraction algorithm for rule extraction. In an attempt to make pedagogical extraction applicable to larger neural networks, Saito and Nakano [SN88], and then Gallant [Gal93], restrict the search space by considering for rule extraction only the truth-values that are referred to in the network's training set. More similar to the ideas presented in this chapter, Thrun's Validity Interval Analysis [Thr94] defines valid intervals of activation to constrain some of the input and output neurons, by performing forward and backward propagations through the network, as in the *Backpropagation* learning algorithm, while Craven and Shavlik's Rule Extraction as Learning method [CS94] builds *Decision Trees* from trained neural networks by querying the network with specific input vectors, starting from the network's training examples and using the constraints imposed by the provisionally built decision tree. Thrun's validity intervals are *valid activation ranges* similar to ABG's fine-tuning and worst case analysis on the values of $A_{\min}$. Craven and Shavlik's *constraints* to guide Rule Extraction as Learning are similar to the constraints imposed by ABG's partial ordering on a network's input vectors. In addition, Craven and Shavlik see the problem of rule extraction from trained networks as a problem of inductive learning (with the extra benefit of having the network as an *oracle* to which one is allowed to make as many *queries* as one desires). The search on the space of input vectors performed by the ABG framework could clearly be seen also as a task of inductive learning. In fact, the different lattice searching algorithms developed for *Inductive Logic Programming* could be adapted for use within the ABG framework.

Although pedagogical rule extraction methods are, in general, applicable to a broader range of networks, they all suffer from scalability problems when the size of the networks increases. Decompositional methods, as discussed in Section 5.1, try to solve this problem by extacting rules from certain types of subnetworks, and then assembling the final rule set. The "Subset" method

[Fu94], the "MofN" method [TS93], "Rulex" [AG95], "Combo" [Kri96], as well as Setiono's extraction method [Set97b, Set97a] all are decompositional. Fu's "Subset" method [Fu94], for instance, attempts to search for subsets of weights of each neuron in the hidden and output layers of $N$, such that the neuron's input potential exceeds its threshold. Each subset that satisfies the above condition is written as a rule. Towell and Shavlik's "MofN" [TS93] clusters and prunes the set of weights of each *BNS* in order to facilitate the extraction of rules. It also uses the *M of N* representation to extract more compact sets of rules. Differently from the Subset method and its variations, Andrews and Geva's "Rulex" [AG95] perform rule extraction from *Radial-Basis Function* (*RBF*) networks by interpreting weight vectors directly as rules, while Krishnan's "Combo" [Kri96] extracts rules in disjunctive normal form from a network by querying its subnetworks with an increasing number of positive literals in each input vector. Setiono's decompositional extraction method [Set97b] applies a *penalty function* for pruning a feedforward neural network during training, and then generates rules from the subnetworks of the pruned network by considering only a small number of activation values at the hidden units. Additionally, the extraction method presented in this chapter [dGBG01, dGBGdS99] and, more recently, Boutsinas and Vrahatis' work [BV01] also take into consideration the fact that neural networks are nonmonotonic systems and, as a result, cater for the extraction of nonmonotonic rules.

# 6. Experiments on Knowledge Extraction

In this chapter, we present the results of an empirical analysis of the ABG system of knowledge extraction from trained networks, using traditional examples and real-world application problems. The implementation of the system has been kept as simple as possible, and does not benefit from all the features of the theory presented in the previous chapter.

Our purpose in this section is to show that the implementation of a sound method of extraction can be efficient, and to confirm the importance of extracting nonmonotonic theories from trained networks. Our intention is not to provide an exhaustive comparative analysis with other extraction methods. Such a comparison could easily be biased, depending on the application at hand, training parameters and testing methodology used. Nevertheless, we also present the results reported in [Fu94, Set97a, TS94a], when available.

We have used three application domains in order to test the extraction algorithm: the Monk's problems [TBB+91], which we briefly describe in the sequel, DNA Sequence Analysis and Power Systems' Fault Diagnosis, which are both described in Chapter 4. The results have shown that a very high fidelity between the extracted set of rules and the network can be achieved.

## 6.1 Implementation

The extraction system consists of three modules: the main module takes a trained neural network (its set of weights and activation functions), searches the set of input vectors, and generates a set of rules accordingly; another module is used to simplify the set of rules; and yet another checks the accuracy of the set of rules against that of the network, given a test set, and the fidelity of the set of rules to the network.[1]

The implementation of the extraction system uses the following systematic way of searching the input vectors' space. Given the maximum element, we order it from left to right w.r.t. the weights associated with each input, such that inputs with greater weights are on the left of inputs with smaller weights.

[1] The system was implemented in ANSI C (5K lines of code).

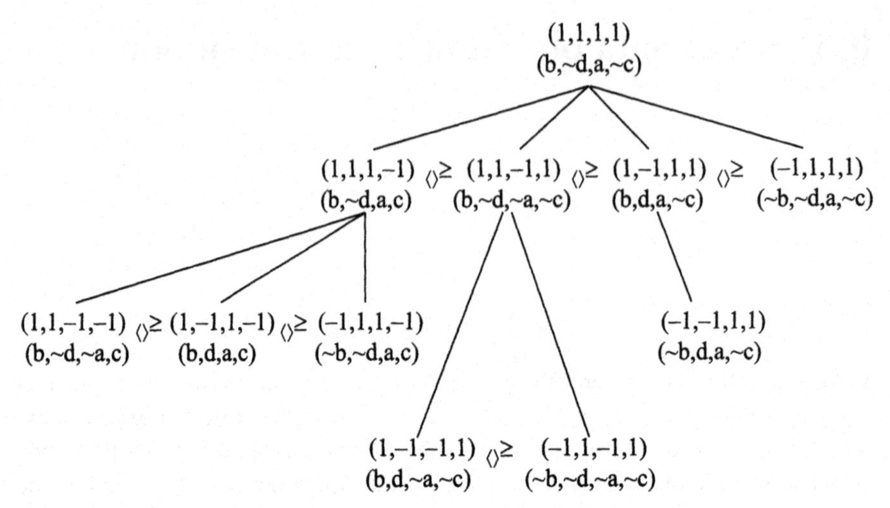


**Fig. 6.1.** Systematically deriving input vectors from $\mathbf{i}_\top$ without repetitions.

Let us use Example 5.3.2 (Chapter 5) to illustrate this process. In Example 5.3.2, $W_{n_i \sim c} \leq W_{n_i a} \leq W_{n_i \sim d} \leq W_{n_i b}$. We rearrange $(a, b, \sim c, \sim d)$ and obtain $(1,1,1,1) = [b, \sim d, a, \sim c]$. The search proceeds by flipping the right-most input, then the second right-most input and so on. At distance 2 from $sup(\mathbf{I})$ and beyond, we only flip the inputs on the left of the left most input $-1$. In this way, we avoid repeating input vectors. Figure 6.1 illustrates this process.

Similarly, starting from the minimum element, we rearrange $(\sim a, \sim b, c, d)$ and obtain $(-1,-1,-1,-1) = [\sim b, d, \sim a, c]$. Figure 6.2 illustrates the search starting from $inf(\mathbf{I})$. Now, at distance 2 from $inf(\mathbf{I})$ and beyond, we only flip the inputs on the left of the left most input 1.

Note the symmetry between Figures 6.1 and 6.2, reflecting, respectively, the use of Pruning Rules 3 and 4 (see Propositions 5.3.3 and 5.3.4 (Chapter 5)). Starting from $sup(\mathbf{I})$, flipping the input with the smallest weight results in the next greatest input, while from $inf(\mathbf{I})$, flipping the input with the smallest weight results in the next smallest input. Note also that the sequence in which the input vectors are generated, according to Figures 6.1 and 6.2, complies with the ordering $\preceq$ on the set of input vectors, shown in Figure 5.16.[2]

We call $dist(\mathbf{i}, inf(\mathbf{I}))$ the *level* of vector $\mathbf{i}$. For example, $(1,-1,-1,-1)$ is at level 1, $(1,1,-1,-1)$ is at level 2, and so on. Each level is divided into *segments* which terminate whenever the left most input is flipped. Thus, level 1 will always contain a single segment and, e.g. in Figure 6.2, level 2 contains

[2] The relation between Figures 6.1 and 6.2, and Figure 5.16 might be more easily recognised if one assumes $(1,1,1,1) = [a, b, c, d]$.

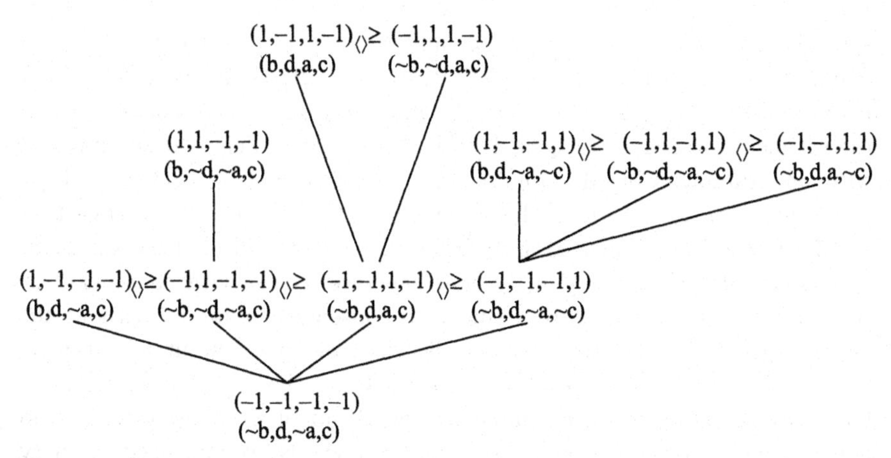


**Fig. 6.2.** Systematically deriving input vectors from $\mathbf{i}_\perp$ without repetitions.

3 segments: *(1)* $(1,-1,-1,1)$, $(-1,1,-1,1)$, $(-1,-1,1,1)$, *(2)* $(1,-1,1,-1)$, $(-1,1,1,-1)$, and *(3)* $(1,1,-1,-1)$. The interest in the definition of a segment is that $\preceq$ is a linear order on the elements of any segment. As a result, our Pruning Rules allow us to jump between segments during the search. For example, take segment *(1)* above. If $(-1,-1,1,1)$ generates a rule then we can generate rules for $(1,-1,-1,1)$ and $(-1,1,-1,1)$ as well, without having to query these input vectors. In the current implementation of the extraction algorithm, we alternate the search between the two trees with $inf(\mathbf{I})$ and $sup(\mathbf{I})$ as root, respectively, such that an element of the latter is always queried after an element of the former and vice versa, starting from $inf(\mathbf{I})$ (this process was illustrated in Example 5.2.6 of the previous chapter). We also perform a breadth first search, i.e. all the segments of a level are covered before we proceed to the next level.

Let us now concentrate on the stopping criteria of the extraction algorithm, namely: *(1)* all elements in the ordering are visited, or *(2)* each element not visited either *(a)* is already represented in the set of rules or *(b)* would not give rise to any new rule. *Case 1* is clearly satisfied when all the elements in both trees are visited. Note that, with the sole purpose of explaining the search process, the elements at level 2 in Figure 6.2 are repeated in Figure 6.1. This does not occur in the actual implementation, where $(1,-1,-1,1)$, $(-1,1,-1,1)$ and $(-1,-1,1,1)$ are generated from $inf(\mathbf{I})$, while $(1,1,-1,-1)$, $(1,-1,1,-1)$ and $(-1,1,1,-1)$ are generated from $sup(\mathbf{I})$. Note also that the elements of the trees are created on the fly, i.e. the trees' structures are not resident in memory.

In order to implement *Case 2*, we simply have to compute the logical *XOR* between the input vectors, and take the sum of the elements of the resulting vector. We define $(x_1, x_2, ..., x_n)$ *XOR* $(y_1, y_2, ..., y_n) = \sum_{i=1}^{n}(x_i \oplus$

$y_i$), where $x_i \oplus y_i = 1$ if $x_i \neq y_i$, and $x_i \oplus y_i = 0$ otherwise. Such an operation is useful because, by Definition 5.2.1, $dist(\mathbf{i}_i, \mathbf{i}_j) = 1$ iff $\mathbf{i}_i$ *XOR* $\mathbf{i}_j = 1$, where $\mathbf{i}_i, \mathbf{i}_j \in \mathbf{I}$. As a result, $\mathbf{i}_i$ is an immediate successor of $\mathbf{i}_j$ w.r.t. the set inclusion ordering $\leq_\mathbf{I}$ on $\mathbf{I}$, iff $\mathbf{i}_i$ *XOR* $\mathbf{i}_j = 1$. This is best explained with an example. Consider Figure 6.2 and let $\mathbf{i}_1 = (-1,-1,-1,1)$, $\mathbf{i}_2 = (-1,-1,1,-1)$, $\mathbf{i}_3 = (-1,1,-1,-1)$ and $\mathbf{i}_4 = (1,-1,-1,-1)$. Assume that $inf(\mathbf{I})$, $\mathbf{i}_1$ and $\mathbf{i}_2$ are queried and do not give rise to rules with consequent $x$. Assume also that querying $\mathbf{i}_3$ does generate a rule with consequent $x$. In this case, we can generate a rule with consequent $x$ for $\mathbf{i}_4$, without the need to query it. The question now is: shall we query input vector $\mathbf{i}_5 = (-1,-1,1,1)$ in the next level of the tree? From the ordering $\leq_\mathbf{I}$ on $\mathbf{I}$, we know that, since $\mathbf{i}_5$ is not an immediate successor of either $\mathbf{i}_3$ or $\mathbf{i}_4$, we can not guarantee that it will activate $x$. In other words, we still need to query $\mathbf{i}_5$. From the point of view of the implementation, however, since we do not have the ordering stored in memory, we have to obtain such an information as the input vectors are generated. We do so by computing the logical *XOR* between $\mathbf{i}_3$, the first input vector to activate $x$ in the preceding level, and $\mathbf{i}_5$, the first input vector to be queried in the current level. We treat '$-1$'s as '0's and obtain $(0,1,0,0)$ *XOR* $(0,0,1,1) = 3$. This tells us that $\mathbf{i}_5$ is not an immediate successor of $\mathbf{i}_3$, and thus $\mathbf{i}_5$ should still be queried. If, however, $\mathbf{i}_1$ were the first input vector to activate $x$, we would have generated rules for $\mathbf{i}_2$, $\mathbf{i}_3$ and $\mathbf{i}_4$, without the need to query these vectors, and would have computed $\mathbf{i}_1$ *XOR* $\mathbf{i}_5$, i.e. $(0,0,0,1)$ *XOR* $(0,0,1,1) = 1$. Similarly, if $\mathbf{i}_2$ were the first vector to activate $x$, we would have generated rules for $\mathbf{i}_3$ and $\mathbf{i}_4$, and computed $\mathbf{i}_2$ *XOR* $\mathbf{i}_5$, obtaining $(0,0,1,0)$ *XOR* $(0,0,1,1) = 1$. In these cases, the result of the *XOR* operation tells us that $\mathbf{i}_5$ is an immediate successor of $\mathbf{i}_1$ and $\mathbf{i}_2$, so that $\mathbf{i}_5$ does not need to be queried.

Before we concentrate on the applications, let us present two simple examples, which will help the reader to recall the sequence of operations contained in the extraction process.

*Example 6.1.1.* (The XOR Problem) A network with $p$ input neurons, $q$ hidden neurons and $r$ output neurons contains $q$ *Input-to-Hidden Basic Neural Structures (BNSs)*, each with $p$ inputs and a single output, and $r$ *Hidden-to-Output BNSs*, each with $q$ inputs and a single output. To each *BNS* we apply a transformation whereby we rename input neurons $x_k$ linked through negative weights to the output, by $\sim x_k$ and replace each weight $W_{lk} \in \Re$ by its modulus. We call the result the *positive form* of the *BNS*. For example, in Figure 6.3, $N_1$ and $N_2$ are the positive forms of the *Input-to-Hidden BNSs* of $N$, while $N_3$ is the positive form of the *Hidden-to-Output BNS* of $N$. We then define the function $\sigma$ mapping input vectors of the positive form into input vectors of the *BNS*. For example, for $N_1$ $\sigma_{[a,\sim b]}(1,1) = (1,-1)$.

Given a $2-ary$ input vector, $\preceq$ is a linear ordering. For $N_1$, $(-1,-1) \preceq (1,-1) \preceq (-1,1) \preceq (1,1)$ and for $N_2$ $(-1,-1) \preceq (-1,1) \preceq (1,-1) \preceq (1,1)$, where $(1,1) = [a, \sim b]$ in both. Querying $N_1$, $\mathrm{h}_0$ is activated for $(1,1)$ only. Thus, by applying $\sigma$ we derive $a \sim b \rightarrow \mathrm{h}_0$. Querying $N_2$, $\mathrm{h}_1$ is not activated for $(-1,-1)$ only. Similarly, we derive $ab \rightarrow \mathrm{h}_1$, $\sim a \sim b \rightarrow \mathrm{h}_1$ and $a \sim b \rightarrow \mathrm{h}_1$. The last two rules can be simplified to obtain $\sim b \rightarrow \mathrm{h}_1$, since $\sim b$ implies $\mathrm{h}_1$ given either $a$ or $\sim a$. Similarly, from $ab \rightarrow \mathrm{h}_1$ and $a \sim b \rightarrow \mathrm{h}_1$ we obtain $a \rightarrow \mathrm{h}_1$.

Considering now *Hidden to Output BNSs*, it is usually assumed that the network's hidden neurons present discrete activation values such as $\{-1,1\}$. We know, however, that this is not the case, and therefore problems may arise from such an assumption. At this point we need to compromise. Either we assume that the activation values of the hidden neurons are in $\{-1, A_{\min}\}$, and then are able to show that the extraction is sound, but incomplete, or we assume that they are in $\{-A_{\min}, 1\}$, obtaining an unsound, but complete, extraction. We have chosen the first approach.[3] For $N_3$ we have $(-1,-1) \preceq (-1, A_{\min}) \preceq (A_{\min}, -1) \preceq (A_{\min}, A_{\min})$, where $(A_{\min}, A_{\min}) = [\sim\mathrm{h}_0, \mathrm{h}_1]$. Taking $A_{\min} = 0.5$, only $(A_{\min}, A_{\min})$ activates o, and we derive the rule $\sim\mathrm{h}_0\mathrm{h}_1 \rightarrow \mathrm{o}$.

Finally, to assemble the rule set of $N$, we take the *completion* of each rule extracted from *Input to Hidden BNSs*. We have $a \sim b \rightarrow \mathrm{h}_0$, $a \rightarrow \mathrm{h}_1$, $\sim b \rightarrow \mathrm{h}_1$ and $\sim\mathrm{h}_0\mathrm{h}_1 \rightarrow \mathrm{o}$. And from $a \sim b \leftrightarrow \mathrm{h}_0$ and $a \vee \sim b \leftrightarrow \mathrm{h}_1$ we obtain $(\sim a \vee b) \wedge (a \vee \sim b) \rightarrow \mathrm{o}$, the $\overline{\mathrm{XOR}}$ function.

*Example 6.1.2.* (EXACTLY 1 OUT OF 5) We train a network with five input neurons $\{a,b,c,d,e\}$, two hidden neurons $\{h_0, h_1\}$ and one output neuron $\{o\}$, on all the 32 possible input vectors. The network's output neuron fires iff exactly one of its inputs fires. Although this is a very simple network, it is not straightforward to verify, by inspecting its weights, that it computes the following rule: "*Exactly* 1 *out of* $\{a,b,c,d,e\}$ *implies* $o$".

Assume the following order on the weights linking the input layer to each hidden neuron $h_0$ and $h_1$: $|W_{h_0 d}| \le |W_{h_0 e}| \le |W_{h_0 c}| \le |W_{h_0 a}| \le |W_{h_0 b}|$ and $|W_{h_1 d}| \le |W_{h_1 e}| \le |W_{h_1 a}| \le |W_{h_1 c}| \le |W_{h_1 b}|$. We split the network into its *BNSs* and apply the extraction algorithm. Taking $\mathcal{I} = [a,b,c,d,e]$ for the *BNS* with output $h_0$, we find out that input $(-1,-1,-1,1,-1)$ activates $h_0$, by querying the *BNS*. Since $|W_{h_0 d}|$ is the smallest weight, from the ordering $\preceq$ on $\mathbf{I}$ and by applying Definitions 5.2.6 and 5.2.8 (Chapter 5), we derive the rule $1(abcde) \rightarrow h_0$. Note that, by Definition 5.2.9, this rule subsumes

[3] Here, we perform a kind of worst case analysis. By choosing activations in $\{-1, A_{\min}\}$, misclassifications occur because of the absence of a rule (incompleteness). Analogously, by choosing $\{-A_{\min}, 1\}$, misclassifications are due to the inappropriate presence of rules in the rule set (unsoundness). In this context, the choice of $\{-1,1\}$ yields unsound and incomplete rule sets.

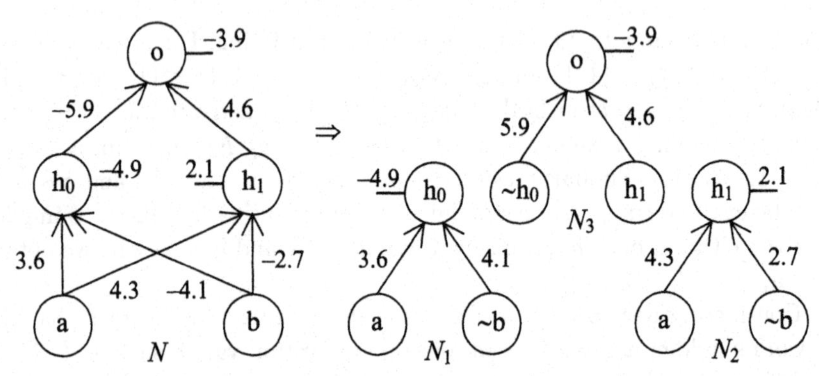


**Fig. 6.3.** The network $N$, having tanh as activation function, computes $\overline{XOR}$. We will extract rules for $h_0$, $h_1$ and $o$ by querying $N_1$, $N_2$ and $N_3$, respectively, and then assemble the set of rules of $N$.

$m(abcde) \rightarrow h_0$, for $m > 1$. Taking again $\mathcal{I} = [a, b, c, d, e]$ but now for the *BNS* with output $h_1$, we find out that input $(-1, -1, -1, 1, 1)$ activates $h_1$. Similarly, from the ordering $\preceq$ on **I** and by applying Definitions 5.2.6 and 5.2.8, we derive the rule $2(abcde) \rightarrow h_1$. Finally, for the *Hidden to Output BNS*, $\mathcal{I} = [h_0, \sim h_1]$. Taking $A_{\min} = 0.5$, $o$ is only activated by $(A_{\min}, A_{\min})$ and we derive the rule $h_0 \sim h_1 \rightarrow o$.

In order to obtain the rules mapping inputs $\{a, b, c, d, e\}$ directly into the output $\{o\}$, since the extraction from *Input to Hidden BNSs* is sound and complete, we can take the *completion* of the rules extracted: $1(abcde) \leftrightarrow h_1$ and $2(abcde) \leftrightarrow h_2$. Therefore, "*Exactly* 1 *out of* $\{a, b, c, d, e\}$ *implies* $o$" is obtained by computing $1(abcde) \wedge \sim 2(abcde) \rightarrow o$, i.e. "*At least* 1 *out of* $\{a, b, c, d, e\}$ AND *At most* 1 *out of* $\{a, b, c, d, e\}$ *implies* $o$".

In both examples above, the value of $A_{\min}$ was chosen empirically, for there was no background knowledge. In the case of knowledge extraction from networks in which *C-IL²P* has been used, the value of $A_{\min}$ is predefined and should be used as the starting point for the extraction. Otherwise, $A_{\min}$ is a parameter of the extraction system, and the choice of $A_{\min}$ should be a design decision. In case of poor accuracy, fine-tuning the value of $A_{\min}$ could improve the results obtained by the extracted set of rules.

For each application below we investigate three parameters:

- The *accuracy* of the rule set against that of the network w.r.t. a test set;
- The *fidelity* of the rule set to the network, i.e. the ability of the rule set to mimic the network's answers; and
- The *readability* of the rule set in terms of the number of rules extracted.

The following example illustrates how we compute the above parameters.

*Example 6.1.3.* Consider a network $N$ with 3 inputs $(i_1, i_2, i_3) \in \{-1, 1\}^3$ and a single output $o_1 \in [-1, 1]$. Assume that the set of rules extracted from $N$ is $R = \{i_1 \sim i_2 \rightarrow o_1; i_3 \rightarrow o_1\}$. $R$ is represented internally as $R_{int} = \{(1, -1, 0), (0, 0, 1)\}_{o_1}$, where 0 indicates "don't care" such that $\{(1, -1, 0)\}_{o_1}$ means $i_1 \sim i_2 \rightarrow o_1$.

Now, let $target\_o_1$ and $net\_o_1$ be, respectively, the target output and the network's output for a given input vector $(i_1, i_2, i_3)$. Table 6.1 shows a hypothetical set of target outputs ($target\_o_1$), network's outputs ($net\_o_1$) and rule set's answers ($rules\_o_1$) for six input vectors $(i_1, i_2, i_3)$. The values of $rules\_o_1$ are obtained from $(i_1, i_2, i_3)$ and $R_{int}$ as follows: If $(i_1, i_2, i_3)$ matches any vector in $R_{int}$ then $rules\_o_1 = 1$. Otherwise, $rules\_o_1 = -1$. For example, $(-1, -1, -1)$ does not match $(1, -1, 0)$ nor $(0, 0, 1)$; $(-1, -1, 1)$ matches $(0, 0, 1)$; $(1, -1, -1)$ matches $(1, -1, 0)$; and so on.

Finally, for each input vector, we compare $rules\_o_1$ and $target\_o_1$ to obtain the rule set accuracy, and $rules\_o_1$ and $net\_o_1$ to obtain the fidelity of the rule set to the network. As usual, $net\_o_1$ and $target\_o_1$ give the network's accuracy. In order to compare $net\_o_1$ with $target\_o_1$ and $rules\_o_1$, if $net\_o_1 > A_{\min}$ then $net\_o_1 := 1$. Otherwise, $net\_o_1 := -1$.

Thus, in this example, the accuracy of the rule set is 4/6 (or 66.6%), in comparison with 3/6 (or 50%) of accuracy obtained by the network. The fidelity of the rule set to the network is 5/6 (or 83.3%). Note that, if the extraction is sound, when $net\_o_1$ is $-1$ then $rules\_o_1$ should never be 1. Dually, if the extraction is complete, when $net\_o_1$ is 1 then $rules\_o_1$ should never be $-1$. In other words, soundness and completeness imply 100% of fidelity.

For a network with $n$ output neurons, the above process should be repeated $n$ times.

**Table 6.1.** Computing accuracy and fidelity

| $i_1$ | $i_2$ | $i_3$ | $target\_o_1$ | $net\_o_1$ | $rules\_o_1$ |
|---|---|---|---|---|---|
| −1 | −1 | −1 | −1 | −0.99 | −1 |
| −1 | −1 | 1 | 1 | 0.99 | 1 |
| 1 | −1 | −1 | 1 | 0.98 | 1 |
| 1 | 1 | −1 | −1 | 0.99 | −1 |
| −1 | 1 | 1 | −1 | 0.98 | 1 |
| 1 | 1 | 1 | −1 | 0.98 | 1 |

In what follows, readability will be measured by the number of rules extracted. Other measures of readability include the average number of antecedents per rule, or simply the size of $R_{int}$.

## 6.2 The Monk's Problems

As a point of departure for testing, we applied the extraction system to the Monk's problems [TBB+91]: three examples which have been used as a benchmark for comparing the performance of a range of symbolic and connectionist machine learning systems. Briefly, in the Monk's problems, robots in an artificial domain are described by six attributes with the following possible values:

1. *head_shape*{round, square, octagon}
2. *body_shape*{round, square, octagon}
3. *is_smiling*{yes, no}
4. *holding*{sword, balloon, flag}
5. *jacket_colour*{red, yellow, green, blue}
6. *has_tie*{yes, no}

Problem 1 trains a network with 124 examples, selected from 432, where $head\_shape = body\_shape \vee jacket\_colour = red$. Problem 2 trains a network with 169 examples, selected from 432, where *exactly two of the six attributes have their first value*. Problem 3 trains a network with 122 examples with 5% noise, selected from 432, where $(jacket\_colour = green \wedge holding = sword) \vee (jacket\_colour \neq blue \wedge body\_shape \neq octagon)$. The remaining examples are used in the respective test sets.

We use the same architectures as Thrun [TBB+91], i.e. single hidden layer networks with three, two and four hidden neurons, for Problems 1, 2 and 3, respectively; 17 input neurons, one for each attribute value, and a single output neuron, for the binary classification task. We use the standard backpropagation learning algorithm [CR95, RHW86]. All networks have been trained for 5000 epochs. Differently from Thrun, we use bipolar activation function, inputs in the set $\{-1, 1\}$, and $A_{\min} = 0$.

Firstly, we present the accuracy and fidelity of the extracted sets of rules. For Problems 1, 2 and 3, the performance of the networks w.r.t. their test sets was 100%, 100% and 93.2%, respectively. The accuracy of the extracted sets of rules, in the same test sets, was 100%, 99.2% and 93.5%. The fidelity of the sets of rules to the networks was 100%, 99.2% and 91%.

Figure 6.4 displays the accuracy of the network, the accuracy of the set of rules, and the fidelity of the set of rules to the network, grouped for each problem. The results show that the accuracy of the sets of rules is very similar to that of the networks. In Problem 1, the rule set matches exactly the behaviour of the network. In Problem 2, the rule set fails to classify correctly two examples, and in Problem 3 the rule set classifies correctly one example wrongly classified by the network. Such discrepancies are due to the incompleteness of the extraction algorithm.

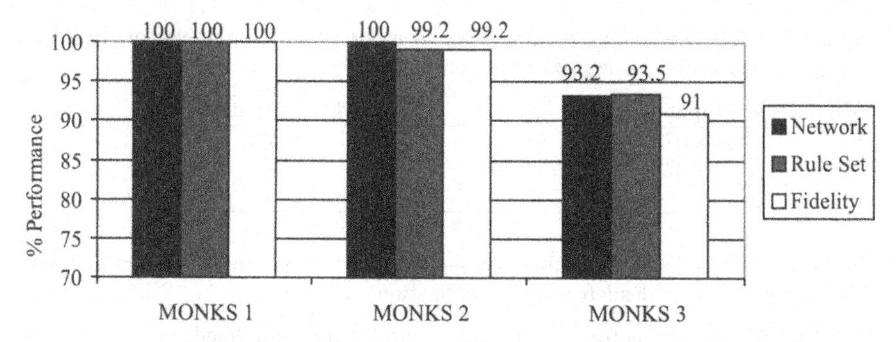


**Fig. 6.4.** The accuracy of the network, the accuracy of the extracted rule set and the fidelity of the rule set to the network w.r.t. the test sets of the Monk's Problems 1, 2 and 3, respectively.

Now, we consider the issue of readability. Tables 6.2, 6.3 and 6.4 present, respectively for Problems 1, 2 and 3, the total number of input vectors, the number of input vectors queried during extraction, the number of rules extracted before simplification, and the number of rules left after simplifications *Complementary Literals* and *Subsumption* are applied. For example, for hidden neuron $h_0$ in Monk's Problem 1 (Table 6.2), 18,724 out of 131,072 input vectors were queried, generating 9,455 rules that, after simplification, were reduced to 2,633 rules.

In general, less than 30% of the set of input vectors are queried and, among these, less than 50% generate rules. Furthermore, *Complementary Literal* and *Subsumption* generally reduce the rule set by 80%. *M of N* and *M of N Subsumption* may further reduce the size of the set of rules.

**Table 6.2.** MONKS 1: The total number of input vectors, input vectors queried, rules extracted and rules left after simplification

| MONKS 1 | Input Vectors | Queried | Extracted | Simplified |
|---|---|---|---|---|
| $h_0$ | 131072 | 18724 | 9455 | 2633 |
| $h_1$ | 131072 | 18598 | 9385 | 536 |
| $h_2$ | 131072 | 42776 | 21526 | 1793 |
| o | 8 | 8 | 2 | 1 |

In particular, the rule set for Problem 1 is presented in Table 6.5. For short, we name each attribute value with a letter from $a$ to $q$ in the sequence presented above, such that $a = (head\_shape = round)$, $b = (head\_shape = square)$, and so on. We use $\bar{a}, \bar{b}, ..., \bar{q}$ to represent $\sim a, \sim b, ..., \sim q$, respectively. We also use the *Integrity Constraints* of the Monk's Problems in order to present a clearer set of rules. For example, we do not present derived rules

**Table 6.3.** MONKS 2: The total number of input vectors, input vectors queried, rules extracted and rules left after simplification

| MONKS 2 | Input Vectors | Queried | Extracted | Simplified |
|---|---|---|---|---|
| $h_0$ | 131072 | 131070 | 58317 | 18521 |
| $h_1$ | 131072 | 43246 | 21769 | 5171 |
| o | 4 | 4 | 1 | 1 |

**Table 6.4.** MONKS 3: The total number of input vectors, input vectors queried, rules extracted and rules left after simplification

| MONKS 3 | Input Vectors | Queried | Extracted | Simplified |
|---|---|---|---|---|
| $h_0$ | 131072 | 18780 | 9240 | 3311 |
| $h_1$ | 131072 | 18618 | 9498 | 794 |
| $h_2$ | 131072 | 43278 | 21282 | 3989 |
| $h_3$ | 131072 | 18466 | 9544 | 1026 |
| o | 16 | 14 | 8 | 2 |

where *has_tie* = *yes* and *has_tie* = *no* appear simultaneously, although the network has generalised to include some of these rules.

By looking at the set of rules extracted and the much simpler description of Monk's Problem 1, it is clear that neural networks do not learn rules in a simple and structured way. Instead, they use a complex and redundant way of encoding rules. Not surprisingly, such a redundant representation is responsible for the network's robustness.

It is interesting that because the rule obtained for the *Hidden-to-Output BNS* of Monk's Problem 1 was $\sim h_1 \sim h_2 \rightarrow o$, and since the set of rules presents 100% accuracy, hidden neuron $h_0$ is not necessary at all, and the problem could have been solved by a network with two hidden neurons only, obtaining the same results. Another interesting exercise is to try to see what the network has generalised, given the set of rules and the classification task learned.

## 6.3 DNA Sequence Analysis

We apply the extraction system to eukaryote promoter recognition and prokaryote splice junction determination, which are large real world problems. These are the same problems used in the experiments with *C-IL²P*, which are described in Chapter 4. Differently from the Monk's Problems, an exhaustive pedagogical extraction (sound and complete) now turns out to be impossible due to the very large number of input neurons (the networks trained in both applications contain more than 200 input neurons).

Although rule extraction is a part of the *C-IL²P* system, we also would like to use the extraction system on networks in which no prior knowledge

**Table 6.5.** Rules extracted for Monk's Problem 1

| Rules for $o$ |
|---|
| $\sim h_1 \sim h_2 \rightarrow o$ |

| Rules for $h_1$ |
|---|
| $\bar{a}bcd\bar{e} \rightarrow h_1$ |
| $bd\bar{e}\bar{l} \rightarrow h_1$ |
| $b\bar{\imath}\bar{l}mn \rightarrow h_1$ |
| $bcd(\bar{l} \vee \bar{e}f) \rightarrow h_1$ |
| $b\bar{e}f(mn \vee mo) \rightarrow h_1$ |
| $\bar{a}bdf(\bar{l} \vee m \vee n) \rightarrow h_1$ |
| $mno(\bar{l} \vee b\bar{e} \vee d\bar{e} \vee bc \vee cd \vee \bar{a}b \vee bf) \rightarrow h_1$ |
| $1(mno)(bc\bar{e}\bar{l} \vee cd\bar{e}\bar{l} \vee \bar{a}bcd \vee bcdf) \rightarrow h_1$ |
| $1(mno)(bd\bar{e} \vee bd\bar{l} \vee b\bar{e}f\bar{l} \vee \bar{a}b\bar{e}\bar{l}) \rightarrow h_1$ |

| Rules for $h_2$ |
|---|
| $abdekl \rightarrow h_2$ |
| $ac\bar{d}em\bar{q} \rightarrow h_2$ |
| $a\bar{b}\bar{d}ef\bar{l} \rightarrow h_2$ |
| $ae\bar{g}jm(n \vee o) \rightarrow h_2$ |
| $\bar{b}e\bar{g}\bar{l}n(a \vee \bar{d}) \rightarrow h_2$ |
| $a\bar{b}\bar{d}e\bar{l}(c \vee \bar{h}) \rightarrow h_2$ |
| $\bar{b}\bar{d}e\bar{g}\bar{l}(m \vee o) \rightarrow h_2$ |
| $a\bar{b}\bar{d}e\bar{l}(j \vee p \vee i) \rightarrow h_2$ |
| $a\bar{b}e\bar{l}\bar{q}(\bar{d} \vee m) \rightarrow h_2$ |
| $a\bar{b}e\bar{g}\bar{l}(\bar{d} \vee m \vee o) \rightarrow h_2$ |
| $aem(\bar{g}n\bar{p} \vee \bar{g}o\bar{p} \vee \bar{h}kn \vee \bar{h}ko) \rightarrow h_2$ |
| $1(mno)(a\bar{b}c\bar{d}\bar{l} \vee \bar{b}c\bar{d}e\bar{l}) \rightarrow h_2$ |
| $1(mno)(ac\bar{d}ef \vee a\bar{b}\bar{d}f\bar{l} \vee \bar{b}\bar{d}ef\bar{l}) \rightarrow h_2$ |
| $1(mno)(a\bar{b}ef\bar{l} \vee a\bar{b}\bar{d}\bar{g}\bar{l} \vee a\bar{b}e\bar{h}\bar{l}) \rightarrow h_2$ |
| $1(mno)1(\bar{b}\bar{g}\bar{h}\bar{l})(a\bar{d}e) \rightarrow h_2$ |
| $1(mno)(a\bar{b}\bar{d}\bar{h}\bar{l} \vee \bar{b}\bar{d}e\bar{h}\bar{l} \vee a\bar{b}ce\bar{l}) \rightarrow h_2$ |

about the learning process and the application domain is provided. We use the experiments on DNA sequence analysis to investigate this case. In the next section, we will use the experiments on Power Systems fault diagnosis to analyse the extraction as a part of the *C-IL²P* framework.

Hence, following [Set97a], we have trained the networks for DNA sequence analysis without adding background knowledge. For the task of promoter recognition[4], we have trained a network with 228 input neurons (57 consecutive DNA nucleotides), a single hidden layer with 16 neurons, and a single output neuron that is responsible for classifying the DNA sequence as promoter or nonpromoter. The training examples consist of 48 promoter and 48 nonpromoter DNA sequences, while the test set contains only 10 examples.

[4] Recall that promoters are short DNA sequences that precede the beginning of genes, and the aim of "*promoter recognition*" is to identify such sequences, the starting location of genes, in long sequences of DNA.

For the task of splice junction determination[5], we have trained a network with 240 input neurons (60 consecutive DNA nucleotides), 26 neurons in a single hidden layer, and two output neurons that are responsible for classifying the DNA sequences into Exon/Intron (E/I) boundaries or Intron/Exon (I/E) boundaries. A third category (neither E/I nor I/E) is assigned to DNA sequences that fail to activate any output neuron. The training set for this task contains 1000 examples, in which approximately 25% are of I/E boundaries, 25% are of E/I boundaries and the remaining 50% are neither. We use a test set with 100 examples. Note that for the splice junction problem, we should not evaluate each output neuron individually. Instead, the combined activation of output neurons E/I and I/E should be considered.

In both applications, due to the intractability of the set of input vectors ($2^{228}$ and $2^{240}$ elements each), we have limited the maximum number of rules generated to 50,000 per hidden neuron. As far as the number of rules derived per output neuron is concerned, the promoter problem had 14,578 rules created out of 29,534 input vectors queried for its output neuron, and the splice junction problem had 83,273 rules, out of 167,316 vectors queried, and 17,919 rules, out of 35,824 vectors queried, created for each output neuron. We have also speeded up the search process by doing the following: we jump, in a kind of binary search, from the ordering's minimum element to a new minimal element in the frontier at which input vectors start to generate rules. This is done as follows: instead of searching from the ordering's maximum and minimum elements, we pick an input vector at distance $n/2$ from them, where $n$ is the number of input neurons, and query it. If it activates the output then it becomes a new maximal element; otherwise, it becomes a new minimal element. We carry on with this process until maximal and minimal elements are at distance 1 from each other. In this case, we have found elements at the *frontier of activation*, and we apply the extraction system as before. Note that, the objective of the search process is, actually, to find the frontier of activation, and a set of rules covering the elements at this frontier would be sufficient to represent all the rules encoded in the *BNS* in question.

Figure 6.5 displays the accuracy of the network, the accuracy of the rule set and the fidelity of the rule set to the network for the promoter recognition and splice junction determination problems. The results reported were obtained using $A_{\min} = 0.5$. In the promoter recognition task, the network classified 9 of the 10 test set examples correctly. The rule set extracted for this task classified the same 9 examples correctly, and thus the fidelity of the rule set to the network was 100%. In the splice junction problem, the

[5] Recall that splice junctions are points on a DNA sequence at which the non-coding regions are removed during the process of protein synthesis. The aim of "*splice junction determination*" is to recognise the boundaries between the part of the DNA retained after splice – called Exons – and the part that is spliced out – the Introns.

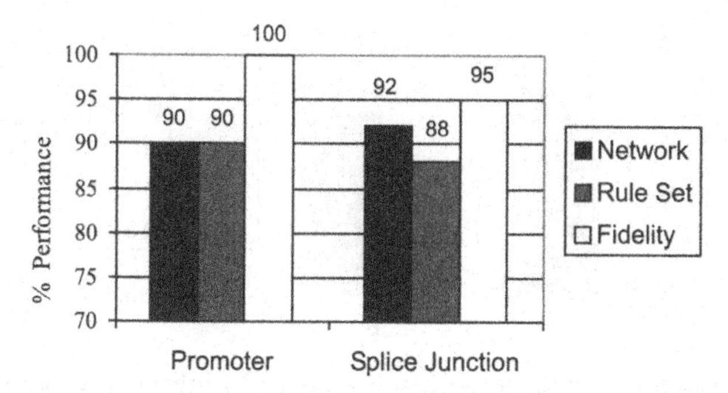


**Fig. 6.5.** The accuracy of the network, the accuracy of the rule set and the fidelity of the rule set to the network for the promoter recognition and splice junction determination problems.

network classified correctly 92 out of 100 examples. The rule set for this task classified 88 out of 100 examples correctly, and 7 of the 8 examples wrongly classified by the network were wrongly classified by the rule set. As a result, the fidelity of the rule set to the network was 95%.

The results obtained for the Promoter problem do not have statistical significance due to the reduced number of examples available for testing. However, the accuracy of the set of rules w.r.t. the network's training set was 90.6%, and thus similar to that obtained for the test set. Unfortunately, it is not easy to compare the results obtained here with the ones obtained by other extraction methods; differences in training and testing methodology are sufficient to preclude comparisons. For example, in [Set97a], Setiono[6] trains a network with three output neurons for the splice junction determination problem, while in [Tow92], Towell uses cross-validation to test the accuracy of the extracted set of rules. To further complicate matters, the figures reported by Towell, concerning the results obtained by the MofN and Subset methods, refer to the training sets of the networks (Towell has pointed out, though, that his results for the networks' test sets were similar). Finally, Towell [Tow92] and Fu [Fu94] extract rules from networks in which a background knowledge had been inserted, while Setiono uses networks trained with no prior knowledge.

Keeping the above discrepancies in mind, in what follows we present the results obtained by the above extraction methods. Towell's *MofN* method [TS93] got 92.3% and 92.8% of accuracy, respectively in the promoter and splice junction problems. Fu's *Subset* [Fu94] got 90.5% and 86%, respectively, and *Setiono*'s method [Set97a] got 93.3% in the splice junction problem. Se-

[6] Setiono's rule extraction algorithm uses a penalty function to prune the weights of a feedforward neural network, and generates rules from the pruned network by considering a small number of activation values at the hidden units.

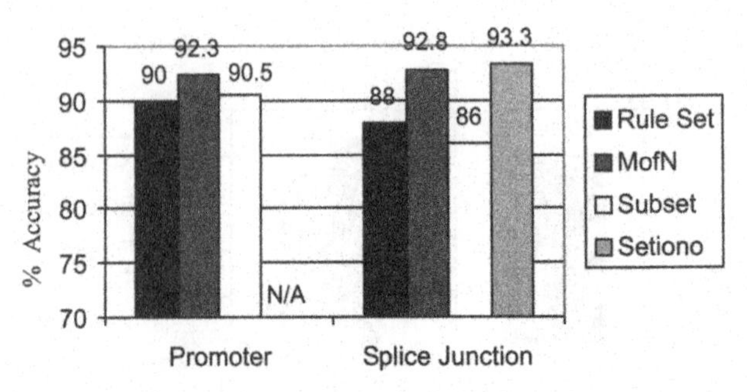


**Fig. 6.6.** Comparison with the accuracy obtained by other extraction methods in the Promoter recognition and splice junction determination problems.

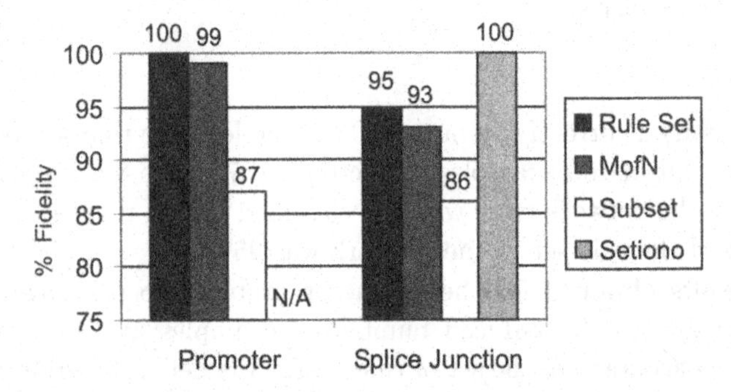


**Fig. 6.7.** Comparison with the fidelity achieved by other extraction methods in the Promoter recognition and splice junction determination problems.

tiono's results in the promoter problem are not available. Figure 6.6 compares the above results with the ones obtained using ABG in both applications. In addition, Figure 6.7 compares the fidelity achieved by these extraction methods, again in both applications. In [Set97a], however, 100% fidelity (which we reproduce here) seems to be assumed from the observation that the accuracy of the set of rules is identical to that of the network. This may not be the case, though, when less than 100% accuracy is obtained.

In spite of the above-mentioned differences in evaluation methodology, one can observe from Figures 6.6 and 6.7 that, apart from the poor fidelity of the Subset method, the ABG extraction method is within a margin of error of less than 5.5% from the results obtained by the remaining methods in both applications.

Finally, a comparison between the sizes of the sets of rules extracted by each of the above methods indicates that a drawback of ABG lies in the much larger size of the set of rules, at least before the simplification of rules is carried out. On the other hand, the above experiments also show that an advantage of ABG is the fact that a provably sound extraction is feasible for very large networks.[7] We will address the problem of readability, and present some alternatives to counteract it, in the discussion at the end of this chapter.

## 6.4 Power Systems Fault Diagnosis

Finally, we apply the extraction algorithm to power systems fault diagnosis (see Section 4.2 for the description of the problem). Power systems applications are an example of safety-critical domains, in which the soundness of the explanations provided by the set of rules extracted is of great importance. In this application, we have used the *C-IL²P* system (see Section 3.4 for the insertion of rules with classical negation into neural networks), and in this section we extract rules from the trained network of Chapter 4 (Figure 4.13).

Recall that we have trained a network, in which 35 rules have been inserted as background knowledge, with 278 examples using standard backpropagation. Each rule of the background knowledge and each training example associated a set of 23 alarms (input neurons) with 32 possible faults (output neurons) of a simplified version of a real power plant. We then used two test sets: one with 92 examples of noisy single faults, and another with 70 examples of noisy multiple faults. The value predefined by the background knowledge for $A_{\min}$ was 0.7.

In what follows, we present, for each output neuron, the accuracy of the network, the accuracy of the rule set, and the fidelity of the rule set to the network, w.r.t. the test set with single faults (Figures 6.8, 6.9 and 6.10). For example, for output neuron Fault 1 at Figure 6.8, the network's accuracy was 95.7% (4 misclassifications in 92 examples), the accuracy of the set of rules extracted was also 95.7%, and the fidelity of the set of rules to the network was 100%, i.e. the network and the set of rules misclassified the same 4 examples.

Figures 6.11, 6.12 and 6.13 show, for each output neuron, the accuracy of the network, the accuracy of the rule set and the fidelity of the rule set to the network, w.r.t. the test set with multiple faults. Apart from faults 24 and 30 in the multiple faults case, the accuracy of the rule set is very good. Similarly, the fidelity of the rule set to the network is excellent in most cases, and in general better than the accuracy of the rule set. This indicates that the extraction algorithm prioritises fidelity over accuracy, i.e. it tries to mimic the network's behaviour, which is a result of soundness and the fact that the extraction is made by querying the network. The graph of Figure 6.14 summarises the results reported above by comparing the mean accuracy and fidelity of the rule set with the mean accuracy of the network for both the single and multiple faults test sets.

[7] We believe that the proof of soundness of the extraction algorithm is a prerequisite for the achievement of a good fidelity in any application domain.

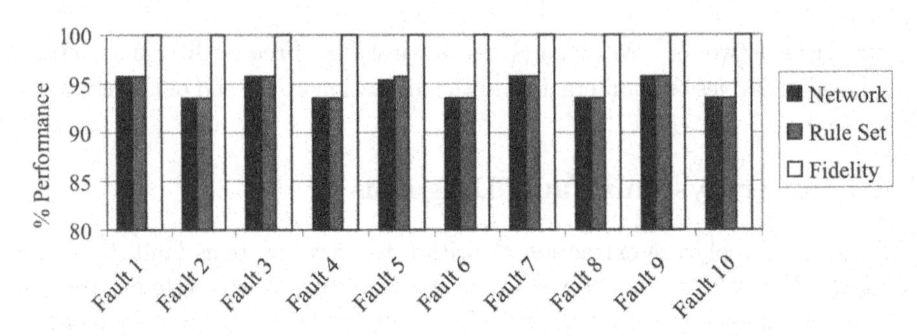


**Fig. 6.8.** Network, Rule Set and Fidelity percentage w.r.t. the single fault test set (outputs 1–10).

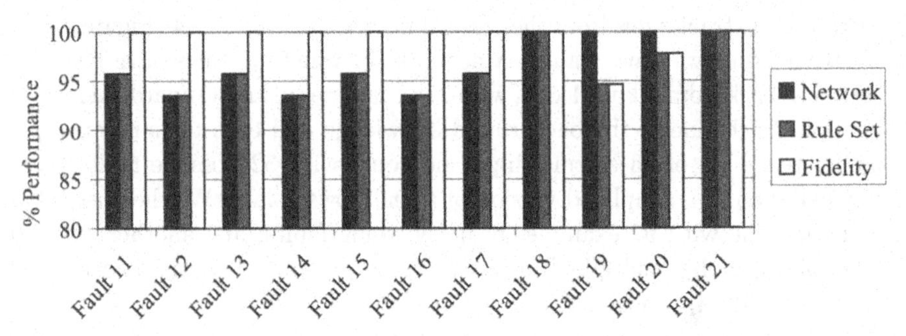


**Fig. 6.9.** Network, Rule Set and Fidelity percentage w.r.t. the single fault test set (outputs 11–21).

As we have seen in Section 4.2, the performance of fault diagnosis systems is typically evaluated not only by determining the percentage of successful diagnosis but also the average size of the ambiguity set (when the system isolates failures from several possible fault modes, but fails to correctly identify the set of faults). In Chapter 4, the network's average sizes of the ambiguity set were 0.5% and 0% of the size of the set of activated faults, respectively, for the single fault and multiple faults test sets. Similarly, using the rule set extracted, the sizes of the ambiguity set were 2.2% and 0% of the size of the set of activated faults, again for the single and multiple faults test sets.

In this application, differently from the ones on DNA sequence analysis, we have not limited the maximum number of rules extracted for each neuron; neither have we used any optimisation in the search process. Tables 6.6 and 6.7 contain the number of input vectors queried during the extraction process and the number of rules created, for each hidden and output neuron, respectively, before simplifications were applied.

**Table 6.6.** Fault diagnosis: the number of input vectors queried and rules extracted out of 8,388,608 input vectors for each hidden neuron ($h_i$)

| | *Queried* | *Extracted* | | *Queried* | *Extracted* |
|---|---|---|---|---|---|
| $h_0$ | 780,696 | 390,145 | $h_{18}$ | 290,998 | 110,659 |
| $h_1$ | 290,548 | 143,230 | $h_{19}$ | 290,996 | 114,642 |
| $h_2$ | 4,096 | 2,045 | $h_{20}$ | 781,280 | 188,813 |
| $h_3$ | 21,806 | 9,764 | $h_{21}$ | 21,768 | 10,728 |
| $h_4$ | 21,766 | 10,808 | $h_{22}$ | 21,774 | 10,708 |
| $h_5$ | 21,806 | 9,855 | $h_{23}$ | 554 | 277 |
| $h_6$ | 4,096 | 2,013 | $h_{24}$ | 4,096 | 1,816 |
| $h_7$ | 21,806 | 9,766 | $h_{25}$ | 21,766 | 10,752 |
| $h_8$ | 21,766 | 9,432 | $h_{26}$ | 21,530 | 9,479 |
| $h_9$ | 290,892 | 135,509 | $h_{27}$ | 21,766 | 9,842 |
| $h_{10}$ | 21,766 | 10,527 | $h_{28}$ | 4,096 | 2,048 |
| $h_{11}$ | 21,806 | 9,568 | $h_{29}$ | 21,754 | 10,910 |
| $h_{12}$ | 21,800 | 10,545 | $h_{30}$ | 4,096 | 2,049 |
| $h_{13}$ | 88,904 | 41,333 | $h_{31}$ | 554 | 277 |
| $h_{14}$ | 21,766 | 10,298 | $h_{32}$ | 554 | 277 |
| $h_{15}$ | 21,806 | 9,736 | $h_{33}$ | 554 | 277 |
| $h_{16}$ | 781,198 | 270,015 | $h_{34}$ | 554 | 277 |
| $h_{17}$ | 290,998 | 100,947 | | | |

**Table 6.7.** Fault diagnosis: the number of input vectors queried and rules extracted out of $3.44 \times 10^{10}$ input vectors for each output neuron ($o_j$)

| | *Queried* | *Extracted* | | *Queried* | *Extracted* |
|---|---|---|---|---|---|
| $o_0$ | 138 | 69 | $o_{16}$ | 138 | 69 |
| $o_1$ | 138 | 69 | $o_{17}$ | 138 | 69 |
| $o_2$ | 138 | 69 | $o_{18}$ | 138 | 69 |
| $o_3$ | 138 | 69 | $o_{19}$ | 138 | 69 |
| $o_4$ | 138 | 69 | $o_{20}$ | 138 | 69 |
| $o_5$ | 138 | 69 | $o_{21}$ | 138 | 69 |
| $o_6$ | 138 | 69 | $o_{22}$ | 138 | 69 |
| $o_7$ | 138 | 69 | $o_{23}$ | 768,334 | 384,172 |
| $o_8$ | 138 | 69 | $o_{24}$ | 138 | 69 |
| $o_9$ | 138 | 69 | $o_{25}$ | 138 | 69 |
| $o_{10}$ | 138 | 69 | $o_{26}$ | 138 | 69 |
| $o_{11}$ | 138 | 69 | $o_{27}$ | 138 | 69 |
| $o_{12}$ | 138 | 69 | $o_{28}$ | 138 | 69 |
| $o_{13}$ | 138 | 69 | $o_{29}$ | 1,316 | 661 |
| $o_{14}$ | 138 | 69 | $o_{30}$ | 1,318 | 661 |
| $o_{15}$ | 138 | 69 | $o_{31}$ | 138 | 69 |

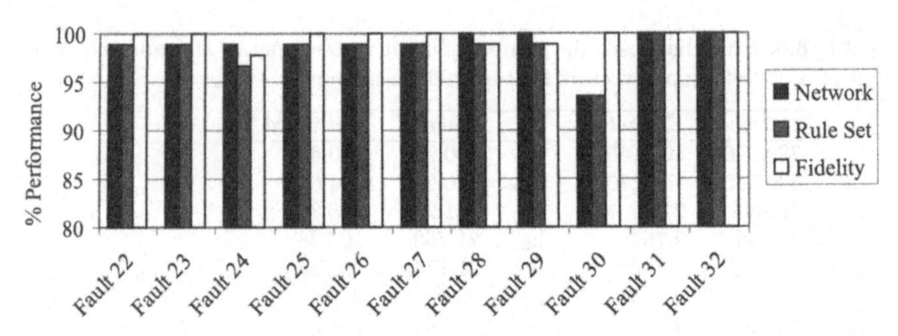


**Fig. 6.10.** Network, Rule Set and Fidelity percentage w.r.t. the single fault test set (outputs 22–32).

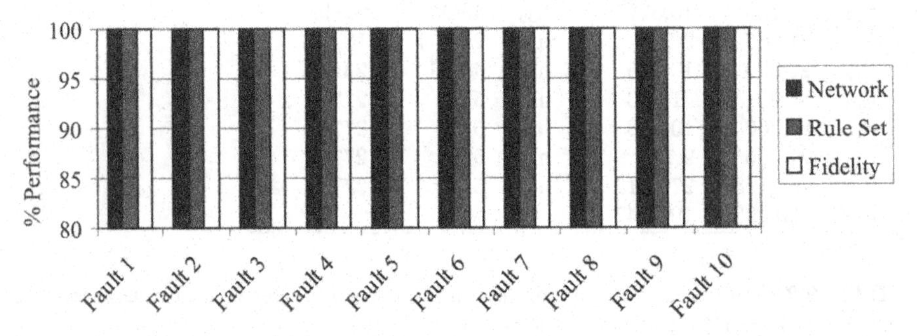


**Fig. 6.11.** Network, Rule Set and Fidelity percentage w.r.t. the multiple faults test set (outputs 1–10).

## 6.5 Discussion

The above experimental results corroborate two important properties of the extraction system: it captures nonmonotonicity and it is sound. Nonmonotonicity is captured by the extraction of rules with default negation, as for example:

$$Fault(overload, *, transformer01, no\_bypass) \leftarrow$$

$$(Alarm(overload, transformer01),$$

$$\sim Alarm(breaker\_open, by\_pass)).$$

which was derived in the experiments with power systems. Soundness is reflected in the high fidelity achieved in the applications, by assuring that any rule extracted is actually encoded in the network, even if such a rule does not comply with the network's test set. The extraction system is, therefore, bound to produce a set of rules that tries to mimic the network, regardless of the network's performance in the training and test sets.

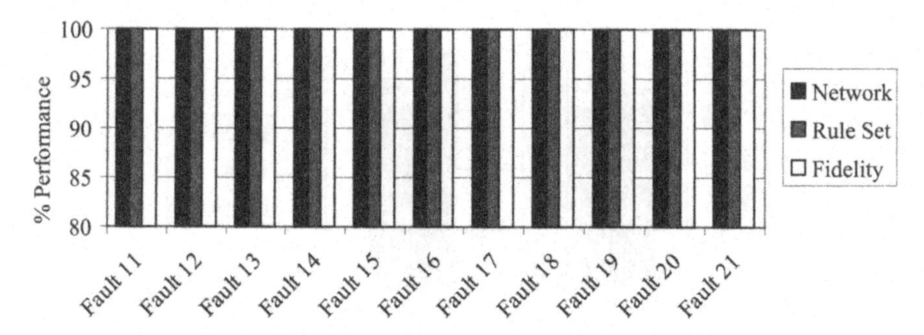


**Fig. 6.12.** Network, Rule Set and Fidelity percentage w.r.t. the multiple faults test set (outputs 11–21).

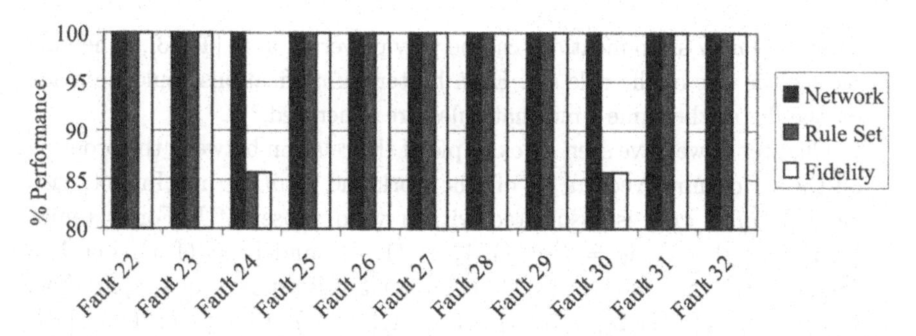


**Fig. 6.13.** Network, Rule Set and Fidelity percentage w.r.t. the multiple faults test set (outputs 22–32).

The above experiments also indicate that the drawback of the extraction system lies in the size of the set of rules. In comparison with [Set97a] and [TS93], in the DNA sequence analysis domain, the number of rules extracted before any simplification is done is considerably bigger than, for example, the number of rules extracted by the *M of N* algorithm (despite the differences in syntax). It seems that less readability is the price one has to pay for soundness. The problem, however, is that we regard the proof of soundness as the minimum requirement of any method of rule extraction. We are, therefore, left with two possible courses of action: *(1)* we can try to enhance readability by manipulating, e.g. simplifying, the extracted set of rules, or *(2)* we can ignore the lack of readability of the set of rules as a whole, and concentrate on providing an explanation for each particular answer of the network.

As far as course of action *(1)* is concerned, there are many possible improvements to be made in the ABG extraction system.

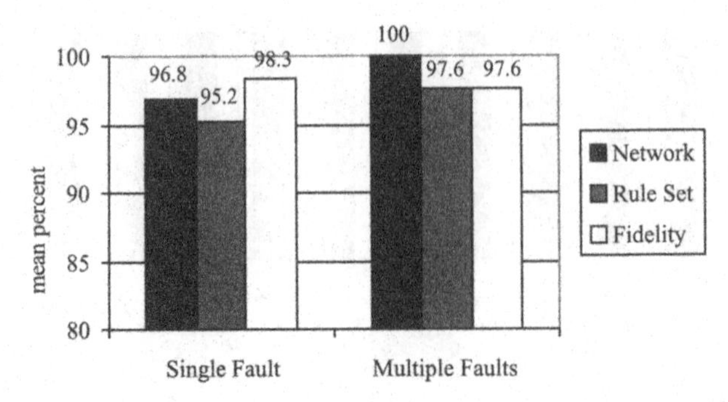


**Fig. 6.14.** Mean network, rule set and fidelity percentage for Faults 1–32.

- Firstly, *M of N* simplifications can be very powerful, as in [TS93], in helping reduce the size of the rule set. Even better, simplifications could be made on the fly, at the same time that rules are generated.[8]
  In Chapter 5, we have seen an example of the relation between the ordering on the set of input vectors ($\mathbf{I}$) of a network and *M of N* rules.In fact, each valid *M of N* rule is associated with a valid subset of $\mathbf{I}$. For example, let $\mathbf{i}_1 = (-1,-1)$, $\mathbf{i}_2 = (-1,1)$, $\mathbf{i}_3 = (1,-1)$ and $\mathbf{i}_4 = (1,1)$. Let $\mathbf{I} = \{\mathbf{i}_1,\mathbf{i}_2,\mathbf{i}_3,\mathbf{i}_4\}$ and $sup(\mathbf{I}) = (1,1)$. There are 5 valid subsets of **I,** apart from $\emptyset$, namely $\{\mathbf{i}_4\}$, $\{\mathbf{i}_4,\mathbf{i}_3\}$, $\{\mathbf{i}_4,\mathbf{i}_2\}$, $\{\mathbf{i}_4,\mathbf{i}_3,\mathbf{i}_2\}$ and $\mathbf{I}$ itself. If $(1,1) = [a,b]$ then each of these subsets corresponds, respectively, to the following *M of N* rules: $2(a,b)$, $1(a)$, $1(b)$, $1(a,b)$, and $0(a,b)$. Any other *M of N* rule is not valid due to the ordering $\preceq$ on $\mathbf{I}$. For example, $1(a,\sim b)$ would require the set $\{\mathbf{i}_4,\mathbf{i}_3,\mathbf{i}_1\}$ to be also a valid subset of $\mathbf{I}$, but this is impossible according to $\preceq$. $1(a,\sim b)$ would be a valid *M of N* rule if, for example, $sup(\mathbf{I}) = (1,1) = [a,\sim b]$. In this case, though, $1(a,b)$ would not be a valid *M of N* rule, for the same reason as described above.[9]
  The relation between *M of N* rules and subsets of $\mathbf{I}$ could facilitate the extraction of more compact sets of rules, thus improving readability. By manipulating *M of N* rules, as in [Mai97], a neater set of rules could also be derived. The characterisation of an algebra for manipulating *M of N* rules is work in progress.
- Improvements could also be made in the optimisation of the system's search process, exploring the ordering on the set of input vectors, and adding some new heuristics to the extraction algorithm. An example is what we have

[8] The idea here is to implement a *buffer* of rules extracted and, whenever a new rule is generated, try to simplify it together with the rules in the buffer. Potentially good rules for simplification, the ones with many *don't cares*, would remain in the buffer for longer periods.

[9] In fact, this is the reason why *M of N* rules ought to be seen as simplifications.

done in the DNA sequence analysis case, when we jump to new minimal elements in the ordering.

The efficiency of the search process could also be enhanced by the implementation of a *time-slice* for each output neuron. This would help the extraction not to get stuck in the generation of thousands of rules about an output, while no rule about the remaining outputs is created. As far as efficiency is concerned, a parallel implementation of the extraction system would be the ultimate goal.

- Finally, the extraction of *metalevel priorities* (in the sense of Section 3.5) directly from the network or, alternatively, the use of metalevel priorities when the rules are assembled to derive the final rule set could result in the enhancement of readability. For example, the set of rules $R_1 = \{ab \rightarrow \neg x, \sim ac \rightarrow x, \sim bc \rightarrow x\}$ could be converted into the following, more compact, set of rules $R_2 = \{r_1 : ab \rightarrow \neg x, r_2 : c \rightarrow x\}$ with $r_1 \succ r_2$.

We will present the current state of development of the above subjects in more detail in Chapter 9.

The idea behind course of action *(2)* is to provide an explanation for individual answers of the network, instead of trying to understand what is computed by it as a whole. When we extract rules from a trained network, we obtain a database, which can be used instead of the network. By querying the database with a particular answer of the network, using, for instance, an automatic theorem prover, we may use the steps of the proof of a literal to provide a symbolic explanation for such an answer of the network. Note that this explanation will only be reliable if the extraction of rules is sound. Of course, when one takes course of action *(2)*, some interesting features of the network as a whole might never be found. On the other hand, in this case, even very large sets of rules are not a major concern.

In this chapter, we have successfully applied a new sound method of rule extraction from trained neural networks in well known traditional examples and real-world application problems. In particular, the experiments on power systems fault diagnosis enabled us to compare the background knowledge inserted in the network with the extracted set of rules, and thus to study the effects of the learning process on the background knowledge. In the next chapter, we will deal with this matter and, more specifically, with the interesting case where inconsistencies between the background knowledge and the training examples arise.

# Part III

# Knowledge Revision in Neural Networks

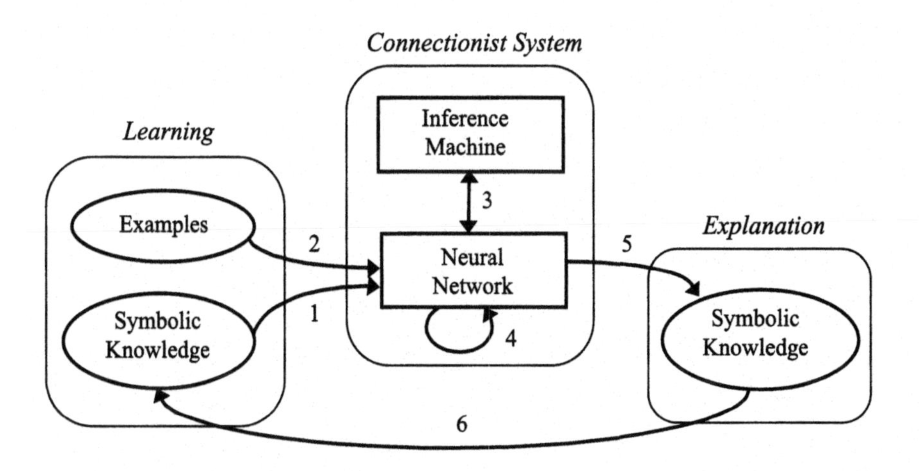


Part III of this book is about *Handling Inconsistencies* in neural networks. It presents the theoretical and practical aspects of process (4) above. It contains a new Learning Algorithm, responsible for fine-tuning trained neural networks. It makes use of the inference machine built in *C-IL²P* networks (3), to fine-tune the network with a minimal loss of information w.r.t. the logical consequences derived by the network. It is useful to solve inconsistencies between the background knowledge and the training examples of a network. Part III concludes with an application of processes (1) to (5) in a real problem of software requirements specification. Attention is focused on process (4), and how to solve inconsistencies in evolving requirements specifications.

# 7. Handling Inconsistencies in Neural Networks

As a nonmonotonic learning system, *C-IL²P* has interesting connections with *Belief Revision* [Gar92] and, more specifically, *Truth Maintenance Systems* (TMS) [Doy79] (see Chapter 2 for an overview of TMSs). When background knowledge is encoded as defeasible into a neural network, the set of training examples may specify a revision of that knowledge. In addition, the set of examples can be inconsistent with the background knowledge, and the resulting network can itself contain inconsistencies. In this chapter, we investigate how to detect and treat *inconsistencies* in the system. We also show the equivalence between *C-IL²P* networks and *TMSs*. As a result, the learning process of *C-IL²P* can be seen as a technique for handling changes of belief in a *TMS*.

## 7.1 Theory Revision in Neural Networks

In Chapter 3, we have seen that the function $\mathbf{f} : \{-1,1\}^r \rightarrow \{-1,1\}^s$ that a neural network $\mathcal{N}$ computes can be the fixed point operator $T_{\mathcal{P}}$ of a logic program $\mathcal{P}$. In addition, $\mathcal{N}$ can be trained from examples to approximate the fixed point operator $T_{\mathcal{P}'}$ of an unknown program $\mathcal{P}'$. According to the *Translation Algorithm* of Chapter 3, $\mathbf{f}$ is initially defined by a background knowledge ($\mathcal{P}$). The learning process, however, adds new information to $\mathcal{N}$, in accordance with a given set of examples ($\mathcal{E}$). Finally, the *Extraction Algorithm* of Chapter 5 tries to find out $\mathcal{P}'$, regardless of $\mathcal{P}$ or $\mathcal{E}$. In what follows, we provide a first account of the effects of the learning process over $\mathcal{N}$ by comparing $\mathcal{P}$ and $\mathcal{P}'$. Briefly, the learning process may:

1. *expand* the background knowledge, when $\mathcal{P} \subset \mathcal{P}'$;
2. *contract* the background knowledge, when $\mathcal{P}' \subset \mathcal{P}$; or
3. *revise* the background knowledge, when an *expansion* is followed by a *contraction*.

Let us exemplify the process of revision with the experiments of Chapters 4 and 6 on Power Systems fault diagnosis. In Chapter 4, we deliberately changed rules 24 and 25 of the background knowledge by replacing

*Alarm(auxiliary, transformer01)* and *Alarm(auxiliary, transformer02)*, respectively, by *Alarm(breaker_open, transformer01)* and *Alarm(breaker_open, transformer02)*. In Chapter 6, however, rules 24 and 25 were both extracted from the trained network. It seems that the network's training examples managed to correct our "mistake". In order to do so, the training process has changed the background knowledge, by changing the network's weights, which is equivalent to firstly expanding and then contracting the background knowledge. Other examples include:

$$Fault(overload, *, transformer01, no\_bypass) \leftarrow (Alarm(overload, transformer01)).$$

which became, after learning,

$$Fault(overload, *, transformer01, no\_bypass) \leftarrow (Alarm(overload, transformer01), \sim Alarm(breaker\_open, by\_pass)).$$

Similarly, *Fault(main_bus, ∗, ∗, ∗)* seems to be more accurately represented by:

$$Fault(main\_bus, *, *, *) \leftarrow (Alarm(breaker\_open, line02), Alarm(breaker\_open, transformer01), \sim Alarm(breaker\_open, by\_pass)).$$

rather than by

$$Fault(main\_bus, *, *, *) \leftarrow (Alarm(breaker\_open, line01), Alarm\ (breaker\_open, line02), Alarm(breaker\_open, transformer01), Alarm(breaker\_open, transformer02)).$$

Finally, *Alarm(breaker_open, by_pass)* became irrelevant to the detection of any fault related to the auxiliary bus. The remaining rules of the background knowledge were extracted unchanged.

### 7.1.1 The Equivalence with Truth Maintenance Systems

The connections between feedforward neural networks and truth maintenance systems become obvious when we look at neural networks as a theory revision tool. More specifically, when we associate neurons with propositional variables and negative weights with default negation, as in *C-IL²P* networks, we find out that each network is, in fact, equivalent to the graphic representation of a *TMS*.

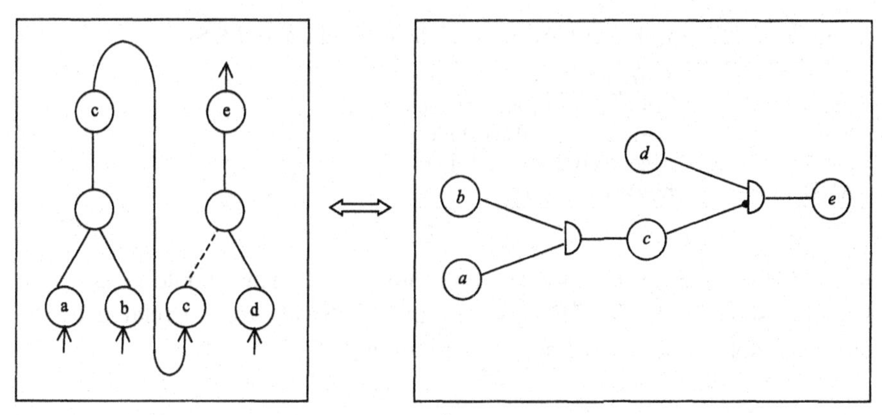


**Fig. 7.1.** The equivalence between *C-IL²P* networks and *TMSs*.

Recall that there are two basic types of entity in a *TMS*: *nodes* (representing propositional *beliefs*) and *justifications* (representing reasons for beliefs). A node can be *in* or *out*, which correspond, respectively, to accepting and not accepting the belief represented by it. A justification is defined as a pair of lists: an inlist ($I$) and an outlist ($O$), together with the node ($n$) that it is a justification for, and it is denoted as $\langle I \mid O \rightarrow n \rangle$, where $I$ contains only nodes that are *in*, $O$ contains only nodes that are *out*, and $n$ is called the consequence of the justification.

Analogously, in a *C-IL²P* network, *input and output neurons* can be associated with propositional beliefs, the activation of the neuron can indicate whether its associated belief is *in* or *out*, and *hidden neurons* can be seen as justifications. Each hidden neuron associates an inlist (positive literals in the input layer of the network) and an outlist (negative literals in the input layer of the network) with a node $n$ (a network's output neuron), which is the consequence of the justification. Hence, we only need to unfold the *C-IL²P* network, creating a multi hidden layer network, to obtain the graphical representation of a *TMS*. Figure 7.1 shows a *C-IL²P* network and its equivalent *TMS*. Note that hidden neurons are converted into AND-gates (which represent justifications), and connections associated with negative weights in the network become connections with a black dot in the *TMS* (which represent negation).[1]

A translation from *C-IL²P* networks to *TMSs* and vice versa can therefore be easily accomplished by performing the association indicated at Table 7.1.

The proof that the dynamics of *C-IL²P* networks, obtained from the *Translation Algorithm* of Chapter 3, is identical to the dynamics of its equiv-

[1] The black dots in the graphical representation of a *TMS* mean default negation (see [GR94]), and this is precisely the meaning of negative weights in *C-IL²P* networks.

**Table 7.1.** The translation between *C-IL2P* networks and *TMSs*

| *TMS* | *C-IL$^2$P* |
|---|---|
| nodes | input and output neurons |
| lines | positively weighted links |
| black dot lines | negatively weighted links |
| AND-Gates | hidden neurons |

alent *TMS* is straightforward from Theorem 3.1.1. This should come as no surprise, since both *C-IL$^2$P* networks and *TMSs* are closely related to the stable model semantics of Logic Programming.

### 7.1.2 Minimal Learning

So far, we have seen that a neural learning process performs theory revision, and that *C-IL$^2$P* networks are equivalent to *TMSs*. However, as we devote the following section to trying to solve inconsistencies in trained networks with a minimal loss of information, we need a method adequate for this task. We call such a method *Minimal Learning.*

Let us use an example to better understand the idea of minimal learning. In what follows, we use vectors as an abstract representation of sets, as in Chapter 5.

*Example 7.1.1.* Consider the network of Figure 7.2. When a training example is supposed to change the output of D, we do not want the output of E to be affected, through connections of either kind 1 or 2 (see Figure 7.2). Moreover, we do not want any output value of D that is associated with an input vector not referred to in the set of examples to be changed. For example, if our training example is $\{[A,B,C]_{input} = (1,1,1), [D,E]_{output} = (1,-1)\}$, and the network's previous outputs for input vectors $[A,B,C]_{input} = (1,1,1)$ and $[A,B,C]_{input} = (1,-1,1)$ are, respectively, $[D,E]_{output} = (-1,-1)$ and $[D,E]_{output} = (-1,1)$, we now want $[A,B,C]_{input} = (1,1,1)$ to output $[D,E]_{output} = (1,-1)$, as trained, but also $[A,B,C]_{input} = (1,-1,1)$ to output $[D,E]_{output} = (-1,1)$, as before.

Minimal learning may be extremely difficult to achieve using, for example, Backpropagation.[2] An alternative is to appropriately set up the new knowledge into the network. This idea originated from Kasabov's work on evolving networks [Kas98, Kas99], in which incremental learning is achieved by the addition of hidden neurons to the network and, in general, by the use of one-shot learning. When evolving in such a way, a network manages to learn new

[2] One possibility is to retrain the whole set of input vectors.

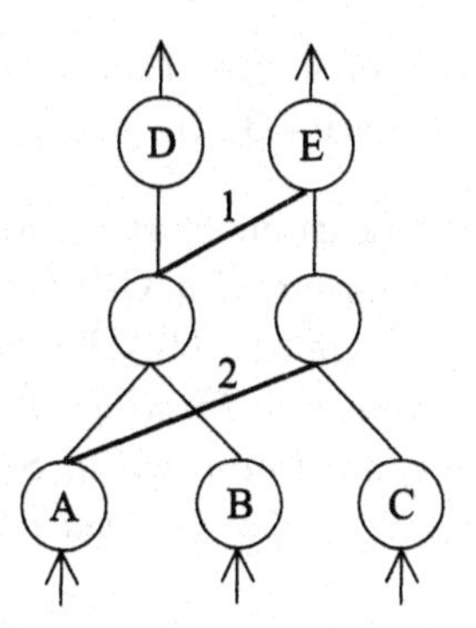


**Fig. 7.2.** Minimal learning.

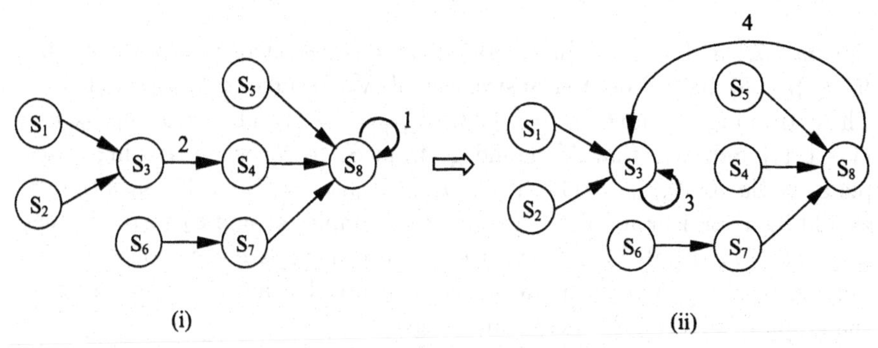


**Fig. 7.3.** Minimal learning and answer sets.

concepts without losing the information previously trained, as long as it is not done too often.[3]

Figure 7.3 illustrates the idea of minimal learning in terms of the answer set computed by a *C-IL²P* network. Let $S_1, S_2, ..., S_8$ be subsets of the set of literals $S$ represented in a network $\mathcal{N}$. Let $S_8$ be the answer set of $\mathcal{N}$ (we write $S_8 = \mathcal{S}(\mathcal{N})$). As a result, $S_1, S_2, ..., S_7$ all converge to $S_8$, as shown in Figure 7.3($i$), where the arrows represent the transitions between the subsets of $S$, computed by $\mathcal{N}$. For example, if $S_1$ is given as input vector of $\mathcal{N}$ then $S_3$ is obtained as output vector of $\mathcal{N}$ (we write $S_3 = \mathcal{N}(S_1)$).

Now, let $S_3$ be the desired answer set of $\mathcal{N}$, according to some set of training examples. The process of minimal learning is the one of converting $S_3$ into the new answer set of $\mathcal{N}$ by *(1)* replacing the arrow from $S_8$ to itself (*arrow* 1) by an arrow from $S_8$ to $S_3$ (*arrow* 4), and *(2)* replacing the arrow from $S_3$ to $S_4$ (*arrow* 2) by an arrow from $S_3$ to itself (*arrow* 3). In addition,

[3] If it is done too often such a process might result in a poor generalisation, as a consequence of overfitting the data into too many hidden neurons.

any other arrow in the graph of Figure 7.3($i$) should not be changed. The expected result is shown in Figure 7.3($ii$).

**Definition 7.1.1.** (Minimal Learning) *Let* $S_1, S_2, ..., S_n \in \wp(\mathcal{I})$*, where* $\mathcal{I}$ *is the set of input and output neurons of a* C-IL$^2$P *network* $\mathcal{N}$*. Let* $S_m$ $(1 \leq m \leq n)$ *be the answer set of* $\mathcal{N}$*, and* $S_k$ $(1 \leq k \leq n)$ *be the answer set of* $\mathcal{N}'$*, where* $\mathcal{N}'$ *is the network obtained by training* $\mathcal{N}$ *with a set of examples* $\mathbf{E}$*. The learning process in* $\mathcal{N}$ *is called minimal iff* $S_k = \mathcal{N}'(S_m)$*,* $S_k = \mathcal{N}'(S_k)$*, and* $\mathcal{N}'(S_l) = \mathcal{N}(S_l)$ *for any* $S_l$ $(1 \leq l \leq n)$ *such that* $S_l \notin \mathbf{E}$*.*

In particular, in order to make $S_k = \mathcal{N}'(S_m)$ and $S_k = \mathcal{N}'(S_k)$, we must have $S_k, S_m \in \mathbf{E}$, and, thus, $l \neq k$ and $l \neq m$.

Let us illustrate the above idea with a simple example.

*Example 7.1.2.* Let $S_j = \{a, b, \neg b\}$ be the answer set of a network $\mathcal{N}$. Let $S_i = \{a, \neg b\}$ be the desired answer set of $\mathcal{N}$. Assume $[a, b, \neg b] = (1, 1, 1)$. Thus, training example $\mathbf{e}_1 = \{[a, b, \neg b]_{input} = (1, 1, 1),\ [a, b, \neg b]_{output} = (1, -1, 1)\}$ indicates that $\mathcal{N}$ should go from $S_j$ to $S_i$, while training example $\mathbf{e}_2 = \{[a, b, \neg b]_{input} = (1, -1, 1),\ [a, b, \neg b]_{output} = (1, -1, 1)\}$ says that $S_i$ should be a stable state of $\mathcal{N}$. If $\mathcal{N}$ learns examples $\mathbf{e}_1$ and $\mathbf{e}_2$ correctly then steps *(1)* and *(2)* of the process of minimal learning are accomplished. The difficult task at this point, however, is to ensure that any other transition in the graph of Figure 7.3 remains unchanged.

Following [Kas99], the idea is to add rules that should *reinforce* the activation of certain output neurons, and *inhibit* the activation of others. In the above example, $\mathbf{e}_1$ indicates that $a, b$ and $\neg b$ should reinforce, i.e. activate, $a$ and $\neg b$, and inhibit, i.e. deactivate, $b$. Similarly, $\mathbf{e}_2$ says that $a, \sim b$ and $\neg b$ should also reinforce $a$ and $\neg b$, and inhibit $b$. We represent such information as follows: $a, b, \neg b \Rightarrow a, \neg b[b]$ and $a, \sim b, \neg b \Rightarrow a, \neg b[b]$, where literals that ought to be inhibited are placed within brackets. These rules can be simplified by a slight variation of the simplification rules *Subsumption* and *Complementary Literals* (see Definitions 5.2.5 and 5.2.6), whereby the sequence $a, \neg b[b]$ is treated as a single literal, to obtain $a, \neg b \Rightarrow a, \neg b[b]$.

Figure 7.4 illustrates the process of evolving a network where $\mathcal{I} = \{a, b, \neg b\}$ to accommodate the rule $a, \neg b \Rightarrow [b]$. This means that output neuron $b$ should be deactivated by any input vector of the form $[a, b, \neg b] = (1, *, 1)$, where $*$ indicates '*don't care*', while the activation values of output neurons $a$ and $\neg b$ should remain unchanged. The question here is how to set up weights $W_3, W_4$ and $W_5$, and the activation function $h(x)$ of the newly added hidden neuron $h_0$, and its threshold $\theta_{h_0}$, such that the intended meaning of this new rule is achieved. We want to satisfy: *(i)* when $h_0$ is not activated, the activation of $b$ is not influenced by $h_0$, and *(ii)* when $h_0$ is activated, the activation of $b$ is smaller than $-A_{\min}$. A straightforward way

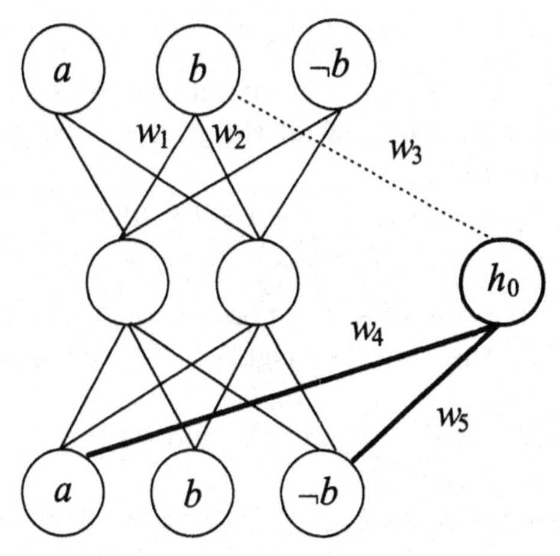


**Fig. 7.4.** Evolving a network to perform minimal learning.

of satisfying Condition *(i)* is to use a non-linear activation function for $h_0$ such that $h_0$ outputs 1 if its input potential is greater than zero, and zero otherwise. In order to satisfy Condition *(ii)*, assume $W_1 > 0$ and $W_2 < 0$. The maximum activation of $b$ is $h(W_1 - W_2 - \theta_b)$. Therefore, the value of $W_3$ ($W_3 < 0$) should be such that $h(W_1 - W_2 + W_3 - \theta_b) < -A_{\min}$. Finally, in order to define the values of $W_4$, $W_5$ and $\theta_{h_0}$, we use a variation of Holldobler and Kalinke's Translation Algorithm for binary threshold networks [HK94], described in Section 3.1. A slight variation is necessary here because we use bipolar inputs $i_j = \{-1, 1\}$. We will describe such a variation in the *Minimal Learning Procedure* below. In this example, $W_4 = 1$, $W_5 = 1$ and $\theta_{h_0} = 1.5$ is sufficient to activate $h_0$ only when $a$ and $\neg b$ are activated.

Let us now present the *Minimal Learning Procedure*. In what follows, the values of the weights associated with reinforcing and inhibiting rules result from the proof of Proposition 7.1.1 below. We use $\{\mathbf{e}_o\}$ to denote the set of examples $\mathbf{e}_o$, $1 \leq o \leq n$.

– *Minimal Learning Procedure* ($\mathcal{N}$, $\{\mathbf{e}_o\}$)

1. For each example $\mathbf{e}_o = \{\mathbf{i}_o, \mathbf{o}_o\}$, where $\mathbf{i}_o = (i_1, ..., i_n)$ and $\mathbf{o}_o = (o_m, ..., o_k)$, find the corresponding rule $r_o$ of the form $L_1, ..., L_n \Rightarrow L_m, ..., L_l[L_{l+1}, ..., L_k]$ such that, for all $L_i$ ($m \leq i \leq l$), $o_i = 1$, and, for all $L_j$ ($l+1 \leq j \leq k$), $o_j = -1$.
2. If possible, simplify rules $r_o$ by using *Complementary Literals* and *Subsumption*.
3. For each rule $r_o$, do:
   a) For each output neuron $L_s$ ($m \leq s \leq l$), do:

Calculate $W_{so} > -\frac{1}{\beta} ln \left( \frac{1-A_{\min}}{1+A_{\min}} \right) + \sum_{i=1}^{y} w_{si} - \sum_{j=y+1}^{z} w_{sj} + \theta_s$, where $\sum_{i=1}^{y} w_{si}$ is the sum of all the positive weights to $L_s$, $\sum_{j=y+1}^{z} w_{sj}$ is the sum of all the negative weights to $L_s$, and $\theta_s$ is the threshold of $L_s$.

b) For each output neuron $L_t$ $(l+1 \leq t \leq k)$, do:
Calculate $W_{to} < -\frac{1}{\beta} ln \left( \frac{1+A_{\min}}{1-A_{\min}} \right) - \sum_{i=1}^{x} w_{ti} + \sum_{j=x+1}^{z} w_{tj} + \theta_t$, where $\sum_{i=1}^{x} w_{ti}$ is the sum of all the positive weights to $L_t$, $\sum_{j=x+1}^{z} w_{tj}$ is the sum of all the negative weights to $L_t$, and $\theta_t$ is the threshold of $L_t$.

c) Insert a hidden neuron $h_o$ into $\mathcal{N}$, and set the step function[4] as the activation function of $h_o$.

d) Connect input neurons $L_i$ $(1 \leq i \leq n)$ to $h_o$. If $L_i$ is a positive literal then set the connection weight $W_{oi}$ to 1; otherwise, set $W_{oi}$ to $-1$. Set the threshold of $h_o$ as $\theta_o = n - 1/2$.[5]

e) Connect output neurons $L_s$ $(m \leq s \leq l)$ to $h_o$ and set the connection weight to $W_{so}$.

f) Connect output neurons $L_t$ $(l+1 \leq t \leq k)$ to $h_o$ and set the connection weight to $W_{to}$.

In Example 3.5.4 (Chapter 3, Section 3.5), we have seen that the threshold $\theta_t$ of any output neuron $L_t$ needs to be changed when more than one hidden neuron has to inhibit the activation of the same output neuron. In other words, if the activation function of the hidden neurons $h_o$ were semi-linear, $\theta_b$ would have to be changed in order to accommodate, for example, rules $a, b \Rightarrow [b]$ and $a, \neg b \Rightarrow [b]$ in the network of Figure 7.4. However, a change in $\theta_b$ could clearly render the above procedure non-minimal, by possibly making $\mathcal{N}'(S_l) \neq \mathcal{N}(S_l)$ (see Definition 7.1.1). The solution to this problem was to make any newly added neuron $h_o$ a binary threshold neuron. This guarantees that neurons $h_o$ do not affect the answer of the network when they are not activated, simply because, in this case, their contribution to the activation of any output neuron is *zero*. Proposition 7.1.1 below shows that the use of binary threshold neurons is sufficient to guarantee that the learning process is minimal.

**Proposition 7.1.1.** *The* Minimal Learning Procedure *complies with* Minimal Learning.

---

[4] $h_o = \begin{cases} 1, \text{if} \sum_i W_{h_o i} i_i - \theta_{h_o} > 0 \\ 0, \text{otherwise} \end{cases}$

[5] In [HK94], $\theta_o = m - 1/2$, where $m$ is the number of positive literals in $L_1, ..., L_n$. Making $\theta_o = n - 1/2$ suffices to cope with inputs in $\{-1, 1\}$, as we will see in the proof of Proposition 7.1.1.

*Proof. We need to show that the Minimal Learning Procedure implements in a network $\mathcal{N}$ exactly the changes intended by the set of examples $\{\mathbf{e}_o\}$, as depicted in Figure 7.3.*

*Firstly, we prove that the variation for bipolar inputs of Holldobler and Kalinke's Translation Algorithm is correct. The only modification here is in the threshold of $h_o$, $\theta_o = n - 1/2$. We need to show that $h_o$ is activated iff $L_1, ..., L_n$ is satisfied. Let $n'$ be the number of positive literals and $n''$ the number of negative literals in $L_1, ..., L_n$. ($\rightarrow$) Assume $L_1, ..., L_n$ is satisfied. Since the activation function of $h_o$ is the step function, we need to satisfy $n'(1 \cdot 1) + n''(-1 \cdot -1) > n - 1/2$. Since $n = n' + n''$, this is clearly true. ($\leftarrow$) When $L_1, ..., L_n$ is not satisfied, in the worst case, either a positive literal has input $-1$ or a negative literal has input 1. Both cases result in the following inequality $n' + n'' - 2 < n - 1/2$, which is also clearly true.*

*Now, from the* Minimal Learning Procedure, step 2, *no rule $r_i$ in $\{r_o\}$ is subsumed by any other rule $r_j$ in $\{r_o\}$, where $\{r_o\}$ is the set of simplified rules. Hence, whenever $h_i = 1$, $h_j = 0$ and, thus, we only need to consider the case where a unique hidden neuron $h_o$ (relative to a single rule $r_o$) is added into $\mathcal{N}$.*

*Let $r_o : L_1, ..., L_n \rightarrow L_m[L_k]$ be a reinforcing/blocking rule to be inserted into $\mathcal{N}$. Let $w_{k1}, ..., w_{kx}$ be the positive weights from the hidden layer of $\mathcal{N}$ to $L_k$, and $w_{k(x+1)}, ..., w_{kz}$ be the negative weights from the hidden layer of $\mathcal{N}$ to $L_k$, prior to the addition of any extra hidden neuron $h_o$ in $\mathcal{N}$. Similarly, let $w_{m1}, ..., w_{my}$ be the positive weights from the hidden layer of $\mathcal{N}$ to $L_m$, and $w_{m(y+1)}, ..., w_{mz}$ be the negative weights from the hidden layer of $\mathcal{N}$ to $L_m$, prior to the addition of any extra hidden neuron $h_o$ in $\mathcal{N}$.*

*Let $W_{ko}$ denote the weight from $h_o$ to $L_k$, and $W_{mo}$ the weight from $h_o$ to $L_m$. $\theta_k$ and $\theta_m$ will denote the thresholds of $L_k$ and $L_m$, respectively. From the variation for bipolar inputs of Holldobler and Kalinke's Translation Algorithm, $h_o$ is activated when $L_1, ..., L_n$ are satisfied, and in this case the activation of $h_o$ is 1. Otherwise, the activation of $h_o$ is zero, due to its non-linear activation function.*

*Case 1 (*blocking rules*): The maximum activation of $L_k$ is obtained when all the hidden neurons $h_1, ..., h_x$ linked to $L_k$ by the positive weights $w_{k1}, ..., w_{kx}$ present activation 1, and all the hidden neurons $h_{x+1}, ..., h_z$ linked to $L_k$ by the negative weights $w_{k(x+1)}, ..., w_{kz}$ present activation $-1$. The activation of $L_k$ has to be blocked when $h_o$ presents activation 1. This yields Equation 7.1, which has to be satisfied.*

$$h\left(\sum_{i=1}^{x} w_{ki} - \sum_{j=x+1}^{z} w_{kj} + W_{ko} - \theta_k\right) < -A_{\min} \tag{7.1}$$

*Solving Equation 7.1 for $W_{ko}$ yields Equation 7.2 below.*

$$W_{ko} < -\frac{1}{\beta} ln \left( \frac{1 + A_{\min}}{1 - A_{\min}} \right) - \sum_{i=1}^{x} w_{ki} + \sum_{j=x+1}^{z} w_{kj} + \theta_k \tag{7.2}$$

*Case 2 (*reinforcing rules*): The minimum activation of $L_m$ is obtained when all the hidden neurons $h_1, ..., h_y$ linked to $L_m$ by the positive weights $w_{m1}, ..., w_{my}$ present activation $-1$, and all the hidden neurons $h_{y+1}, ..., h_z$ linked to $L_m$ by the negative weights $w_{m(y+1)}, ..., w_{mz}$ present activation 1. The activation of $L_m$ has to be enforced when $h_o$ presents activation 1. This yields Equation 7.3, which has to be satisfied.*

$$h(-\sum_{i=1}^{y} w_{mi} + \sum_{j=y+1}^{z} w_{mj} + W_{mo} - \theta_m) > A_{\min} \tag{7.3}$$

*Solving Equation 7.3 for $W_{mo}$ yields Equation 7.4 below.*

$$W_{mo} > -\frac{1}{\beta} ln \left( \frac{1 - A_{\min}}{1 + A_{\min}} \right) + \sum_{i=1}^{y} w_{mi} - \sum_{j=y+1}^{z} w_{mj} + \theta_m \tag{7.4}$$

*As a result, if Equation 7.2 is satisfied by $W_{ko}$ and Equation 7.4 is satisfied by $W_{mo}$ then $h_o$ blocks $L_k$ and reinforces $L_m$. This completes the proof, since Equations 7.2 and 7.4 are clearly satisfied in the* Minimal Learning Procedure. □

## 7.2 Solving Inconsistencies in Neural Networks

In this book, we consider two *sources of inconsistency* in neural networks. Inconsistencies either occur in the *answer set* computed by a neural network, which is the *standard* notion of inconsistency, or at any point during the computation of $T_{\mathcal{P}}$, which we call *strong* inconsistency. For example, let $P = \{\neg a;$ $\neg a \rightarrow b;\ \sim b \rightarrow a\}$ with answer set $\{\neg a, b\}$. As a result, the network $N$ corresponding to $P$ is not inconsistent according to the standard notion above. However, a possible computation of $T_{\mathcal{P}}$ includes $[a, \neg a, b]_{input} = (-1, 1, -1)$, $[a, \neg a, b]_{output} = (1, 1, 1)$, which shows that $N$ exhibits strong inconsistency. In this chapter, we consider standard inconsistencies. In Chapter 8, we will deal with strong inconsistencies as well.

Inconsistencies in neural networks may arise (*i*) directly from the background knowledge, (*ii*) within the set of training examples, or (*iii*) between the background knowledge and the set of examples. *C-IL²P* networks are not sensitive to inconsistencies, for there is no special mechanism in $\mathcal{N}$ stating that $x$ should imply $\neg\neg x$. As in a *TMS*, one could add rules of the form $x, \neg x \rightarrow \perp$, where $\perp$ should be a new output neuron to indicate that $x$ and $\neg x$ are not desirable, and hence that a revision process must be carried out.[6]

[6] More generally, one could use $x, y \rightarrow \perp$ to flag any integrity constraint.

*C-IL²P* networks are *paraconsistent*, i.e. inconsistencies are locally propagated without trivialising the theory[7] (see [dC74, BS89, KKT94]). As a result, one should be able to derive some plausible conclusions from a network's inconsistent answer set. However, although paraconsistency supports inconsistencies, this does not mean that it avoids belief revision, since, for example, one may want to reason about conflicting beliefs and their conclusions. "The presence of two mutually contradictory beliefs occasions in us the recognition of a quandary, and we treat those beliefs differently from others, until we are able to resolve the difficulty." [Per96] This is precisely what we intend to do here.

In this section, we use *minimal learning* to solve inconsistencies in trained networks. The idea is to change the preference relation encoded in the network when its answer set is inconsistent, thus changing its stable state. We start with an example.

*Example 7.2.1.* Consider the birds example. Let $P = \{penguin;\ bird \rightarrow flies;\ penguin \rightarrow bird;\ penguin \rightarrow \neg flies\}$, and let $N$ be the equivalent neural network. The unique stable state of $N$ (or the unique answer set of $P$) is, therefore, $S_1 = \{bird,\ penguin,\ flies,\ \neg flies\}$.

Now, suppose that the following information is made available: rule $penguin \rightarrow \neg flies$ should have higher priority than rule $bird \rightarrow flies$. This tells us that we prefer $\neg flies$ over $flies$, and thus that we want the following consistent answer set for $N$: $S_2 = \{bird,\ penguin,\ \neg flies\}$.

As a result, we only need to (minimally) train $N$ to move from the inconsistent stable state $S_1$ to the new, consistent, stable state $S_2$, analogously to the changes shown in Figure 7.3. Table 7.2 shows examples $\mathbf{e}_1$ and $\mathbf{e}_2$ which are sufficient to convert $N$ into a consistent network by applying minimal learning. Example $\mathbf{e}_1$ says that, from state $S_1 = (1,1,1,1)$, $N$ should move to state $S_2 = (1,1,-1,1)$, while example $\mathbf{e}_2$ stresses the fact that $S_2$ should be the new stable state of $N$.

Finally, the *Minimal Learning Procedure* of Section 7.1.2 would convert $\mathbf{e}_1$ and $\mathbf{e}_2$ into the simplified rule $r_o = bird,\ penguin,\ \neg flies \Rightarrow bird,\ penguin,\ \neg flies\ [flies]$, and define the weights of a neuron newly added into $N$ such that the network computes $S_2$.

**Table 7.2.** Solving inconsistencies through learning

| | *bird* | *penguin* | *flies* | $\neg$*flies* | *bird* | *penguin* | *flies* | $\neg$*flies* |
|---|---|---|---|---|---|---|---|---|
| $\mathbf{e}_1$ | 1 | 1 | 1 | 1 | 1 | 1 | −1 | 1 |
| $\mathbf{e}_2$ | 1 | 1 | −1 | 1 | 1 | 1 | −1 | 1 |

[7] "*to trivialise the theory*" means: to derive the trivial inferences that follow the *ex falso quodlibet* proof rule ($\alpha, \neg\alpha \models \beta$) which holds in classical logic.

Note that, in the above example, the whole process of solving inconsistencies in $\mathcal{N}$ requires only the current answer set ($S_j$) and the desired answer set ($S_i$) of $\mathcal{N}$ to be given. No information about the knowledge ($P$) encoded in the network is needed. In other words, the process of revision does not depend on the extraction of rules from the trained network, which means that $\mathcal{P}$ can be unknown. We assume, though, that $\mathcal{N}$ computes an acceptable program, and thus that it has a unique stable state. We call this process *Outcome Revision*, for we are exclusively concerned about converting the outcome of $\mathcal{N}$ (its associated answer set) into a consistent one, regardless of the changes performed in the knowledge actually encoded in $\mathcal{N}$. The definition of examples $\mathbf{e}_1$ and $\mathbf{e}_2$, in turn, depends on the revision policy used (see Chapter 2, Section 2.6). In what follows, we concentrate on the compromise and foundational policies, and, more specifically, on how to define examples $\mathbf{e}_1$ and $\mathbf{e}_2$ such that these cases are catered for by the minimal learning procedure above. Throughout, we refer to input and output neurons of $\mathcal{N}$ and literals of $\mathcal{P}$ interchangeably.

### 7.2.1 Compromise Revision

In order to choose between $x$ and $\neg x$, and between what is derived from $x$ and $\neg x$, new information is necessary. Such information basically needs to define a new priority relation between $x$ and $\neg x$, but it also needs to define which of the conclusions of $x$ and $\neg x$ should be retained or rejected. This will depend on the revision policy adopted, according to a given definition of *Minimal Change*, as the following example illustrates.

**Definition 7.2.1.** (Minimal Change) *Let $\mathcal{S}$ be the* inconsistent answer set *of a program $\mathcal{P}$ computed by a network $\mathcal{N}$. Let $X \subseteq \mathcal{S}$. Let $\mathcal{P}'$ be the program obtained by removing from $\mathcal{P}$ all the rules in which any literal $x \in X$ occurs in the head. Let $\mathcal{S}'$ be the answer set of $\mathcal{P}'$. The change in $\mathcal{N}$ is called* minimal *iff the new $\mathcal{N}$ computes $\mathcal{S}' \cup \{\mathcal{S} - X\}$.*

*Example 7.2.2.* Consider the *Nixon Diamond* problem. Let $P$ = {*Quaker (Nixon)*; *Republican(Nixon)*; *Quaker(Nixon)* → *Pacifist (Nixon)*; *Republican (Nixon)* → ¬ *Pacifist (Nixon)*; ¬*Pacifist (Nixon)* → *Footballfan (Nixon)*}. Assume that the stable state of a neural network $N$ is $S_1$ = {*Quaker(Nixon), Republican(Nixon), Pacifist(Nixon),* ¬*Pacifist(Nixon), Footballfan(Nixon)*}.[8] Now suppose that, as a matter of fact, Nixon is a pacifist. If we want to minimally change $S_1$, in the sense of Definition 7.2.1, we need to make $N$

[8] Recall that the original definition of Answer Sets [GL91] would assign the entire set of literals in the language as the answer set of an inconsistent program. Differently, we use a paraconsistent approach when faced with an inconsistent answer set (see [BE99] and Chapter 3, Section 3.4).

compute $S_2$ = {*Quaker(Nixon), Republican(Nixon), Pacifist(Nixon), Footballfan(Nixon)*}.

But this is the Compromise approach to revision of Gabbay [Gab99]. If *Footballfan(Nixon)* is a consequence of $\neg$*Pacifist(Nixon)*, which we have just retracted from $N$, but *Footballfan(Nixon)* is not itself a source of inconsistency in $N$, then we have no reason to retract *Footballfan(Nixon)* from $N$ as well. As we have seen in Section 7.1.2, we may do so by evolving the network from $S_1$ to $S_2$, using examples $\mathbf{e}_1$ and $\mathbf{e}_2$ of Table 7.3.[9]

**Table 7.3.** Minimal change

| | $q$ | $r$ | $p$ | $p'$ | $f$ | $q$ | $r$ | $p$ | $p'$ | $f$ |
|---|---|---|---|---|---|---|---|---|---|---|
| $\mathbf{e}_1$ | 1 | 1 | 1 | 1 | 1 | 1 | 1 | 1 | −1 | 1 |
| $\mathbf{e}_2$ | 1 | 1 | 1 | −1 | 1 | 1 | 1 | 1 | −1 | 1 |

Hence the concept of *Minimal Change* (Definition 7.2.1), together with the process of *Minimal Learning* (Section 7.1.2), implements a Compromise Revision in a neural network.

### 7.2.2 Foundational Revision

Let us now consider the case of Foundational revision. In the Nixon Diamond problem of Example 7.2.2, if $\neg$*Pacifist(Nixon)*$\rightarrow$ *Footballfan(Nixon)* is the only rule in $N$ with *Footballfan(Nixon)* in it, and we retract $\neg$*Pacifist(Nixon)* from $N$ then we should also retract *Footballfan(Nixon)*, for there is no more reason in our database to believe in it. In other words, *Footballfan(Nixon)* is no longer a *supported* belief.

**Definition 7.2.2.** *A belief* $\mathrm{B}_0$ *is called* supported *in a propositional belief set* $\mathcal{B}$ *iff for* $\mathrm{B}_1, ..., \mathrm{B}_m, \sim \mathrm{B}_{m+1}, ..., \sim \mathrm{B}_n \rightarrow \mathrm{B}_0$ *in* $\mathcal{B}$ *we have that* $\mathcal{B} \models \mathrm{B}_1, ..., \mathrm{B}_m, \sim \mathrm{B}_{m+1}, ..., \sim \mathrm{B}_n$.

Note that, by definition, $x \in \mathcal{S}_{\mathcal{P}}$ iff $x$ is supported in $\mathcal{P}$, where $\mathcal{S}_{\mathcal{P}}$ is the answer set of $\mathcal{P}$.

**Definition 7.2.3.** *(*Weaker Minimal Change*) Let* $\mathcal{S}$ *be the inconsistent answer set of a program* $\mathcal{P}$ *computed by a network* $\mathcal{N}$. *Let* $X \subseteq \mathcal{S}$. *Let* $\mathcal{P}'$ *be the program obtained by removing from* $\mathcal{P}$ *all the rules in which any literal* $x \in X$ *occurs in the head. Let* $\mathcal{S}'$ *be the answer set of* $\mathcal{P}'$. *The change in* $\mathcal{N}$ *is called* weakly minimal *iff* $\mathcal{N}$ *computes* $\mathcal{S}'$.

[9] In Table 7.3, we use $q$, $r$, $p$, $p'$ and $f$ to abbreviate, respectively, *Quaker(Nixon), Republican(Nixon), Pacifist(Nixon),* $\neg$*Pacifist(Nixon)* and *Footballfan(Nixon).*

*Example 7.2.3.* (Example 7.2.2 continued) Assume $S_3 = \{Quaker\ (Nixon),$ $Republican(Nixon),\ Pacifist(Nixon)\}$ and $X = \{\neg Pacifist\ (Nixon)\}$. It is not difficult to verify that the change in $N$ is weakly minimal iff $N$ computes $S_3$. Thus, as opposed to $S_2$ in the Compromise case (see Example 7.2.2), $S_3$ is the desired answer set of $S_1 - \{\neg Pacifist(Nixon)\}$ when a foundational approach is to be used. Table 7.4 below shows examples $\mathbf{e}_1$ and $\mathbf{e}_2$, which should be used by the minimal learning algorithm in this case.[10]

Note that, if *Pacifist(Nixon)* were to be retracted from $S_1$, $S_4 = \{Quaker$ $(Nixon),\ Republican(Nixon),\ \neg Pacifist(Nixon),\ Footballfan(Nixon)\}$ would be the desired answer set of $N$ for both the Compromise and Foundational approaches.

**Table 7.4.** Weaker minimal change

| | $q$ | $r$ | $p$ | $p'$ | $f$ | $q$ | $r$ | $p$ | $p'$ | $f$ |
|---|---|---|---|---|---|---|---|---|---|---|
| $\mathbf{e}_1$ | 1 | 1 | 1 | 1 | 1 | 1 | 1 | 1 | −1 | −1 |
| $\mathbf{e}_2$ | 1 | 1 | 1 | −1 | −1 | 1 | 1 | 1 | −1 | −1 |

This time, the concept of *Weaker Minimal Change* (Definition 7.2.3), together with the process of *Minimal Learning* (Section 7.1.2), implements a Foundational Revision in a neural network.

The problem now is how to find out the set of supported beliefs in $\mathcal{N}$ once a given belief is retracted from it.[11] We start by giving two simple examples in which $\mathcal{N}$ does not contain default negation ($\sim$). Let $\mathcal{I} = \{a, b, c, d\}$, $S = \{a, b, c\}$ and $[a, b, c, d] = (1, 1, 1, 1)$. In what follows, when we write *query* $\mathcal{N}$ *with* $S$ *to obtain* $S^*$, one should read *present input vector* $[a, b, c, d] = (1, 1, 1, -1)$ *to* $\mathcal{N}$ *and obtain, after one pass through* $\mathcal{N}$, *the corresponding output vector of* $\mathcal{N}$, *say,* $[a, b, c, d] = (-1, 1, 1, 1)$, *which is equivalent to* $S^* = \{b, c, d\}$. As before, we use $S^* = \mathcal{N}(S)$ to denote that $S^*$ is the set obtained from querying $\mathcal{N}$ with $S$.

*Example 7.2.4.* Let $P = \{a; a \rightarrow b; a \rightarrow \neg b; b \rightarrow c; \neg b \rightarrow d\}$. The answer set of $P$ is $S_1 = \{a, b, \neg b, c, d\}$. Suppose we get to know that $b$ is preferred over $\neg b$, and thus we want to solve the inconsistency in $S_1$ by retracting $\neg b$ as well as any conclusions that are no longer supported in $S_1 - \{\neg b\}$. We assume though that we do not know $P$, but have only a network $N$ that computes

[10] As before, in Table 7.4 we use $q$, $r$, $p$, $p'$ and $f$ to abbreviate, respectively, *Quaker(Nixon)*, *Republican(Nixon)*, *Pacifist(Nixon)*, *¬Pacifist(Nixon)* and *Footballfan(Nixon)*.

[11] Recall that we assume that the program $\mathcal{P}$ encoded in $\mathcal{N}$ is unknown.

$S_1$.[12] We are left with the option of querying $N$. We proceed as follows: Query $N$ with $S_1 - \{\neg b\}$, i.e. present input vector $[a, b, \neg b, c, d] = (1, 1, -1, 1, 1)$ to $N$. The output obtained, $[a, b, \neg b, c, d] = (1, 1, 1, 1, -1)$, corresponds to the set $S_1^* = \{a, b, \neg b, c\}$. Now, let $\overline{S}_1 = S_1 - S_1^*$. In this example, $\overline{S}_1 = \{d\}$. Then, query $N$ with $S_1^*$ to obtain $S_2^* = \{a, b, \neg b, c, d\}$. Since $S_2^* = S_1$, we stop querying $N$. If $\neg b$ is to be retracted from $N$ then $d$ also needs to be taken into account.

*Example 7.2.5.* Let $P = \{a; a \rightarrow b; a \rightarrow \neg b; b \rightarrow c; c \rightarrow d; \neg b \rightarrow e\}$ and let $N$ be a network equivalent to $P$. $S_1 = \{a, b, \neg b, c, d, e\}$ is the answer set of $P$. Assume we query $N$ with $S_1 - \{b\}$ and obtain $S_1^* = \{a, b, \neg b, d, e\}$ and, thus, $\overline{S}_1 = \{c\}$. Then, we query $N$ with $S_1^*$ and obtain $S_2^* = \{a, b, \neg b, c, e\}$ and, thus, $\overline{S}_2 = \{d\}$. Finally, we query $N$ with $S_2^*$ and obtain $S_3^* = \{a, b, \neg b, c, d, e\}$ and $\overline{S}_3 = \emptyset$. The set $D = \overline{S}_1 \cup \overline{S}_2 \cup \overline{S}_3 = \{c, d\}$ will contain the literals in $S_1$ that are of our interest when $b$ is to be retracted from $N$.

Let us define precisely the literals that must belong to $D$.

**Definition 7.2.4.** *Let $\mathcal{S}$ be the answer set of a program $\mathcal{P}$ computed by a network $\mathcal{N}$. Let $x \in \mathcal{S}$. Let $\mathcal{P}'$ be the program obtained by removing from $\mathcal{P}$ all the rules in which $x$ occurs in the head, and let $\mathcal{S}'$ be the answer set of $\mathcal{P}'$. We say that $y$ ($y \in \mathcal{S}$)* results *from $x$ ($x \neq y$) iff $y \notin \mathcal{S}'$. We say that $z$ ($z \in B_{\mathcal{P}}$)* results *from $\sim x$ ($x \neq z$) iff $z \notin \mathcal{S}$ and $z \in \mathcal{S}'$.*

A literal $y$ should belong to $D$ iff $y$ results from $x$. This is so because we want to find out all (and only) the literals that are supported in $\mathcal{P}$ and become not supported in $\mathcal{P}'$. For example, let $P = \{a;\ b;\ a \rightarrow c;\ b \rightarrow c;\ b \rightarrow d;\ d \rightarrow e;\ \sim a \rightarrow f;\ p \rightarrow q\}$ and, thus, $S_1 = \{a, b, c, d, e\}$. The literal $c$ does not result from $a$ because it is supported by $b$. Both $d$ and $e$ result from $b$ because they depend only on $b$ to be supported. Similarly, $f$ results from $\sim a$. Finally, although $q$ depends on $p$, $p$ is not supported, and thus $q$ does not result from $p$.

Let us consider firstly the case of monotonic networks, i.e. networks in which default negation ($\sim$) is not present. We want to revise a network $\mathcal{N}$ by rejecting a set $X \subseteq \mathcal{S}(\mathcal{N})$, where $\mathcal{S}(\mathcal{N})$ is the answer set of $\mathcal{N}$, and using weaker minimal change. The following algorithm should return the set $D$ of literals in $\mathcal{S}(\mathcal{N})$ that result from the literals in $X$.

- *Dependency Finding Algorithm (Monotonic Networks)* $(\mathcal{N}, X)$

1. $S_1 = \mathcal{S}(\mathcal{N})$; // Calculate the answer set of $\mathcal{N}$.
2. $S_1^* = \mathcal{N}(S_1 - X)$; // query $\mathcal{N}$ with $S_1 - X$ and obtain $S_1^*$.

[12] This complies with the idea of *Outcome Revision*, in which we are not concerned about the actual database and the changes performed in it to restore consistency, but only in obtaining a maximally consistent answer set.

3. $\overline{S}_1 = S_1 - S_1^*$; // $\overline{S}_1$ contains the literals that result directly from $X$.
4. $i = 1$;
5. While $\overline{S}_i \neq \emptyset$ do: // find the set of literals that result from $X$ after $i$ iterations.
   a) $S_{i+1}^* = \mathcal{N}(S_i^*)$; // query $\mathcal{N}$ with $S_i^*$ and obtain $S_{i+1}^*$.
   b) $\overline{S}_{i+1} = S_1 - S_{i+1}^*$; // $\overline{S}_{i+1}$ contains the literals that result from $X$ after $i$ iterations.
   c) $i = i + 1$;
6. Return $D = \bigcup_{j=1}^{i} \overline{S}_j$. // $D$ contains the literals that result from $X$.

In order to satisfy weaker minimal change, we need to show that $D$ is the set of all the literals that result from some literal in $X$.

**Proposition 7.2.1.** *Let $\mathcal{P}$ be an acceptable extended logic program without negation as failure ($\sim$). Let $\mathcal{N}$ be a neural network that computes the fixed point operator $T_{\mathcal{P}}$ of $\mathcal{P}$, $D$ be obtained from the* Dependency Finding Algorithm (Monotonic Networks), *$\mathcal{S}(\mathcal{N})$ be the answer set of $\mathcal{P}$, and $X \subseteq \mathcal{S}(\mathcal{N})$. Then, $d \in D$ iff $\exists x \in X$ such that $d$* results *from $x$.*

*Proof.* *By assumption, $\mathcal{P}$ is an acceptable program. As a result, $\mathcal{N}$ converges to a stable state $\mathcal{S}(\mathcal{N})$, starting from any initial state. Therefore, for any $X$, there exists an $i \in \aleph$ such that, starting from $\mathcal{S}(\mathcal{N}) - X$, $\mathcal{N}$ converges to $\mathcal{S}(\mathcal{N})$ after $i$ iterations, and hence the above* Dependency Finding Algorithm *always terminates. This value of $i$ will be used in the rest of the proof.*

*We need to show that $D \subseteq \mathcal{S}(\mathcal{N})$. From the* Dependency Finding Algorithm, *$D = \{\mathcal{S}(\mathcal{N}) - S_1^*\} \cup \{\mathcal{S}(\mathcal{N}) - S_2^*\} \cup ... \cup \{\mathcal{S}(\mathcal{N}) - S_i^*\}$. By definition, $\{\mathcal{S}(\mathcal{N}) - S\} \subseteq \mathcal{S}(\mathcal{N})$, for any set $S$, and the union of subsets of $\mathcal{S}(\mathcal{N})$ is also a subset of $\mathcal{S}(\mathcal{N})$. Thus, $D \subseteq \mathcal{S}(\mathcal{N})$.*

*By Definition 7.2.2, $y \in \mathcal{S}(\mathcal{N})$ iff $y$ is supported in $\mathcal{P}$. From $D \subseteq \mathcal{S}(\mathcal{N})$, $y \in D$ only if $y \in \mathcal{S}(\mathcal{N})$ and, thus, $y \in D$ only if $y$ is supported in $\mathcal{P}$.*

*Now, let $x \in X, y \in \mathcal{S}(\mathcal{N})$. We need to show that $y \in \bigcup_{j=1}^{i}\{\mathcal{S}(\mathcal{N}) - S_j^*\} \leftrightarrow results(y, x)$, where $results(u, v)$ should read $u$ results from $v$.*

*($\leftarrow$) "If $results(y, x)$ then $y \in \bigcup_{j=1}^{i}\{\mathcal{S}(\mathcal{N}) - S_j^*\}$". Assume $results(y, x)$. Note that $\bigcup_{j=1}^{i}\{\mathcal{S}(\mathcal{N}) - S_j^*\} = \mathcal{S}(\mathcal{N}) - \bigcap_{j=1}^{i} S_j^*$. By assumption, $y \in \mathcal{S}(\mathcal{N})$. Thus, we only need to show that $y \notin \bigcap_{j=1}^{i} S_j^*$. It suffices to show that $y \notin S_1^*$. The proof is by contradiction. Assume $y \in S_1^*$. From the* Dependency Finding Algorithm, *$S_1^* = \mathcal{N}(\mathcal{S}(\mathcal{N}) - X) = T_{\mathcal{P}}(\mathcal{S}(\mathcal{N}) - X)$. From the definition of $T_{\mathcal{P}}$, there is a rule of the form $L_1, ..., L_n \rightarrow y$, $n \geq 0$, in $\mathcal{N}$ s. t. $L_1, ..., L_n \in \mathcal{S}(\mathcal{N}) - X$. Hence, by Definition 7.2.4, $\neg results(y, x)$, $x \in X$. This contradicts the assumption that $results(y, x)$ and, thus, $y \notin S_1^*$.*

*($\rightarrow$) "If $y \in \bigcup_{j=1}^{i}\{\mathcal{S}(\mathcal{N}) - S_j^*\}$ then $results(y, x)$". If $y \in \bigcup_{j=1}^{i}\{\mathcal{S}(\mathcal{N}) - S_j^*\}$ then $y \in \mathcal{S}(\mathcal{N})$ and $y \notin \bigcap_{j=1}^{i} S_j^*$. Assume $\neg results(y, x)$. By Definition 7.2.4, there is a rule of the form $L_1, ..., L_n \rightarrow y$, $n \geq 0$, in $\mathcal{N}$ s. t. $L_1, ..., L_n \in$*

$\mathcal{S}(\mathcal{N}) - X$. *From the* Dependency Finding Algorithm, $S_j^* = \mathcal{N}(S_{j-1}^*) = \mathcal{N}(...(\mathcal{N}(\mathcal{S}(\mathcal{N}) - X)))$ *and, since* $\mathcal{S}(\mathcal{N})$ *is a stable state of* $\mathcal{N}$, $y \in S_j^*$ *for* $1 \leq j \leq i$, *that is,* $y \in \bigcap_{j=1}^{i} S_j^*$. *This contradicts the assumption that* $y \notin \bigcap_{j=1}^{i} S_j^*$ *and, thus, results*$(y, x)$. □

Once the set $D$ is obtained, we simply need to apply *Minimal Learning*, training the network not to derive any literal in $D$, in addition to the ones in $X$. In Example 7.2.5 above, if we want to reject $b$ from the answer set of $N$, we should also train the network not to derive $c$ or $d$. Table 7.5 contains the training examples for performing a foundational revision in this case.

**Table 7.5.** Weaker minimal change (revisited)

| | $a$ | $b$ | $\neg b$ | $c$ | $d$ | $e$ | $a$ | $b$ | $\neg b$ | $c$ | $d$ | $e$ |
|---|---|---|---|---|---|---|---|---|---|---|---|---|
| $\mathbf{e}_1$ | 1 | 1 | 1 | 1 | 1 | 1 | 1 | −1 | 1 | −1 | −1 | 1 |
| $\mathbf{e}_2$ | 1 | −1 | 1 | −1 | −1 | 1 | 1 | −1 | 1 | −1 | −1 | 1 |

In what follows, we present an example with some problematic cases, which might help clarify the idea behind the above algorithm.

*Example 7.2.6.* Let $P = \{a; b; a \to \neg b; \neg b \to a\}$. $S_1 = \mathcal{S}(N) = \{a, b, \neg b\}$. Let $X = \{b\}$. $S_1^* = N(S_1 - X) = \{a, b, \neg b\}$. $\overline{S}_1 = S_1 - S_1^* = \emptyset$. As expected, no literal results from $b$. Now let $X = \{\neg b\}$. $S_1^* = N(S_1 - X) = \{a, b, \neg b\}$ and $\overline{S}_1 = \emptyset$. Although $\neg b \to a$, $a$ does not result from $\neg b$ and, in this case, $D$ is empty.

Let $P = \{\neg a; \neg b; \neg a \to b; a \to x; b \to x\}$. $S_1 = \mathcal{S}(N) = \{\neg a, b, \neg b, x\}$. If $S_1^* = N(S_1 - \{b\}) = \{\neg a, b, \neg b\}$ then $\overline{S}_1 = \{x\}$. $S_2^* = N(S_1^*) = \{\neg a, b, \neg b, x\}$ and $\overline{S}_2 = \emptyset$. Hence, $x$ results from $b$. This is so because, although $a \to x$, this rule is never used, and thus $x$ is only supported in $P$ by $b$. If, however, one adds $a$ as a fact into $P$ then $x$ will depend on $a$ or $b$ to be supported. In this case, $x$ will not result from $b$. Finally, if one replaces $a \to x$ and $b \to x$ by $ab \to x$ then, as expected, $x$ will result from $b$, as long as $a \in S_1$.

Note that the Dependency Algorithm is not only necessary when we want to perform foundational revision, but also when we want to draw some plausible conclusions from an inconsistent answer set. In Example 7.2.2, for instance, suppose that we do not have any extra information about the pacifism of Nixon. We are not able, then, to solve the inconsistency for now, but if we believe that Nixon is a football fan because he is non-pacifist, and we know that there is a conflict about Nixon's pacifism, then we should lower our belief that Nixon is a football fan. On the other hand, if we believe that Nixon is a football fan because he is republican, and there is no conflict about this, then we should have no reason not to believe that Nixon is indeed a football fan. We will come back to this matter later on in this chapter.

Let us now concentrate on the general case, in which the network contains negation by default.

### 7.2.3 Nonmonotonic Theory Revision

Nonmonotonic networks allow the derivation of new facts as a result of the rejection of other facts. Take, for example, $P = \{a; \neg a; \sim a \rightarrow b\}$. $S_1 = \mathcal{S}(N) = \{a, \neg a\}$. If $S_1^* = N(S_1 - \{a\})$ then $S_1^* = \{a, \neg a, b\}$. In other words, we have lost the property that $S_j^* \subseteq \mathcal{S}(N)$ for all $j \in \aleph$. Now, if we want to revise $P$ by rejecting $a$, we also need to decide what should be done about $b$.

An immediate result of $S_j^* \nsubseteq \mathcal{S}(N)$ is that inconsistencies might no longer be solvable in a single learning step. This is so because, now, the revision of $P$ might add new inconsistent conclusions to its answer set. Hence the processes of finding dependencies and minimal learning might need to be applied repeatedly until a consistent answer set is obtained.

As expected, nonmonotonicity makes things more complicated. Nevertheless, it is not difficult to extend the Dependency Finding Algorithm to nonmonotonic theories. Take, for example, $P = \{a; a \rightarrow b; b \rightarrow \neg a; \sim a \rightarrow c\}$, and thus $S_1 = \mathcal{S}(N) = \{a, \neg a, b\}$. Let $S_1^* = N(S_1 - \{a\}) = \{a, \neg a, c\}$. If, as before, we take $S_1 - S_1^* = \{b\}$, we find out that $b$ results from $a$. However, it makes sense now to also compute $S_1^* - S_1 = \{c\}$, which informs us that $c$ results from $\sim a$. When revising $P$ by $a$, we would want to keep $b$, but also to add $c$, which is now nonmonotonically derived. Returning to the example, and taking, as before, $S_2^* = N(S_1^*)$, we obtain $S_2^* = \{a, b\}$. Thus, $S_1 - S_2^* = \{\neg a\}$ and $S_2^* - S_1 = \emptyset$. $S_1 - S_2^* \neq \emptyset$ tells us that we need to carry on. We compute $S_3^* = N(S_2^*) = \{a, \neg a, b\}$ and $S_1 - S_3^* = \emptyset$. Now, $S_3^* - S_1 = \emptyset$ as well, and we can stop. The results obtained are $D_a = \{\neg a, b\}$ and $D_{\sim a} = \{c\}$, indicating, respectively, the literals that result from $a$ and $\sim a$. As before, $D_a = \bigcup_{j=1}^{i} \overline{S}_j$, where $\overline{S}_j = S_1 - S_j^*$. We simply need to define $D_{\sim a} = \bigcup_{j=1}^{i} \overline{\overline{S}}_j$, where $\overline{\overline{S}}_j = S_j^* - S_1$, in order to extend the dependency algorithm to the general, nonmonotonic, case.

$D_x$ and $D_{\sim x}$ contain all the information necessary for performing either Compromise or Foundational revision. In the Compromise case, only $D_{\sim x}$ is relevant. In the Foundational case, we need to retract the literals in $D_x$ and add the ones in $D_{\sim x}$. In the above example, given $D_a = \{\neg a, b\}$ and $D_{\sim a} = \{c\}$, the training examples for minimal learning can be worked out for the Compromise and Foundational revisions, as shown in Tables 7.6 and 7.7, respectively.

In order to obtain the above training examples, we have proceeded as follows: in both the Compromise and Foundational cases, $S_1$ provides the input vector $\mathbf{i}_1$ of $\mathbf{e}_1$. In the Compromise case, from $\mathbf{i}_1$, we flip from 1 to $-1$ the input being revised ($a$), and we flip from $-1$ to 1 the inputs in $D_{\sim a}$, i.e.

**Table 7.6.** Nonmonotonic compromise revision

| | $a$ | $\neg a$ | $b$ | $c$ | $a$ | $\neg a$ | $b$ | $c$ |
|---|---|---|---|---|---|---|---|---|
| $\mathbf{e}_1$ | 1 | 1 | 1 | −1 | −1 | 1 | 1 | 1 |
| $\mathbf{e}_2$ | −1 | 1 | 1 | 1 | −1 | 1 | 1 | 1 |

**Table 7.7.** Nonmonotonic Foundational Revision

| | $a$ | $\neg a$ | $b$ | $c$ | $a$ | $\neg a$ | $b$ | $c$ |
|---|---|---|---|---|---|---|---|---|
| $\mathbf{e}_1$ | 1 | 1 | 1 | −1 | −1 | −1 | −1 | 1 |
| $\mathbf{e}_2$ | −1 | −1 | −1 | 1 | −1 | −1 | −1 | 1 |

*c*. All remaining inputs are kept as in $\mathbf{i}_1$. This gives the output vector $\mathbf{o}_1$ of $\mathbf{e}_1$, which should be identical to both the input and output vectors $\mathbf{i}_2$ and $\mathbf{o}_2$ of $\mathbf{e}_2$; the new stable state of the network. In the Foundational case, however, from $\mathbf{i}_1$, we flip from 1 to −1 the input being revised ($a$) and the inputs in $D_a$ ($\neg a$ and $b$). We flip from −1 to 1 the inputs in $D_{\sim a}$ ($c$). Again, all the remaining inputs should be kept as in $\mathbf{i}_1$. This process gives $\mathbf{o}_1$, which should also be identical to $\mathbf{i}_2$ and $\mathbf{o}_2$.

We need to perform some simple changes to the Dependency Finding Algorithm in order to extend it to the general case. As before, we want to revise a network $\mathcal{N}$ by rejecting a set $X \subseteq S_1$, where $S_1 = \mathcal{S}(\mathcal{N})$. The following algorithm returns two sets, $D_X$ and $D_{\sim X}$, containing, respectively, the literals of $\mathcal{N}$ that result from the literals in $X$ and the literals of $\mathcal{N}$ that result from the (default) negation of the literals in $X$. Throughout, we use $\sim X$ to indicate the set $\{\sim L_1, \sim L_2, ..., \sim L_n\}$, obtained from $X = \{L_1, L_2, ..., L_n\}$.

– *Dependency Finding Algorithm (Nonmonotonic Networks)* $(\mathcal{N}, X)$

1. $S_1 = \mathcal{S}(\mathcal{N})$; // $S_1$ is the answer set of $\mathcal{N}$.
2. $S_1^* = \mathcal{N}(S_1 - X)$; // Query $\mathcal{N}$ with $S_1 - X$ and obtain $S_1^*$.
3. $\overline{S}_1 = S_1 - S_1^*$; // $\overline{S}_1$ contains the literals that result directly from $X$.
4. $\overline{\overline{S}}_1 = S_1^* - S_1$; // $\overline{\overline{S}}_1$ contains the literals that result directly from $\sim X$.
5. $i = 1$;
6. While $\overline{S}_i \neq \emptyset$ or $\overline{\overline{S}}_i \neq \emptyset$ do: // find the set of literals that result from $X$ and $\sim X$ after $i$ iterations.
   a) $S_{i+1}^* = \mathcal{N}(S_i^*)$; // Query $\mathcal{N}$ with $S_i^*$ and obtain $S_{i+1}^*$.
   b) $\overline{S}_{i+1} = S_1 - S_{i+1}^*$; // $\overline{S}_{i+1}$ contains the literals that result from $X$ after $i$ iterations.
   c) $\overline{\overline{S}}_{i+1} = S_{i+1}^* - S_1$; // $\overline{\overline{S}}_{i+1}$ contains the literals that result from $\sim X$ after $i$ iterations.
   d) $i = i + 1$;
7. Return $D_X = \bigcup_{j=1}^{i} \overline{S}_j$ and $D_{\sim X} = \bigcup_{j=1}^{i} \overline{\overline{S}}_j$. // $D_X$ and $D_{\sim X}$ contain the literals that result from $X$ and $\sim X$, respectively.

Let us see how the above algorithm works in a larger example.

*Example 7.2.7.* Let $P_1 = \{a; a \rightarrow b; a \rightarrow \neg b; b \rightarrow c; \neg b \rightarrow d; \sim c \rightarrow x; \sim d \rightarrow y; p \rightarrow q\}$. Let $P_2 = P_1 \cup \{\sim a \rightarrow p\}$ and $P_3 = P_2 \cup \{\sim a \rightarrow \neg q\}$. $N_1$, $N_2$ and $N_3$ will denote, respectively, the neural networks that compute $P_1$, $P_2$ and $P_3$.

Let $S_1 = \mathcal{S}(N_1) = \{a, b, \neg b, c, d\}$. Let $S_1^* = N_1(S_1 - \{b\}) = \{a, b, \neg b, d\}$. Thus, $\overline{S}_1 = S_1 - S_1^* = \{c\}$ and $\overline{\overline{S}}_1 = S_1^* - S_1 = \emptyset$. $S_2^* = N_1(S_1^*) = \{a, b, \neg b, c, d, x\}$. Thus, $\overline{S}_2 = S_1 - S_2^* = \emptyset$ and $\overline{\overline{S}}_2 = S_2^* - S_1 = \{x\}$. $S_3^* = N_1(S_2^*) = \{a, b, \neg b, c, d\}$, and $\overline{S}_3 = S_1 - S_3^* = \emptyset$, $\overline{\overline{S}}_3 = S_3^* - S_1 = \emptyset$, that is, $S_3^* = S_1$. This example shows that the algorithm can terminate only when both $\overline{S}_j$ and $\overline{\overline{S}}_j$ are empty sets; in other words, when $S_j^* = S_1$.[13]

Now, let $S_1 = \mathcal{S}(N_2) = \{a, b, \neg b, c, d\}$. Assume we want to retract $a$ and $b$ from $S_1$ (e.g. assume $a, c \rightarrow \perp$ is an integrity constraint). We could apply the above procedure twice, for instance, firstly for $a$, and, if necessary, for $b$. However, if we want to try to solve the problem in one goal, we might simply assume $S_1^* = N_2(S_1 - \{a, b\})$. In this case, we obtain $S_1^* = \{a, d, p\}$ and, hence, $\overline{S}_1 = \{b, \neg b, c\}$ and $\overline{\overline{S}}_1 = \{p\}$. Then, $S_2^* = N_2(S_1^*) = \{a, b, \neg b, x, q\}$, and $\overline{S}_2 = \{c, d\}$, $\overline{\overline{S}}_2 = \{x, q\}$. $S_3^* = N_2(S_2^*) = \{a, b, \neg b, c, d, x, y\}$, and $\overline{S}_3 = \emptyset$, $\overline{\overline{S}}_3 = \{x, y\}$. Finally, $S_4^* = N_2(S_3^*) = \{a, b, \neg b, c, d\} = S_1$, and we can stop. As a result, $D_{ab} = \{b, \neg b, c, d\}$ and $D_{\sim a \sim b} = \{p, q, x, y\}$. Although we are not able to distinguish the literals that result from $a$ from the ones that result from $b$, the information in $D_{ab}$ and $D_{\sim a \sim b}$ is sufficient to perform the revision of $P_2$ by learning. The training examples for Compromise and Foundational revision could be defined exactly as before.

Finally, let $S_1 = \mathcal{S}(N_3) = \{a, b, \neg b, c, d\}$. Assume, as before, that $a, c \rightarrow \perp$ and take $S_1^* = N_3(S_1 - \{a, b\})$. Thus, $S_1^* = \{a, d, p, \neg q\}$, $\overline{S}_1 = \{b, \neg b, c\}$ and $\overline{\overline{S}}_1 = \{p, \neg q\}$. As before, $S_2^* = N_3(S_1^*) = \{a, b, \neg b, x, q\}$, $\overline{S}_2 = \{c, d\}$ and $\overline{\overline{S}}_2 = \{x, q\}$. And, $S_3^* = N_3(S_2^*) = \{a, b, \neg b, c, d, x, y\}$, $\overline{S}_3 = \emptyset$ and $\overline{\overline{S}}_3 = \{x, y\}$. Finally, $S_4^* = S_1$. Now, $D_{ab} = \{b, \neg b, c, d\}$ and $D_{\sim a \sim b} = \{p, q, \neg q, x, y\}$. This example illustrates the case in which inconsistencies can not be eliminated through a single learning step, because $D_{\sim a \sim b}$ is itself inconsistent. When retracting $a$ from the network, both $q$ and $\neg q$ were nonmonotonically derived. The same revision process could be repeated when a preference relation between $q$ and $\neg q$ is made available.

---

[13] By definition, $S_j^* \subseteq S_1$ implies $S_j^* - S_1 = \emptyset$. Similarly, if $S_1 \subseteq S_j^*$ then $S_1 - S_j^* = \emptyset$. Thus, if $S_1 - S_j^* = \emptyset$ and $S_j^* - S_1 = \emptyset$ then $S_j^* = S_1$.

As a result, $S_j^* = S_1$ is the stopping criterion for the general case Dependency Finding Algorithm. This is the same stopping criterion of the algorithm for monotonic networks, but, since $S_j^* \subseteq S_1$ in that case, $S_1 - S_j^* = \emptyset$ is sufficient to guarantee that $S_j^* = S_1$.

**Proposition 7.2.2.** *Let $\mathcal{P}$ be an acceptable extended logic program. Let $\mathcal{N}$ be a neural network that computes the fixed point operator $T_{\mathcal{P}}$ of $\mathcal{P}$ and $X \subseteq \mathcal{S}(\mathcal{N})$. Let $\sim X$ be obtained from $X$ as described above and $\mathcal{S}(\mathcal{N})$ be the answer set of $\mathcal{P}$. Then, $d \in D_X$ iff $\exists x \in X$ such that $d$* results *from $x$, and $d \in D_{\sim X}$ iff $\exists x \in \sim X$ such that $d$* results *from $x$, where $D_X$ and $D_{\sim X}$ are obtained from the* Dependency Finding Algorithm (Nonmonotonic Networks).

*Proof. By assumption, $\mathcal{P}$ is an acceptable program. As a result, $\mathcal{N}$ converges to a stable state $\mathcal{S}(\mathcal{N})$, starting from any initial state. Therefore, for any $X$, there exists an $i \in \aleph$ such that, starting from $\mathcal{S}(\mathcal{N}) - X$, $\mathcal{N}$ converges to $\mathcal{S}(\mathcal{N})$ after $i$ iterations, and hence the above* Dependency Finding Algorithm *always terminates. This value of $i$ will be used in the rest of the proof.*

*Note that Proposition 7.2.1 generalises to* Nonmonotonic Networks *and, therefore, if $S_j^* = \mathcal{S}(\mathcal{N})$ then for all $d \in D_X$ there exists $x \in X$ such that $d$* results *from $x$. Also, if $d$* results *from $x \in X$ then $d \in D_X$. Hence, we only need to worry about $D_{\sim X}$. Let $x \in X$, $y \in B_{\mathcal{P}}$, where $B_{\mathcal{P}}$ is the Herbrand base of $\mathcal{P}$. We need to show that $y \in \bigcup_{j=1}^{i}\{S_j^* - \mathcal{S}(\mathcal{N})\} \leftrightarrow results(y, \sim x)$.*

*($\leftarrow$) "If $results(y, \sim x)$ then $y \in \bigcup_{j=1}^{i}\{S_j^* - \mathcal{S}(\mathcal{N})\}$". Note that $\bigcup_{j=1}^{i}\{S_j^* - \mathcal{S}(\mathcal{N})\} = \bigcup_{j=1}^{i} S_j^* - \mathcal{S}(\mathcal{N})$. Assume $results(y, \sim x)$. By Definition 7.2.4, $y \notin \mathcal{S}(\mathcal{N})$. Thus, it suffices to show that $y \in \bigcup_{j=1}^{i} S_j^*$. From the* Dependency Finding Algorithm, *$S_1^* = \mathcal{N}(\mathcal{S}(\mathcal{N}) - X) = T_{\mathcal{P}}(\mathcal{S}(\mathcal{N}) - X)$. From the definition of $T_{\mathcal{P}}$, there is a rule of the form $L_1, ..., L_n, \sim x \rightarrow y$, $n \geq 0$, in $\mathcal{N}$ s. t. $L_1, ..., L_n \in \mathcal{S}(\mathcal{N})$. Since $x \in X$ and $S_1^* = T_{\mathcal{P}}(\mathcal{S}(\mathcal{N}) - X)$, $y \in S_1^*$. Thus, $y \in \bigcup_{j=1}^{i} S_j^*$.*

*($\rightarrow$) "If $y \in \bigcup_{j=1}^{i}\{S_j^* - \mathcal{S}(\mathcal{N})\}$ then $results(y, \sim x)$". If $y \in \bigcup_{j=1}^{i}\{S_j^* - \mathcal{S}(\mathcal{N})\}$ then $y \in \bigcup_{j=1}^{i} S_j^*$ and $y \notin \mathcal{S}(\mathcal{N})$. Assume $\neg results(y, \sim x)$. By Definition 7.2.4, there is no rule of the form $L_1, ..., L_n, \sim x \rightarrow y$ in $\mathcal{N}$ s.t. $L_1, ..., L_n \in \mathcal{S}(\mathcal{N}) - X$. From the* Dependency Finding Algorithm, *$S_j^* = \mathcal{N}(S_{j-1}^*) = \mathcal{N}(...(\mathcal{N}(\mathcal{S}(\mathcal{N}) - X)))$. Thus, $y \notin S_j^*$, $1 \leq j \leq i$. This contradicts the assumption that $y \in \bigcup_{j=1}^{i} S_j^*$ and, thus, $results(y, \sim x)$.* □

We now give some examples of the process of revision. In what follows, we use standard backpropagation, instead of minimal learning. By using backpropagation, a network may loose previous information when trained with new examples. Nevertheless, backpropagation can still be useful to exemplify the process. Take, for example, the well-known birds example. Let $P = \{b \rightarrow f; p \rightarrow b; p \rightarrow \neg f; p\}$. We translate $P$ into a network $N$ using the Translation Algorithm of Chapter 3. We then obtain $S(N) = \{b, p, f, \neg f\}$. At this point, we want to solve the inconsistency regarding $f$. Let us say that $\neg f$ is the result of an observation and, thus, has priority over $f$. We use the Dependency Finding Algorithm to obtain $D_f = \emptyset$ and $D_{\sim f} = \emptyset$, from which we derive the training examples of Table 7.8. We then train $N$ on these two examples using standard backpropagation to obtain $N'$. The following program was extracted from $N'$ by an exhaustive pedagogical extraction: $P' = \{p; b \sim p \rightarrow f; bpf \rightarrow b; bpf \rightarrow \neg f; p \sim f \rightarrow b; p \sim f \rightarrow \neg f\}$. As desired, the answer set of $P'$ is $S(N') = \{b, p, \neg f\}$.

**Table 7.8.** The birds example (retracting $f$)

| | $b$ | $p$ | $f$ | $\neg f$ | $b$ | $p$ | $f$ | $\neg f$ |
|---|---|---|---|---|---|---|---|---|
| $\mathbf{e}_1$ | 1 | 1 | 1 | 1 | 1 | 1 | −1 | 1 |
| $\mathbf{e}_2$ | 1 | 1 | −1 | 1 | 1 | 1 | −1 | 1 |

We now consider the Nixon Diamond problem. We translate $P = \{n; n \rightarrow q; n \rightarrow r; q \rightarrow p; r \rightarrow \neg p\}$ into a network $N$. If we assume that religious beliefs are stronger than political affiliations then we would like to retract $\neg p$ from the answer set of $P$. We do so by training the examples of Table 7.9. The knowledge extracted from the trained network was $P' = \{n; n \rightarrow q; n \rightarrow r; q \rightarrow p; \sim n \sim qr \rightarrow \neg p; \sim n \sim qr \sim p \rightarrow \neg p\}$, whose answer set is $S(N') = \{n, q, r, p\}$. Alternatively, when retracting $p$ from $P$, the knowledge extracted after training was $P'' = \{n; n \rightarrow q; n \rightarrow r; r \rightarrow \neg p; \sim nq \sim r \rightarrow p\}$, whose answer set is $S'' = \{n, q, r, \neg p\}$.

**Table 7.9.** The Nixon Diamond problem (retracting $\neg p$)

| | $n$ | $q$ | $r$ | $p$ | $\neg p$ | $n$ | $q$ | $r$ | $p$ | $\neg p$ |
|---|---|---|---|---|---|---|---|---|---|---|
| $\mathbf{e}_1$ | 1 | 1 | 1 | 1 | 1 | 1 | 1 | 1 | 1 | −1 |
| $\mathbf{e}_2$ | 1 | 1 | 1 | 1 | −1 | 1 | 1 | 1 | 1 | −1 |

We are finally in a position to describe the general procedure for solving inconsistencies in trained networks.

– *Revision Procedure* ($\mathcal{N}$, $X$)

1. Apply the *Dependency Finding Algorithm* ($\mathcal{N}$, $X$) to obtain $D_X$ and $D_{\sim X}$;
2. From $D_X$ and $D_{\sim X}$, find $\{\mathbf{e}_1, \mathbf{e}_2\}$, according to the revision policy adopted; and
3. Apply the *Minimal Learning Procedure* ($\mathcal{N}$, $\{\mathbf{e}_1, \mathbf{e}_2\}$).

**Theorem 7.2.1.** *The* Revision Procedure *complies with* Minimal Change.

*Proof. From Definition 7.2.1 and Proposition 7.2.2,* $\{\mathbf{e}_1, \mathbf{e}_2\}$ *complies with minimal change. From Definition 7.2.3 and Proposition 7.2.2,* $\{\mathbf{e}_1, \mathbf{e}_2\}$ *complies with weaker minimal change. From Definition 7.1.1 and Proposition*

*7.1.1, the Minimal Learning Procedure performs exactly the changes intended by* $\{\mathbf{e}_1, \mathbf{e}_2\}$*, as depicted in Figure 7.3, that is, Minimal Learning. Hence the Revision Procedure complies with Minimal Change in the Compromise framework and with Weaker Minimal Change in the Foundational framework.* □

Let us now apply the complete *Revision Procedure* in a final example.

*Example 7.2.8.* Let $\mathbf{I} = \mathbf{O} = [a, b, c, d, \neg d]$ and $S(N) = \{1, 1, -1, 1, 1\}$, i.e. $S(N)$ is the inconsistent answer set of an unknown program $P$ computed by a trained neural network $N$. Let $d \succ \neg d$, i.e. $d$ has priority over $\neg d$. Assume that, by applying the Dependency Finding Algorithm and using the Foundational policy, we obtain examples $\mathbf{e}_1 = \{\mathbf{i}_1 = (1, 1, -1, 1, 1), \mathbf{o}_1 = (1, -1, 1, 1, -1)\}$ and $\mathbf{e}_2 = \{\mathbf{i}_2 = (1, -1, 1, 1, -1), \mathbf{o}_2 = (1, -1, 1, 1, -1)\}$. In other words, $b$ results from $\neg d$, and $c$ results from $\sim \neg d$. From $\mathbf{e}_1$ and $\mathbf{e}_2$, we obtain $r_1 = a, b, \sim c, d, \neg d \Rightarrow a, c, d[b, \neg d]$ and $r_2 = a, \sim b, c, d, \sim \neg d \Rightarrow a, c, d[b, \neg d]$. $r_1$ and $r_2$ can not be simplified. These rules can be inserted into $N$ by the addition of two hidden neurons ($h_4$ and $h_5$), as depicted in Figure 7.5.

Let $A_{\min} = 0.5$ and $\beta = 1$. Let $w_{a(h_1,h_2,h_3)} = (0.5, 1.5, -3)$, $w_{b(h_1,h_2,h_3)} = (1, -2, 5)$, $w_{c(h_1,h2,h3)} = (-4, 3, 2)$, $w_{d(h_1,h_2,h_3)} = (0.5, 2, -1.5)$, $w_{\neg d(h_1,h2,h3)} = (-1, -5, 4)$.[14] Let also $\theta_a = 1.5$, $\theta_b = -0.2$, $\theta_c = 0.4$, $\theta_d = 2$, $\theta_{\neg d} = 1.2$. We calculate $W_{a\{h_4,h_5\}} > -ln\left(\frac{1-0.5}{1+0.5}\right) + \sum_{i=1}^{2} w_{ah_i} - \sum_{j=3}^{3} w_{ah_j} + \theta_a = 7.6$, and similarly, $W_{c\{h_4,h_5\}} > 10.5$, $W_{d\{h_4,h_5\}} > 7.1$, and $W_{b\{h_4,h_5\}} < -9.3$, $W_{\neg d\{h_4,h_5\}} < -9.9$ (see the Minimal Learning Procedure).[15]

Taking $W_{a\{h_4,h_5\}} = 8$, $W_{b\{h_4,h_5\}} = -10$, $W_{c\{h_4,h_5\}} = 11$, $W_{d\{h_4,h_5\}} = 8$, and $W_{\neg d\{h_4,h_5\}} = -10$, we finally add hidden neurons $h_4$ and $h_5$ into $N$, set $W_{h_4\{a,b,d,\neg d\}} = 1$, $W_{h_4 c} = -1$, $W_{h_5\{a,c,d\}} = 1$, $W_{h_5\{b,\neg d\}} = -1$, set $\theta_{h_4} = 3.5$ and $\theta_{h_5} = 2.5$, and connect $h_4$ and $h_5$ to the output layer of $N$ with the weights calculated above.

According to Theorem 7.2.1, the network of Figure 7.5, with newly added neurons $h_4$ and $h_5$, computes the consistent set $\{a, c, d\}$, which is a maximally consistent set according to the Foundational policy.

Finally, note that, as discussed in Remark 3.2.1 (Chapter 3), the network's input and output layers are not required to be identical.

*Remark 7.2.1.* The Dependency Finding Algorithm can also be used to filter the answer given by a trained network when its answer set is inconsistent. We give an example. Let $P = \{n; n \rightarrow q; n \rightarrow r; q \rightarrow p; r \rightarrow \neg p; r \rightarrow ff; p \rightarrow am; ff \rightarrow \neg am\}$, where $ff$ stands for *football-fan*, $am$ stands for *anti-military*, and the remaining literals have the same meaning as in the Nixon

[14] We use $w_{x(y_1,y_2,y_3)} = (z_1, z_2, z_3)$ as a short representation for $w_{xy_1} = z_1$, $w_{xy_2} = z_2$ and $w_{xy_3} = z_3$.

[15] We use $W_{x\{y_1,y_2\}} \gtrless z$ as a short representation for $W_{xy_1} \gtrless z$ and $W_{xy_2} \gtrless z$.

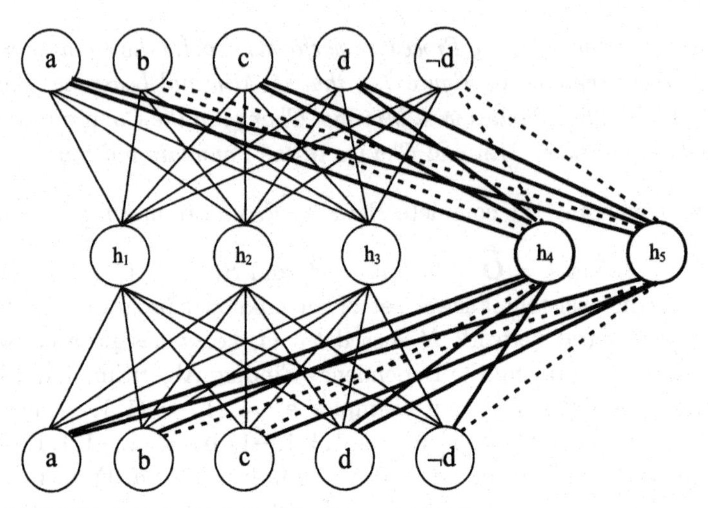


**Fig. 7.5.** Evolving a network to solve inconsistency.

Diamond problem. The answer set of $P$ is $S(P) = \{n, q, r, p, \neg p, ff, am, \neg am\}$. We have two inconsistencies to solve. However, even if we do not yet know how to solve them (e.g. if we do not know how to choose between $p$ and $\neg p$), we still would like to be able to derive some plausible conclusions from $P$.

When a network $N$ encodes an inconsistent program $P$, differently from Gelfond and Lifschitz's answer set semantics, $N$ does not derive $Lit$, the set of literals in $P$, as its answer set. As a result, $N$ may still provide reasonable answers even in the presence of inconsistency. By tracking down the dependencies in $N$, we may inform the user whether a conclusion is plausible or not. We borrow the unary connectives $\Box A$ (box $A$) and $\Diamond A$ (diamond $A$) from Modal Logic to indicate *necessarily* $A$ and *possibly* $A$, respectively.[16]

[16] The intended meaning of $\Box$ and $\Diamond$ may vary. In a *possible world semantics*, necessity ($\Box$) is understood as truth in all possible worlds and possibility ($\Diamond$) as truth in at least one possible world.

A *modal Kripke frame* $\mathcal{F} = \langle W, R \rangle$ consists of a non-empty set of worlds $W$ and an arbitrary binary (*accessibility*) relation $R$ on $W$. If $R(x, y)$ we say that world $y$ is accessible from world $x$. A *Kripke model* of a propositional modal language $\mathcal{L}$ is a pair $\mathcal{M} = \langle \mathcal{F}, v \rangle$ where $\mathcal{F}$ is a frame and $v$ a valuation in $\mathcal{F}$. A truth-relation $(\mathcal{M}, x) \models \varphi$ ($\varphi$ is *true* at world $x$ in model $\mathcal{M}$) is defined inductively by taking:

$(\mathcal{M}, x) \models \Box\psi \Leftrightarrow \forall y \in W \cdot R(x, y), (\mathcal{M}, y) \models \psi$;

$(\mathcal{M}, x) \models \Diamond\psi \Leftrightarrow \exists y \in W \cdot R(x, y), (\mathcal{M}, y) \models \psi$.

In addition, every world is governed by the laws of Classical Logic: an atomic proposition is either *true* or *false* and the remaining non-modal propositions are determined by the usual *truth table* (see [CZ97] and [Che80] if interested in the subject of Modal Logic).

From the Dependency Finding Algorithm, we obtain $D_p = \{am\}, D_{\neg p} = \emptyset$, $D_{am} = \emptyset$ and $D_{\neg am} = \emptyset$. Since $am$ results from $p$ and we do not know whether $p$ should be retracted from $P$, we conclude that $\Diamond am$ ($am$ is *possibly true*). Similarly, since $ff$ does not appear in any of the dependency sets, we conclude $\Box ff$ ($ff$ is *necessarily true*). The (paraconsistent) answer set of $P$ is, therefore, $S_\perp(P) = \{\Box n, \Box q, \Box r, \Diamond p, \Diamond\neg p, \Box ff, \Diamond am, \Diamond\neg am\}$.[17] Note that if we also had, say, $am \to x$ in $P$ then $\Diamond x$ would belong to $S_\perp(P)$ as well.

A method to obtain $S_\perp$ is as follows. Whenever $S$ is inconsistent, find the dependencies of each inconsistent literal $A_i$ in $S$, $D_{A_i} = \{X_1, ..., X_j\}$. Let $S^\Diamond_{A_i} = \{D_{A_i} \cup \{A_i\}\}$ and $S^\Box_{A_i} = S - S^\Diamond_{A_i}$. Then, $S^\Diamond = \bigcup_i S^\Diamond_{A_i}$ and $S^\Box = \bigcup_i S^\Box_{A_i}$. Finally, let $S_\perp = \{\Diamond Y_1, ..., \Diamond Y_m, \Box Z_1, ..., \Box Z_n\}$ where $Y_{k(1 \le k \le m)} \in S^\Diamond$ and $Z_{l(1 \le l \le n)} \in S^\Box$.

## 7.3 Summary of the Chapter

In this chapter, we have investigated how to detect and treat *inconsistencies* in the *C-IL²P* system. We have done so by defining a procedure called *Minimal Learning*, which complies with the concept of *minimal change* either in a Foundational or Compromise Revision framework. In addition, a *Dependency Finding Algorithm* was defined, and it can be used both for establishing the set of examples that should be trained during minimal learning, and for reasoning about inconsistent answer sets.

By showing that *C-IL²P* networks and *Truth Maintenance Systems* of Belief Revision [Doy79] are, in fact, equivalent, we have realised that a *revision* of the background knowledge might occur during learning and, more importantly, that *inconsistencies* between the background knowledge and the training examples may arise. In this case, there are two courses of action. Firstly, even when a network computes an inconsistent answer set, we would like to be able to derive some plausible conclusions from it. Secondly, we would like to be able to solve the inconsistencies in the network, with a minimal loss of information, whenever a new preference relation between conflicting literals is made available. In order to perform the latter, we have used *Minimal Learning* for changing the stable state of the network to a new state with a minimal loss of information. The definition of such a new state, and the

[17] When we have $A$ and $\neg A$ in an answer set $S$, we want to say that $A$ is *true* in a possible world $v \in W$ and that $\neg A$ is *true* in a possible world $u \in W$, but also that $v \neq u$. As a result, using Modal Logic $K$ (see [CZ97]), $S = \{\Diamond A, \Diamond\neg A\}$ should be interpreted as $\Diamond A \wedge \Diamond\neg A \wedge \neg\Diamond(A \wedge \neg A)$. The last term $\neg\Diamond(A \wedge \neg A)$ guarantees that $A$ and $\neg A$ are not *true* in the same possible world. Note, from the definition of $\Diamond$, that $\Diamond A \wedge \Diamond\neg A \nrightarrow \Diamond(A \wedge \neg A)$, even though the converse is true, i.e., $\Diamond(A \wedge \neg A) \to \Diamond A \wedge \Diamond\neg A$.

measure of the loss of information, however, depend on the particular revision policy used. The *Dependency Finding Algorithm* was then applied for finding the new stable state of the network, according to the revision policy adopted: either Foundational Revision [GR94] or Compromise Revision [Gab99]. The same algorithm can also be applied for reasoning about inconsistent answer sets, when no extra information is available to adjudicate a conflict, thus enabling the derivation of plausible conclusions from inconsistent networks.

# 8. Experiments on Handling Inconsistencies

In this chapter, we apply $C\text{-}IL^2P$ in a real problem of software requirements specifications. We use a partial specification of an automobile cruise control system to illustrate all phases of $C\text{-}IL^2P$ (rule insertion, refinement and extraction). Particularly relevant to the area of requirements specifications is the problem of handling inconsistencies. It has been the focus of an increasing amount of work in recent years [dGRNK01, vLDL98, HN98, GN99, SZ01, FGH+94]. As a result, we concentrate on how to treat *inconsistencies* in requirements specifications, by applying the ideas of theory revision put forward in Chapter 7.

## 8.1 Requirements Specifications Evolution as Theory Refinement

This chapter is about how neural-symbolic learning systems can be applied to the process of rationally changing *requirements specifications* of software systems. "The development of most large systems involves many people, each with their own perspectives on the system defined by their knowledge, responsibilities and commitments... [As] the different perspectives intersect, [they give rise to] the possibility of inconsistency [in the specification]" [FGH+94] Nevertheless, "models for reasoning about current alternatives and future plausible changes have received relatively little attention to date, even though such reasoning should be at the heart of the requirements engineering process" [vL00].

### 8.1.1 Analysing Specifications

Following [MW01], in the sequel we see the process of evolving requirements specifications, either with the objective of solving inconsistencies [GN99] or in order to accommodate new information [dGRNK01], as a problem of theory refinement. We concentrate on requirements specifications composed of deterministic, state transition based *system descriptions* and *global system properties*, such as safety properties.

To illustrate the process, we provide a simple example derived from [RMNK00]. Consider an electric circuit consisting of a single light bulb and two switches (SwitchA and SwitchB), all connected in series. Let us assume that a (possibly incorrect) description $D$ of our electric circuit includes the following rules $r_1$ to $r_4$, formalised using logic programming and the "prime" notation often used in formal specifications, where unprimed literals $c$ denote that $c$ is *true* at the *current state*, and primed literals $c'$ denote that $c$ is *true* at the *next state* [AFB+92, HKL96]. For example, rule $r_1$ can be read as "if, in the current state, switch A is not on, the light is not on and A is flicked then, in the next state, the light will be on".

$$\begin{array}{lr}
\sim\text{SwitchA-On} \wedge \sim\text{Light-On} \wedge \text{FlickA} \rightarrow \text{Light-On}' & (r_1) \\
\sim\text{SwitchB-On} \wedge \sim\text{Light-On} \wedge \text{FlickB} \rightarrow \text{Light-On}' & (r_2) \\
\sim\text{SwitchA-On} \wedge \text{FlickA} \rightarrow \text{SwitchA-On}' & (r_3) \\
\sim\text{SwitchB-On} \wedge \text{FlickB} \rightarrow \text{SwitchB-On}' & (r_4)
\end{array}$$

In order to model actions such as FlickA, we need to reason in time, and not in a static world. If we consider a feedforward neural network, for example, we need to add the concept that the input is *true* at an arbitrary time $t$, and the output is *true* at time $t+1$. In the case of rule $r_1$, for instance, if ~SwitchA-On and ~Light-On are *true* and the event FlickA happens at time $t$ then Light-On must be *true* at time $t+1$. Here, as we try to specify the effects of an action in time (or event), we face the well-known *Frame Problem* [PMG98], since it could be impractical to specify the effects of the action on all concepts of the specification. To illustrate, when we apply rule $r_1$, we do not want to have to specify the effect of FlickA on the truth-value of SwitchB-On. Rather, we would like to simply assume that each concept preserves its truth-value, unless it is explicitly changed by the occurrence of some action. We refer to this as the *no change* or *default assumption.* As we will see in the sequel, in a *C-IL*$^2$*P* network, the *no change* assumption will be implemented by connecting each input neuron to its corresponding output neuron with the lowest possible priority. In the light bulb example, this would require the addition of the following rules. (Note that FlickA and FlickB are exogenous events, and therefore not supposed to be represented by *no change* rules.)

$$\begin{array}{lr}
\text{SwitchA-On} \rightarrow \text{SwitchA-On}' & (r_5) \\
\text{SwitchB-On} \rightarrow \text{SwitchB-On}' & (r_6) \\
\text{Light-On} \rightarrow \text{Light-On}' & (r_7)
\end{array}$$

We would like the above description $D$ to satisfy a number of the system's properties, such as "if the light is on then both switches A and B are on":

$$P = \text{Light-On}' \rightarrow \text{SwitchA-On}' \wedge \text{SwitchB-On}'$$

We write $P$ as a number of *integrity constraints* (clauses of the form $\neg P \rightarrow \bot$)[1]. In this example, $\neg P = \neg P_1 \vee \neg P_2$, where:

$$\neg P_1 = \text{Light-On}' \wedge \neg\text{SwitchA-On}'$$
$$\neg P_2 = \text{Light-On}' \wedge \neg\text{SwitchB-On}'$$

The above description $D$ represents the (partial) specification of the system, while property $P$ represents one of the specification's requirements.[2] Ultimately, we would like the specification to satisfy the requirement, i.e. $D \vdash P$. When this is not the case, there will exist an input vector $\mathbf{i}$ such that the network containing $D$ and $P$ will eventually activate $\bot$ in the output [3]. This would prompt the need to revise $D$. Here is where theory refinement is brought to bear in the process of software requirements specifications evolution.

We start by describing what we mean by a *state transition*. We call an input vector $\mathbf{i}$ a *current state*, and an output vector $\mathbf{o}$ a *next state*, where $\mathbf{i}$ contains unprimed literals and $\mathbf{o}$ contains primed literals. We call the pair $\Delta = (\mathbf{i}, \mathbf{o})$ a *state transition*.

Now, continuing with the example, assume that the *Translation Algorithm* of Chapter 3 is applied over $D$. Assume further that both switches A and B are not *on* at the current state, and the light is not *on* when one flicks B. From $r_2$, the network derives that the light must be *on* in the next state, and from $r_4$, the network derives that switch B must be *on* in the next state. From the fact that no rule activates switch A, the network derives by default that switch A must remain not *on* in the next state. Hence we have identified the following *state transition*:

$$\Delta = ((\sim \mathit{SwitchA\text{-}On}, \sim \mathit{SwitchB\text{-}On}, \sim \mathit{Light\text{-}On}, \sim \mathit{FlickA}, \mathit{FlickB}),$$
$$(\neg \mathit{SwitchA\text{-}On}', \mathit{Switch\text{-}On}', \mathit{Light\text{-}On}'))$$

Finally, we do not quite need to add $\neg P \rightarrow \bot$ to the network to realise that state transition $\Delta$ violates $P$, since Light-On′ and ¬SwitchA-On′ is in $\Delta$ and $\neg P_1 = \text{Light-On}' \wedge \neg\text{SwitchA-On}'$.[4] Note that we need to *hypothesise* current states when trying to identify state transitions $\Delta$ that could violate the property, according to the description. An alternative to hypothesising is

[1] Note that $(\neg P \rightarrow \bot) = P$ since, by definition, $\neg P = P \rightarrow \bot$ and $\neg\neg P = P$.

[2] A requirements specification is a description of the *required behaviour* of a system or component.

[3] If an input vector $\mathbf{i}$ causes the network $\mathcal{N}$ to activate $\bot$ then $\exists \mathbf{i}. D \cup \{\mathbf{i}\} \vdash \neg P$. If, in addition, $\mathbf{i}$ is consistent with $D$, i.e. the network containing $D$ but not $P$ does not activate $\bot$, then $\exists \mathbf{i}. D \cup \{\mathbf{i}\} \nvdash P$, which is equivalent to $\neg \forall \mathbf{i}. D \cup \{\mathbf{i}\} \vdash P$, and hence the problem of showing that $D \vdash P$ is reduced to the problem of showing that there does not exist an $\mathbf{i}$ such that $\mathcal{N}$ activates $\bot$.

[4] In this example, we assume that $\sim A \rightarrow \neg A$ for any literal $A$.

to use *abduction* and apply backward reasoning. Abduction is a form of hypothetical reasoning in which assumptions are proposed to explain observations [FK00, KKT94]. It has been used in the context of requirements engineering to analyse specifications for consistency [Men96, RMNK01].

### 8.1.2 Revising Specifications

Evolving incomplete descriptions $D$, according to desirable system properties $P$, can be viewed as a learning process. It involves changing $D$ into a new description $D'$, based on new training examples $\{\mathbf{e}_1, ..., \mathbf{e}_k\}$ (desirable state transitions of system executions) and negative examples $\Delta_i = (\mathbf{i}_i, \mathbf{o}_i)$ (undesirable state transitions of system executions). As a result, we would like to obtain $D' \vdash \mathbf{e}_i$ and $D' \nvdash \Delta_i$. The task of training $D' \vdash \mathbf{e}_i$ will be carried out by applying the algorithms described in Chapter 3. Guaranteeing that $D' \nvdash \Delta_i$, though, will require the use of the ideas put forward in Chapter 7, as we will exemplify in the sequel.

In a process of requirements elicitation, training examples can be seen as an initial set of available *scenarios* [vLW98], which may be complemented (possibly with new contradicting *scenarios*) later on. This would allow the use of *successive refinements* in the process of requirements elicitation. *Successive refinements* are supported in the *C-IL²P* system by *Minimal Learning* (Section 7.1.2), in which a one-shot learning algorithm is used whenever minor adjustments – resulting in a minimal loss of information – become necessary in the process of data collection, analysis and training.

Negative examples $\Delta_i$ tell us that a given *current state* $\mathbf{i}_i$ should not produce a given *next state* $\mathbf{o}_i$ in the new system's description $D'$. If we make sure that $\mathbf{i}_i$ produces a different *next state* $\mathbf{o}_j$ in the refined network then this will be the case. We need, however, to find an admissible (and appropriate) state $\mathbf{o}_j$ such that the new state transition $(\mathbf{i}_i, \mathbf{o}_j)$ is *valid* w.r.t. the system's properties.

**Definition 8.1.1.** *[dGRNK01] (*Valid State Transition*) Let D be a deterministic system description, $\{P_i\}$ a set of system properties, and $\Delta_i = (\mathbf{i}_i, \mathbf{o}_i)$ a negative example. $\Delta'_i = (\mathbf{i}_i, \mathbf{o}_j)$, obtained from $\Delta_i$, is a* valid state transition *iff (i) $\mathbf{o}_j \neq \mathbf{o}_i$ and (ii) $\{P_i\} \cup \{\mathbf{o}_j\} \nvdash \perp$.*[5]

Given $\mathbf{o} = (o_1, ..., o_k)$, there exist $2^k - 1$ possible state transitions to be checked for validity. Following [dGRNK01], we start by flipping the truth-values in $\mathbf{o}$ one at a time, then two at a time and so on, until a *valid state transition* is obtained, in which case we stop. For example, in the electric circuit example above, if we flip the truth-value of SwitchA-On′ from *false* to *true*, obtaining $\Delta'_1 = (\mathbf{i}, (\text{SwitchA-On}', \text{SwitchB-On}', \text{Light\_On}'))$, we might

[5] Eventually, $D$ is to be revised, while $\{P_i\}$ is assumed to be invariant.

derive an inconsistency from $D \cup \{\Delta'_1\} \cup \{P_i\}$, by observing that switch A has changed its position without having been flicked (we assume that domain properties such as ¬SwitchA-On ∧ ¬FlickA → ¬SwitchA-On′ are given in $\{P_i\}$). Similarly, if we try to flip the truth-value of SwitchB-On′ from *true* to *false*, we may derive an inconsistency from the observation that switch B has not changed its position, despite having been flicked. However, flipping the truth-value of Light-On′ from *true* to *false* does not seem to violate any *domain property*, in which case it would have produced the following valid state transition: $\Delta'_2 = (\mathbf{i}, (\neg\text{SwitchA-On}', \text{SwitchB-On}', \neg\text{Light-On}'))$.

We are now in a position to derive a revised description $D'$ from $\Delta'$. Thus far, *C-IL²P's Translation Algorithm* (Section 3.1) has been used to insert rules $r_1 - r_7$ of the (partial) description $D$ into the initial architecture of a neural network $\mathcal{N}$. Recall that it does so by mapping each rule $(r_i)$ from the input layer to the output layer of $\mathcal{N}$, through a hidden neuron $N_i$. For example, rule $r_1$ = ∼SwitchA_On ∧ ∼Light_On ∧ Flick_A → Light_On′ is mapped into $\mathcal{N}$ by simply: (*a*) connecting input neurons representing the concepts SwitchA_On, Light_On and Flick_A to a hidden neuron $N_1$, (*b*) connecting hidden neuron $N_1$ to an output neuron representing the concept Light_On′, and (*c*) setting the weights of these connections in such a way that the output neuron representing the concept Light_On′ is activated (or *true*) if the input neurons representing SwitchA_On, Light_On and Flick_A are, respectively, deactivated (or *false*), deactivated (*false*) and activated (*true*), thus reflecting the information provided by rule $r_1$.

Figure 8.1 shows part of the neural network $\mathcal{N}$ obtained from the above description $D$ of the electric circuit example, containing the translation of rules $r_1, ..., r_4$. It also shows the extension of $\mathcal{N}$, via hidden neuron $N_{\Delta'}$, which is used to accommodate $\Delta'$ by applying the *Minimal Learning Procedure* of Section 7.1.2. The procedure defines the weights of the connections to and from newly added neurons, in this example $N_{\Delta'}$. The resulting network accommodates $\Delta'$ *minimally,* according to Definition 7.1.1. Using the notation introduced in Chapter 7, one could think of $\Delta'$ as the following metarule for minimal learning: ∼SwitchA-On, ∼SwitchB-On, ∼Light-On, FlickB ⇒ SwitchB-On′ [SwitchA-On′, Light-On′], which tells us that the input vector (SwitchA-On, SwitchB-On, Light-On, FlickA, FlickB) $= [-1, -1, -1, -1, 1]$ must activate output neuron SwitchB-On′ and block the activation of output neurons SwitchA-On′ and Light-On′. It should be clear by now that *C-IL²P* networks are being used here to evolve the description $D$ into $D'$, by learning new state transitions. This is precisely the idea of performing *Theory Revision in Neural Networks,* as illustrated in Figure 7.3 (Chapter 7).

Finally, we apply *C-IL²P's Extraction Algorithm* (Section 5.3) to obtain the new knowledge from the network. As expected, in the case of the electrical circuit example, the extraction algorithm derived a new rule $r'_2$ =

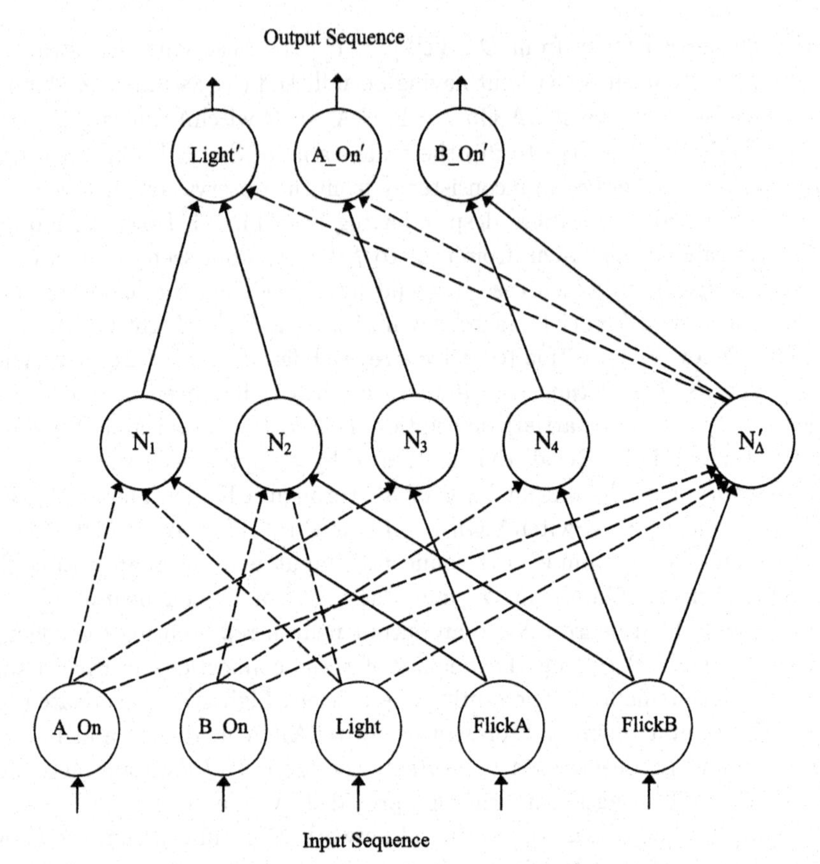


**Fig. 8.1.** Part of the network $\mathcal{N}$ obtained from description $D$ of the electric circuit example, and its extension to accommodate $\Delta'$.

SwitchA_On $\wedge$ ~SwitchB_On $\wedge$ ~Light_On $\wedge$ Flick_B $\rightarrow$ Light_On′, as well as the previous rules $r_1$, $r_3$ and $r_4$. In other words, the learning process has changed rule $r_2$ into rule $r_2'$, without having changed the remaining rules. Clearly, rule $r_2$ was under-specifying the system, and the suggestion of $C$-$IL^2P$ to the requirements engineer, as a result of learning $\Delta'$, was to add to $r_2$ the condition that switch A also needs to be *on* for the light to come *on* once switch B is flicked to *on*.

The revision of $D$ into $D'$ guarantees that negative examples do not hold in the trained network. However, it does not guarantee that $D'$ will not violate $P$. This is why we regard the process of evolving specifications as one of theory refinement, in which the specification is refined successively, until the *system properties* are satisfied.

## 8.2 The Automobile Cruise Control System

We have applied *C-IL²P* to evolve the specification of an automobile cruise control system [Kir87], which is described next. In this case study, the system must be in one of four possible modes at any given time: *off*, *inactive*, *cruise* and *override*. Several variables, such as the position of the ignition switch, the position of the cruise control lever, and the automobile's speed, are considered, and changes in the values of these variables may cause changes in the system's mode. For example, when the system is *inactive*, if the ignition is on, the engine is running, and the brake is off, changing the cruise control lever to the *activate* position is supposed to take the system into *cruise*.

Different specifications of the *Cruise Control System* have been presented in the literature (see [Sha95]), one of which uses the *Software Cost Reduction* (SCR) requirements method [PM95].[6] In the SCR method, state transitions are defined using tables from which rules similar to the ones given in the above *light bulb example* can be obtained. For example, in the *Automobile Cruise Control System*, rules such as $off \wedge ignited \rightarrow inactive'$ are obtained[7].

Consider, for example, rules $cruise \wedge \sim toofast \wedge brake \rightarrow override'$ and $override \wedge \sim brake \wedge activate \rightarrow cruise'$, which are part of a given system *specification* and are depicted in the state transitions of Figure 8.2(a). This figure shows that, among other conditions, the event of pressing the *brake* takes the system from *cruise* to *override,* provided that the automobile is not going *too fast.* Similarly, the event of moving the lever into *activate* takes the system from *override* back to *cruise*, provided that the *brake* is not engaged[8].

A *requirement* of the cruise control system is that the event of pressing the *brake* should *override* the automobile's *cruise* control. This is not completely reflected in the transitions of Figure 8.2(a), since, by applying the *no change* assumption, when the system is in mode *cruise* and the automobile *is* going *too fast*, the system will remain in *cruise* even if the *brake* is applied. In this case, the automobile continues to accelerate. This clearly violates the requirement that pressing the *brake* should *override* the automobile's *cruise*

[6] The SCR method is based on Parnas's *Four Variable Model*, which describes a required system behaviour as a set of mathematical relations between monitored and controlled variables, and input and output data items. It has been used to detect serious errors in the requirements specifications of a range of large-scale real-world systems [AFB+92].

[7] A complete formalisation of the SCR method would require the use of a $happens(x)$ predicate to clearly distinguish actions from conditions. In this case, the rule just given would be written as (1) $off \wedge \neg ignited \wedge happens(ignited) \rightarrow inactive'$ and (2) $\neg ignited \wedge happens(ignited) \rightarrow ignited'$ [HKL96, RMNK01]. While these rules explicitly model the occurrences of actions, we have chosen to concentrate on their effects for the sake of simplicity of this case study.

[8] In this case study, as before, we assume that $\sim A \rightarrow \neg A$ for any literal $A$.

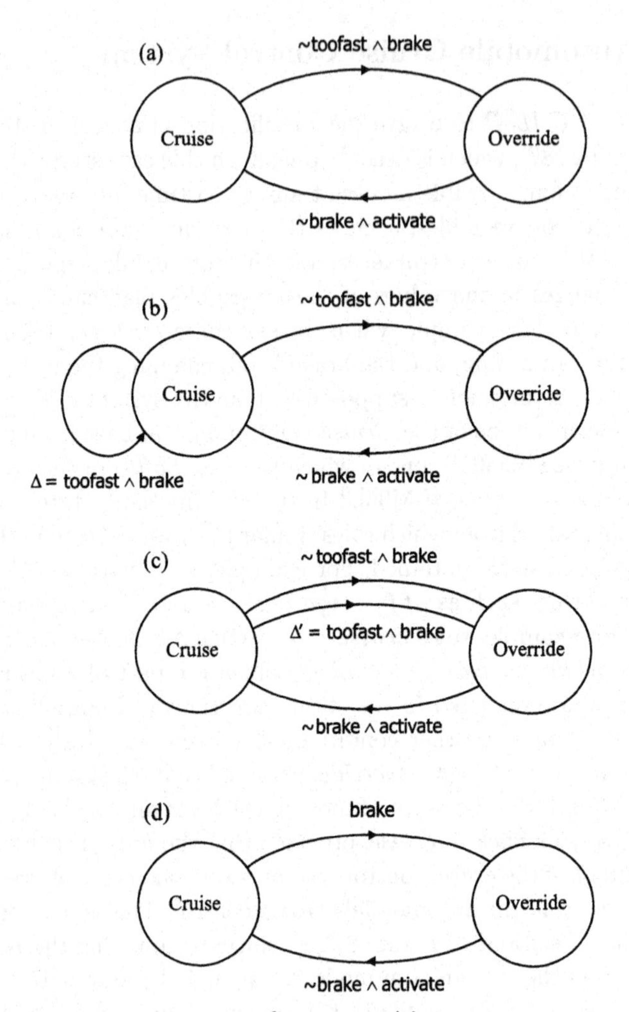


**Fig. 8.2.** Analysis and revision of state transitions.

control system. Such an incorrect transition is depicted in Figure 8.2(b) as the negative example $\Delta$.

Figure 8.2(c) shows a new transition $\Delta'$ as an alternative to transition $\Delta$, in which the system moves from mode *cruise* to mode *override*, when the *brake* is engaged and the automobile is going *too fast*. Transition $\Delta'$ is obtained from $\Delta$ by applying Definition 8.1.1 above. Finally, we need to accommodate $\Delta'$ consistently into the original specification. In this example, this could be achieved just by simplifying the conditions on the two transitions from *cruise* to *override*, as illustrated in Figure 8.2(d) (compare with Figures 8.2(a) and 8.2(c)).

Sections 8.2.1, 8.2.2 and 8.2.3, respectively, will discuss the tasks of rule insertion, revision (including traditional and minimal learning) and rule extraction for the automobile cruise control system. In this case study, safety properties $P_1$ to $P_5$ below will be considered (we use | to represent *exclusive or*)[9].

$$
\begin{array}{ll}
off' \mid inactive' \mid cruise' \mid override' & (P_1) \\
off' \leftrightarrow \neg\, ignited' & (P_2) \\
inactive' \rightarrow ignited' \wedge (\neg running' \vee \neg activated') & (P_3) \\
cruise' \rightarrow ignited' \wedge running' \wedge \neg brake' & (P_4) \\
override' \rightarrow ignited' \wedge running' & (P_5)
\end{array}
$$

Each property $P_i$ is converted into a set of *integrity constraints* (*ic*). For example, property $P_4$ gives rise to three integrity constraints, namely, $ic_{4.1} = \{cruise' \wedge \neg ignited' \rightarrow \bot\}$, $ic_{4.2} = \{cruise' \wedge \neg running' \rightarrow \bot\}$ and $ic_{4.3} = \{cruise' \wedge brake' \rightarrow \bot\}$. Similarly, property $P_5$ produces two integrity constraints: $ic_{5.1} = \{override' \wedge \neg ignited' \rightarrow \bot\}$ and $ic_{5.2} = \{override' \wedge \neg running' \rightarrow \bot\}$. The complete set of integrity constraints derived from $P_1 - P_5$ is given in Table 8.4 (Appendix 8.4).

### 8.2.1 Knowledge Insertion

Assume that the following rules compose our background knowledge[10]:

$$
\begin{array}{lll}
r_1 & : & off \wedge ignited \rightarrow inactive' \\
r_2 & : & inactive \wedge ignited \wedge running \wedge \sim brake \wedge activate \wedge \\
 & & \sim deactivate \wedge \sim resume \rightarrow cruise' \\
r_3 & : & cruise \wedge ignited \wedge running \wedge \sim toofast \wedge brake \rightarrow \\
 & & override'
\end{array}
$$

together with a *no change* (default) assumption for each system mode ($off$, *inactive*, *cruise* and *override*) and variables (*ignited*, *running*, $toofast$, *brake*, *activate*, *deactivate* and *resume*)

[9] The requirement that "pressing the *brake* should *override* the automobile's *cruise* control system" is reflected in safety property $P_4$, where having both *cruise* and *brake* is not allowed.

[10] Note that rule $r_3$ defines completely the state transition from *cruise* to *override* that is only partially illustrated in the diagrams of Figure 8.2.

$r_4 : off \rightarrow off'$ $\quad$ $r_{10} : toofast \rightarrow toofast'$
$r_5 : inactive \rightarrow inactive'$ $\quad$ $r_{11} : brake \rightarrow brake'$
$r_6 : cruise \rightarrow cruise'$ $\quad$ $r_{12} : activate \rightarrow activate'$
$r_7 : override \rightarrow override'$ $\quad$ $r_{13} : deactivate \rightarrow deactivate'$
$r_8 : ignited \rightarrow ignited'$ $\quad$ $r_{14} : resume \rightarrow resume'$
$r_9 : running \rightarrow running'$

and the following preference relations: $r_1 \succ r_4$, $r_1 \succ r_2 \succ r_5$, and $r_2 \succ r_3 \succ r_6$.

The above preference relations are used to guarantee the desired behaviour of the *no change* assumptions. For example, $r_1 \succ r_4$ indicates that the system should remain in mode $off$ (rule $r_4$), unless it moves from $off$ to mode *inactive* due to the application of rule $r_1$. In this case, the application of rule $r_1$ should block the outcome of $r_4$, i.e. $r_1 \succ r_4$.

Let us see how to construct the initial architecture of a neural network $\mathcal{N}$ from the background knowledge $\mathcal{P} = \{r_1, ..., r_{14}\}$ above. Intuitively, the network's input layer is supposed to represent a current state of affairs $S_t$, while its output layer represents the next state of affairs $S_{t+1}$, given $S_t$. It suffices to consider unprimed and primed literals $L$ and $L'$ as distinct, so that the application of the *Translation Algorithm* of Section 3.1 becomes straightforward, as illustrated below.

Firstly, we calculate $MAX_{\mathcal{P}}(\boldsymbol{k}, \boldsymbol{\mu}) = 7$, $A_{\min} > 3/4$ and, taking $A_{\min} = 0.8$, $W > 5.49$. Using $W = 6.0$, we calculate the thresholds $\theta_1, ..., \theta_{14}$ of the hidden neurons $N_1, ..., N_{14}$, which correspond, respectively, to rules $r_1, ..., r_{14}$ of the background knowledge. We obtain: $\theta_1 = 5.4$, $\theta_2 = 32.4$, $\theta_3 = 21.6$, $\theta_4, ..., \theta_{14} = 0.0$. Then, we calculate the thresholds $\theta_{inactive'} = \theta_{cruise'} = \theta_{override'} = -5.4$ of the output neurons corresponding to concepts *inactive*, *cruise* and *override*, respectively. All the remaining output neurons have threshold 0.0. Finally, we need to calculate the necessary negative weights to be added from the hidden to the output layer of $\mathcal{N}$ so that the preference relations $r_1 \succ r_4$, $r_1 \succ r_2 \succ r_5$, and $r_2 \succ r_3 \succ r_6$ are implemented. We do so by applying *Metalevel Priorities Algorithm 1* to implement $r_1 \succ r_4$ and *Metalevel Priorities Algorithm 3* to implement $r_1 \succ r_2 \succ r_5$ and $r_2 \succ r_3 \succ r_6$ (see Section 3.5).

For $r_1 \succ r_4$, since $W_{off,r_4} = 6.0$, taking $\beta = 1$, we calculate $3.4 < \theta_{off} < 86.0$, and taking $\theta_{off} = 5.0$ we obtain $0.66 < \delta < 1.26$ for $n = 1$. We use $\delta = 1.0$, and therefore obtain $W_{off,r_1} = -6.0$. For the remaining preference relations, taking $\varepsilon = 0.01$, from $W_{inactive,r_5} = 6.0$, we get $W_{inactive,r_2} = -5.99$ and $W_{inactive,r_1} = 12$, and from $W_{cruise,r_6} = 6.0$, we get $W_{cruise,r_3} = -5.99$ and $W_{cruise,r_2} = 12$. Again, taking $\beta = 1$, we calculate $-6.195 < \theta_1 < -4.605$ for $j = 3$, and taking $\theta_1 = -5.0$, we change the thresholds of output neurons *inactive* and *cruise* to $\theta_{inactive} = \theta_{cruise} = \theta_1$.

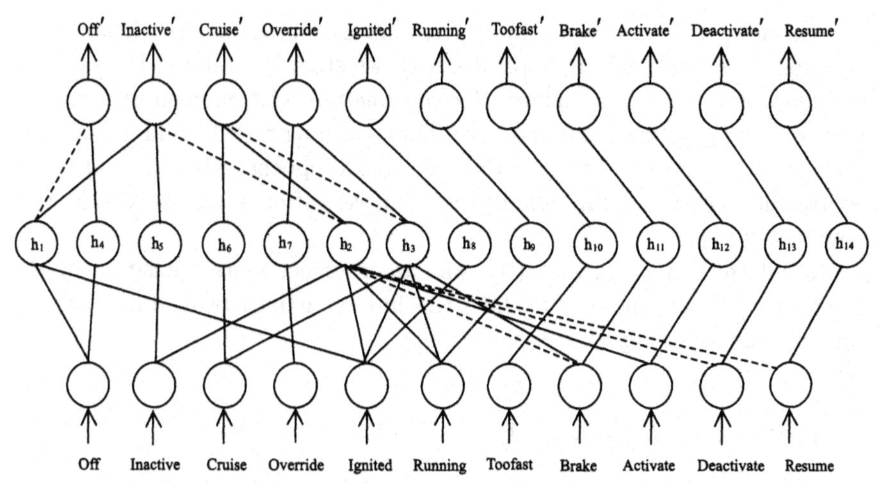


**Fig. 8.3.** The initial neural network for the automobile cruise control system (implementing rules $r_1, ..., r_{14}$ and $r_1 \succ r_4$, $r_1 \succ r_2 \succ r_5$, and $r_2 \succ r_3 \succ r_6$).

Note that, in this case study, when we apply *Metalevel Priorities Algorithm 3*, instead of considering only complementary literals $x$ and $\neg x$, we consider more general integrity constraints, such as *the system should be in a single mode at each time point.* As a result, we cannot allow, for example, output neurons *cruise* and *override* to fire concomitantly, which is identical to having the constraint $cruise \wedge override \rightarrow \perp$ to indicate that an inconsistency should be derived from *cruise* and *override*.

We are now in a position to build the initial neural network for the cruise control system. It is given in Figure 8.3, where hidden neuron $h_i$ represents rule $r_i$.

### 8.2.2 Knowledge Revision: Handling Inconsistencies

The network of Figure 8.3 can be trained with a number of training examples given in the form of input and output patterns. For instance, training example $(i, o)$, where:

$$\begin{aligned} i &= (-1, 1, -1, -1, 1, 1, -1, -1, 1, -1, -1) \\ o &= (-1, -1, 1, -1, 1, 1, -1, -1, 1, -1, -1) \end{aligned}$$

indicates that, from a current mode *inactive*, if the engine is *ignited* and *running*, and the cruise control system is *activated* then the system should move into a new mode, *cruise*. In addition, the *no change* (default) assumption is reflected in this particular training example when, from $i$ (current state), concepts such as *ignited*, *running* and *activated* are supposed to maintain their truth-value at $o$ (next state), unless stated otherwise.

Assuming 10 training examples from the given system specification are available (see Table 8.5 in Appendix 8.4), we start by training the network of Figure 8.3. We use standard *Backpropagation* with momentum, bipolar inputs $\{-1, 1\}$, activation function $tanh(x)$, learning rate 0.1 and momentum constant 0.4. The network is trained for 1,000 epochs, where one epoch is equivalent to one pass through all the examples, after which the $RMS$ error approaches *zero*.

Recall that $\Delta$ (Figure 8.2(b)) was an incorrect state transition of the given specification. In order to check this in the partially trained network, we consider the following current state:

$$i_{\perp} = (-1, -1, 1, -1, 1, 1, 1, 1, -1, -1, -1)$$

which produces the following next state as the network's output:

$$o_{\perp} = (-1, 1, -1, 1, 1, 1, 1, 1, -1, 1, -1)$$

The pair $(i_{\perp}, o_{\perp})$ shows that if one assumes *cruise*, *ignited*, *running*, $toofast$ and *brake* to be *true* (see $i_{\perp}$), one obtains *inactive*, *override*, *ignited*, *running*, $toofast$, *brake* and *deactivated* ($o_{\perp}$) from the trained network, showing that transition $\Delta$ has been eliminated. Since $\Delta'$ was not given explicitly as one of the 10 training examples (see Appendix 8.4), it is clear that the network trained with *Backpropagation* has generalised to avoid the incorrect transition $\Delta$. However, the same set of 10 training examples has produced another incorrect state transition in the network, namely $(i_{\perp}, o_{\perp})$, which clearly violates property $P_1$, via $ic_{1,5}$ : $inactive \wedge override \rightarrow \perp$, since system modes are supposed to be exclusive.

One might want to solve the above inconsistency straight away, without necessarily having to resort to new training examples and a new training phase. At this point, the techniques and algorithms of *Theory Revision* introduced in Chapter 7 are brought to bear. Recall that the answer set computed by a *C-IL²P* network is a subset $S_i$ of the set of literals $S$ represented in the network. Given the initial state $S_1 = \{cruise, ignited, running, toofast, brake\}$, corresponding to $i_{\perp}$, the trained cruise control network converges to the following stable state (or answer set)[11]:

$$S_4 = \{inactive, cruise, override, ignited, running, toofast, brake, resume\}$$

having passed through states $S_2 = \{inactive,\ override,\ ignited,\ running,\ toofast,\ brake,\ deactivated\}$, which corresponds to $o_{\perp}$, and $S_3 = \{inactive,$

[11] Note that the network that implements the cruise control system may contain multiple answer sets.

*override*, *ignited*, *running*, *toofast*, *brake*, *deactivated*, *resume*}, in this order. Recall from Chapter 7 that we write $S_2 = \mathcal{N}(S_1)$, $S_3 = \mathcal{N}(S_2)$, $S_4 = \mathcal{N}(S_3)$, and $S_4 = \mathcal{N}(S_4)$, where $\mathcal{N}$ is the trained network in question.

States $S_2$, $S_3$ and $S_4$ clearly violate $ic_{1,4}$ : *inactive* $\wedge$ *cruise* $\rightarrow \bot$, $ic_{1,6}$ : *cruise* $\wedge$ *override* $\rightarrow \bot$ and $ic_{1,5}$ : *inactive* $\wedge$ *override* $\rightarrow \bot$. In addition, stable state $S_4$ violates $ic_{4,3}$ : *cruise* $\wedge$ *brake* $\rightarrow \bot$. In Chapter 7, only inconsistencies in stable states were considered (standard inconsistencies). In this case study, inconsistencies in non-stable states (strong inconsistencies) need to be considered as well, since safety properties are supposed to hold in any system state. As a result, in what follows, we use the information provided by stable state $S_4$ to find a new (consistent) stable state $S_5$ such that $S_5 = \mathcal{N}(S_1)$.

Considering $S_4$, if we choose to keep *override*, we need to retract both *inactive* and *cruise*. In fact, this happens to be the *desired behaviour* of the system when $S_1$ is the initial state, as discussed in Section 8.2. We are now faced with the situation considered in Section 7.2 in which $X = \{inactive, cruise\}$, $X \subseteq S$, is to be retracted from $\mathcal{N}$. Taking $S = S_4$, we apply the *Dependency Finding Algorithm* (general case) of Section 7.2.3. It computes $S_1^* = \mathcal{N}(S - X) = \{inactive, cruise, override, ignited, running, toofast, brake\}$, $\overline{S}_1 = S - S_1^* = \{resume\}$ and $\overline{\overline{S}}_1 = S_1^* - S = \varnothing$. Then, it computes $S_2^* = \mathcal{N}(S_1^*)$, $\overline{S}_2 = S - S_2^* = \varnothing$, and $\overline{\overline{S}}_2 = S_2^* - S = \varnothing$, i.e. $S_2^* = S$. In this case, the algorithm stops. From Definition 7.2.4, we conclude that $D_X = \bigcup_{j=1}^{i} \overline{S}_j = \{resume\}$ contains the literals that *result* from $X = \{inactive, cruise\}$, while $D_{\sim X} = \bigcup_{j=1}^{i} \overline{\overline{S}}_j = \varnothing$ contains the literals that *result* from $\sim X = \{\sim inactive, \sim cruise\}$. This information will be used to define the pair of training examples that must be applied during *Minimal Learning*, either in a *Compromise* or *Foundational* revision process. Remember that in a *Compromise* approach (Section 7.2.1), we may choose to keep *resume* once it has been derived, even when $X$ is retracted, whereas in the *Foundational* case (Section 7.2.2), *resume* must be retracted together with $X$, since its derivation would otherwise be unfounded.

Using a *Compromise* revision policy, training examples $(i_1, o_1)$ and $(i_2, o_2)$ of Figure 8.4 are obtained. Input $i_1$ represents state $S_1$. Output $o_1$ reflects the necessary retraction of $X$ (the outcomes to be revised), and our *compromise* of keeping *resume*, even though it *results* from $X$. Input $i_2$ and output $o_2$ are equal to $o_1$, and simply reinforce the fact that $o_1$ is supposed to be a stable state of $\mathcal{N}$.

In a *Foundational* revision, exactly the same idea of having two training examples, namely $(i_1, o_1)$ and $(i_2, o_2)$, is used. Again, $i_1$ represents state $S_1$, but $o_1$ must reflect not only the necessary retraction of $X$ (the outcomes to be revised), but also the fact that any outcome in $D_X$ must be *false* since

| | off | inactive | cruise | override | ignited | running | toofast | brake | activate | deactivate | resume |
|---|---|---|---|---|---|---|---|---|---|---|---|
| $i_1$ | –1 | –1 | 1 | –1 | 1 | 1 | 1 | 1 | –1 | –1 | –1 |
| $o_1$ | –1 | –1 | –1 | 1 | 1 | 1 | 1 | 1 | –1 | –1 | 1 |
| $i_2$ | –1 | –1 | –1 | 1 | 1 | 1 | 1 | 1 | –1 | –1 | 1 |
| $o_2$ | –1 | –1 | –1 | 1 | 1 | 1 | 1 | 1 | –1 | –1 | 1 |

Outcomes to be revised

Compromise

**Fig. 8.4.** Cruise control training examples for compromise revision.

| | off | inactive | cruise | override | ignited | running | toofast | brake | activate | deactivate | resume |
|---|---|---|---|---|---|---|---|---|---|---|---|
| $i_1$ | –1 | –1 | 1 | –1 | 1 | 1 | 1 | 1 | –1 | –1 | –1 |
| $o_1$ | –1 | –1 | –1 | 1 | 1 | 1 | 1 | 1 | –1 | –1 | –1 |
| $i_2$ | –1 | –1 | –1 | 1 | 1 | 1 | 1 | 1 | –1 | –1 | –1 |
| $o_2$ | –1 | –1 | –1 | 1 | 1 | 1 | 1 | 1 | –1 | –1 | –1 |

Outcomes to be revised

Unfounded Outcome

**Fig. 8.5.** Cruise control training examples for foundational revision.

such outcomes are unfounded; in this example, *resume*. As before, $i_2 = o_1$ and $o_2 = o_1$, indicating that $o_1$ is to be a stable state of $\mathcal{N}$.

We are finally in a position to apply the *Minimal Learning Procedure* of Section 7.1.2 to $\mathcal{N}$, given examples $\mathbf{e}_o = (i_o, o_o), 1 \leq o \leq 2$, in order to try to solve the inconsistencies identified in the network, with a minimum loss of information. In what follows, we adopt a *Foundational* policy, and therefore use the examples of Figure 8.5.

We start by generating a rule $\mathbf{r}_o$ of the form $L_1, ..., L_n \Rightarrow L_1, ..., L_l$ $[L_{l+1}, ..., L_n]$ from each example $\mathbf{e}_o$ such that, for each $L_i$ $(1 \leq i \leq l)$ in the head of $\mathbf{r}_o$, $o_i = 1$, and, for each $L_j$ $(l+1 \leq j \leq n)$ in the head of $\mathbf{r}_o$, $o_j = -1$. As a result, we obtain rules $\mathbf{r}_1$ and $\mathbf{r}_2$ below, respectively from $(i_1, o_1)$ and $(i_2, o_2)$ shown in Figure 8.5 (we use $o$, $in$, $cr$, $ov$, $ig$, $rn$, $tf$, $br$, $a$, $d$ and $r$ as shorthand for $off$, *inactive*, *cruise*, *override*, *ignited*, *running*, $toofast$, *brake*, *activate*, *deactivate* and *resume*, respectively).

$$\begin{aligned}\mathbf{r}_1 \quad = \quad & \sim o, \sim in, cr, \sim ov, ig, rn, tf, br, \sim a, \sim d, \sim r \Rightarrow \\ & ov, ig, rn, tf, br[o, in, cr, a, d, r]\end{aligned}$$

$$\mathbf{r}_2 = \sim o, \sim in, \sim cr, ov, ig, rn, tf, br, \sim a, \sim d, \sim r \Rightarrow ov, ig, rn, tf, br[o, in, cr, a, d, r]$$

For each output neuron $L_s$, $1 \leq s \leq l$, we calculate $W_{so} > -\frac{1}{\beta} ln\left(\frac{1-A_{\min}}{1+A_{\min}}\right) + \sum_{i=1}^{y} w_{si} - \sum_{j=y+1}^{z} w_{sj} + \theta_s$, where $\sum_{i=1}^{y} w_{si}$ is the sum of all the positive weights to $L_s$, $\sum_{j=y+1}^{z} w_{sj}$ is the sum of all the negative weights to $L_s$, and $\theta_s$ is the threshold of $L_s$ (see Section 7.1.2). We obtain, for $L_s \in \{ov, ig, rn, tf, br\}$, $W_{ov} > 10.9$, $W_{ig} > 14.8$, $W_{rn} > 15.7$, $W_{tf} > 25.5$ and $W_{br} > 13.9$, respectively. Similarly, for each output neuron $L_t$, $l+1 \leq t \leq n$, we calculate $W_{to} < -\frac{1}{\beta} ln\left(\frac{1+A_{\min}}{1-A_{\min}}\right) - \sum_{i=1}^{x} w_{ti} + \sum_{j=x+1}^{z} w_{tj} + \theta_t$, where $\sum_{i=1}^{x} w_{ti}$ is the sum of all the positive weights to $L_t$, $\sum_{j=x+1}^{z} w_{tj}$ is the sum of all the negative weights to $L_t$, and $\theta_t$ is the threshold of $L_t$. We obtain, now for $L_t \in \{o, in, cr, a, d, r\}$, $W_o < -19.5$, $W_{in} < -24.2$, $W_{cr} < -21.3$, $W_a < -16.2$, $W_d < -18.7$, and $W_r < -38.3$, respectively.

Finally, hidden neurons $h_{15}$ and $h_{16}$ with binary activation function and thresholds $\theta_{15} = \theta_{16} = 10.5$, are added to the trained network. Neuron $h_{15}$ is used to implement rule $\mathbf{r}_1$, and neuron $h_{16}$ to implement rule $\mathbf{r}_2$. As a result, considering the antecedent of rule $\mathbf{r}_1$, we connect input neurons *cr*, *ig*, *rn*, *tf* and *br* to $h_{15}$ using weight 1, and input neurons *o*, *in*, *ov*, *a*, *d* and *r* using weight −1. Taking the antecedent of rule $\mathbf{r}_2$, we connect input neurons *ov*, *ig*, *rn*, *tf*, and *br* to $h_{16}$ using weight 1, and input neurons *o*, *in*, *cr*, *a*, *d*, and *r* using weight −1. The result is that $h_{15}$ (respectively $h_{16}$) will fire if and only if the antecedent of $\mathbf{r}_1$ (respectively $\mathbf{r}_2$) is satisfied. Figure 8.6 illustrates the addition of hidden neurons $h_{15}$ and $h_{16}$. As before, dotted lines are used to indicate negative weights. Now, the weight from $h_{15}$ to any output neuron is supposed to be the same as the weight from $h_{16}$ to such an output neuron. This is so because rules $\mathbf{r}_1$ and $\mathbf{r}_2$ will always have the same consequent (a stable state). Thus, based on the calculation of $W_{so}$ and $W_{to}$ above, we have chosen $W_{ov} = 11$, $W_{ig} = 15$, $W_{rn} = 16$, $W_{tf} = 26$, and $W_{br} = 14$, as well as $W_o = -20$, $W_{in} = -25$, $W_{cr} = -22$, $W_a = -17$, $W_d = -19$, and $W_r = -39$. These became the weights from both $h_{15}$ and $h_{16}$ to the output layer of the network.[12]

### 8.2.3 Knowledge Extraction

The network of Figure 8.6, when presented with input state $S_1 = \{$*cruise*, *ignited*, *running*, *toofast*, *brake*$\}$, produces output state $S_5 = \{$*override*,

[12] In order to implement the *compromise approach* to revision, it would suffice in this case to connect hidden neurons $h_{15}$ and $h_{16}$ with a *large positive weight* to output neuron *resume*′.

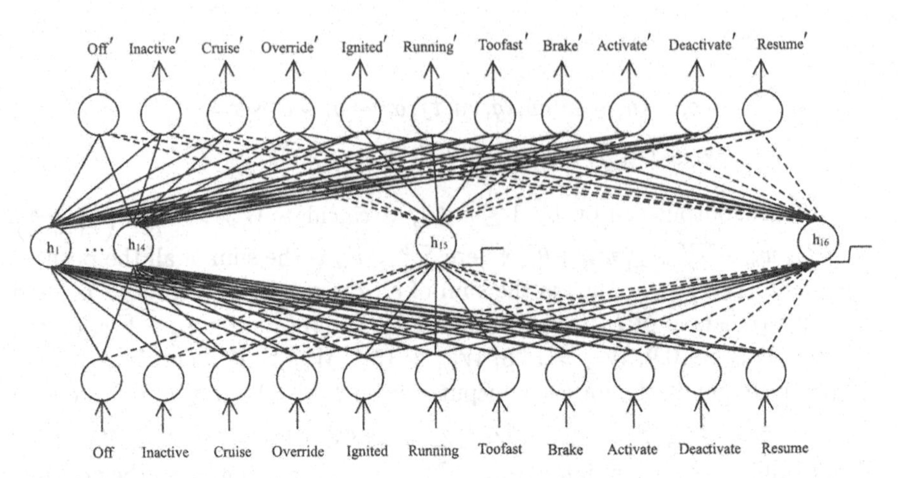


**Fig. 8.6.** The trained neural network for the automobile cruise control system, extended with binary threshold hidden neurons $h_{15}$ and $h_{16}$ to implement minimal learning.

*ignited*, *running*, *toofast*, *brake*}, as opposed to the sequence $(S_2, S_3)$ leading to $S_4 = \{$*inactive*, *cruise*, *override*, *ignited*, *running*, *toofast*, *brake*, *resume*}. This is precisely the expected behaviour of the network after minimal learning. In fact, whenever neuron $h_{15}$ or neuron $h_{16}$ is activated, the behaviour of the network will be to produce stable state $S_5$, according to Theorem 7.2.1, such that the examples given for minimal learning are satisfied. However, unravelling the behaviour of the network as a whole, would require the use of a rule extraction method. Particularly interesting in this case study, would be to verify whether the inconsistency detected before minimal learning had been eliminated.

As we have seen in Section 5.3, decompositional extraction methods may be incomplete. An immediate result of incompleteness is that one may not be able to verify the consistency of the network w.r.t. a safety property $P$ by simply inspecting its extracted set of rules. This would be the case if all the rules that violate $P$ and are encoded in the network were not extracted.

However, in the cruise control network, performing a pedagogical extraction is not prohibitive, since it does not contain a large number of input neurons. Hence, by performing a (sound and complete) pedagogical extraction, we can verify whether minimal learning had solved the inconsistency in the network, for the sake of this case study.

We have performed a pedagogical *Rule Extraction* on the trained cruise control network, both before and after minimal learning had been applied. In what follows, we present the results obtained for the network after minimal learning. For the sake of completeness, we include results obtained for the network before minimal learning in Appendix 8.4.

Table 8.1 below shows, for each output neuron, the number of input vectors queried during the extraction process, the number of rules created, and the number of rules left after simplifications *Complementary Literals* and *Subsumption* were applied (see Section 5.2 for the definitions of the simplification rules). Table 8.6, given in Appendix 8.4, contains the same information for the trained network before minimal learning.

**Table 8.1.** Cruise control system after minimal learning: number of input vectors queried, rules extracted and rules left after simplification (except MofN) for each output neuron

| | *Queried* | *Extracted* | *Simplified* |
|---|---|---|---|
| $o$ | 2048 | 886 | 146 |
| $in$ | 2048 | 1039 | 167 |
| $cr$ | 2048 | 1018 | 225 |
| $ov$ | 2048 | 1053 | 143 |
| $ig$ | 2048 | 1029 | 10 |
| $rn$ | 2048 | 1028 | 10 |
| $tf$ | 2048 | 856 | 106 |
| $br$ | 2048 | 1026 | 10 |
| $a$ | 2048 | 941 | 58 |
| $d$ | 2048 | 691 | 95 |
| $r$ | 2048 | 1427 | 200 |

Below, we present the set of rules derived from the cruise control network for output neurons $off$, *inactive*, *cruise* and *override* (Tables 8.2 and 8.3). Simplification rules *MofN* and *MofN Subsumption* were applied on the *simplified* set of rules of Table 8.1, in order to derive this relatively compact set of rules. As in Chapter 6, we use $m(a_1, ..., a_n)$ to represent $MofN$ rules, and $\overline{o}, \overline{in}, ..., \overline{r}$ to represent $\sim o, \sim in, ..., \sim r$, respectively. The set of rules derived from the network before minimal learning for outputs $off$, *inactive*, *cruise* and *override* is given in Appendix 8.4 (Tables 8.7 and 8.8).

We can use the above set of rules (Tables 8.2 and 8.3) to verify, as guaranteed by *Minimal Learning* (Theorem 7.2.1), that the cruise control network contains the transitions depicted in Figure 8.7, where $S_1$ to $S_5$ are the same as above, and $S_6 = \{in, ov, ig, rn, tf, br\}$. Recall from Chapter 7 that a state $S_i$ is a subset of the set of literals $S$ represented in the network, and that any literal in $S$ that is not in $S_i$ will be assigned truth-value $false$ in this state ($S_i$). For example, state $S_6$ is a short representation for the tuple $(\overline{o}, in, \overline{cr}, ov, ig, rn, tf, br, \overline{a}, \overline{d}, \overline{r})$. Finally, due to the completeness of the extraction used here, one may also verify that a given state does not produce certain outputs (e.g. that state $S_5$ does not produce outputs $o$, $in$ and $cr$).

**Table 8.2.** Rules extracted after Minimal Learning (off and inactive)

| Rules for *off* |
|---|
| $o, \overline{ig}, \overline{tf} \rightarrow o'$ |
| $o, \overline{ig}, \overline{rn}(tf, d \vee \overline{br} \vee \overline{a} \vee \overline{d}) \rightarrow o'$ |
| $o, \overline{ig}, rn(\overline{br} \vee \overline{a} \vee \overline{d} \vee \overline{r}) \rightarrow o'$ |
| $o, rn, \overline{tf}, \overline{br}, \overline{a}, d, \overline{r} \rightarrow o'$ |
| $in, ig, \overline{rn}, \overline{br}, \overline{a}, d, \overline{r} \rightarrow o'$ |
| $in, \overline{rn}, \overline{tf}(\overline{br}(\overline{a} \vee \overline{d}) \vee r(\overline{a} \vee \overline{d})) \rightarrow o'$ |
| $in, \overline{tf}, \overline{br}, \overline{a}, \overline{d}, \overline{r} \rightarrow o'$ |
| $in, \overline{ig}(\overline{rn} \vee \overline{tf} \vee 2(\overline{br}, \overline{a}, \overline{d}) \vee \overline{r}) \rightarrow o'$ |
| $cr, \overline{ig}, \overline{tf}(\overline{rn} \vee \overline{br} \vee \overline{r}) \rightarrow o'$ |
| $cr, \overline{ig}, \overline{rn}, \overline{r} \rightarrow o'$ |
| $ov, \overline{ig}(\overline{rn} \vee \overline{tf} \vee \overline{r}) \rightarrow o'$ |
| $ov, \overline{tf}, \overline{br}(rn, \overline{a}, \overline{d} \vee \overline{rn}, d), \overline{r} \rightarrow o'$ |
| $ov, ig, \overline{rn}, \overline{tf}, \overline{br}, d, \overline{r} \rightarrow o'$ |
| $1(in, ov), \overline{ig}(\overline{br}, 1(\overline{a}, \overline{d}, \overline{r}) \vee 2(\overline{a}, \overline{d}, \overline{r})) \rightarrow o'$ |
| $1(in, cr, ov), \overline{ig}(\overline{tf}, 2(\overline{br}, \overline{a}, \overline{d}) \vee \overline{rn}, \overline{br}, 1(\overline{a}, \overline{d}) \vee \overline{br}, \overline{r}(\overline{a} \vee \overline{d})) \rightarrow o'$ |
| $1(in, cr), \overline{ig}, \overline{rn}, 1(\overline{a}, \overline{d}) \rightarrow o'$ |
| $1(o, in), \overline{ig}, \overline{br}, \overline{rn} \rightarrow o'$ |
| $1(o, in, cr), \overline{ig}, \overline{tf}, 2(\overline{br}, \overline{a}, \overline{d}) \rightarrow o'$ |
| **Rules for *inactive*** |
| $o, ig(\overline{rn} \vee tf \vee rn, \overline{tf}, \overline{br}, \overline{d}, r \vee \overline{tf}, br, \overline{d} \vee \overline{a}, \overline{d}) \rightarrow in'$ |
| $o, \overline{ig}, \overline{rn}, tf(br, \overline{d} \vee \overline{br}, \overline{d}, r) \rightarrow in'$ |
| $in, ig, rn, tf, br, d, \overline{r} \rightarrow in'$ |
| $in, ig, tf(d \vee r) \rightarrow in'$ |
| $in, ig, \overline{rn}(tf \vee \overline{tf}, br, a \vee \overline{tf}, br, \overline{d} \vee \overline{tf}, br, \overline{r} \vee \overline{a}, \overline{r}) \rightarrow in'$ |
| $in, \overline{ig}, \overline{rn}, tf(br, \overline{d}, \overline{r} \vee a, \overline{r}) \rightarrow in'$ |
| $cr, ig, rn, tf(a \vee \overline{a}, d \vee \overline{a}, \overline{d}, r) \rightarrow in'$ |
| $cr, ig, \overline{tf}, br(a \vee \overline{d}), \overline{r} \rightarrow in'$ |
| $cr, tf(ig, \overline{br} \vee \overline{ig}, \overline{rn}, a) \rightarrow in'$ |
| $cr, ig, \overline{rn} \rightarrow in'$ |
| $ov, ig, rn, tf(a, \overline{d} \vee \overline{a}, d, \overline{r} \vee \overline{a}, \overline{d}, r) \rightarrow in'$ |
| $ov, ig, tf(br(a \vee \overline{a}, d) \vee \overline{br}(\overline{a}, \overline{r} \vee \overline{d})) \rightarrow in'$ |
| $ov, ig, \overline{rn} \rightarrow in'$ |
| $ov, \overline{ig}, \overline{rn}, tf, a, \overline{d}, \overline{r} \rightarrow in'$ |
| $1(o, in, cr, ov), ig, \overline{rn}, \overline{a}, \overline{d} \rightarrow in'$ |

**Table 8.3.** Rules extracted after Minimal Learning (cruise and override)

| Rules for *cruise* |
|---|
| $o, ig, rn, \overline{tf}, \overline{br}, a, \overline{d} \rightarrow cr'$ |
| $o, rn, tf, br, r \rightarrow cr'$ |
| $o, ig, rn(\overline{tf}, a, 2(\overline{br}, \overline{d}, \overline{r}) \lor \overline{br}, a, \overline{d}, \overline{r}) \rightarrow cr'$ |
| $o, \overline{ig}, \overline{rn}, \overline{tf}, br, \overline{a}, r \rightarrow cr'$ |
| $o(rn \lor \overline{rn}, tf), \overline{a}, r \rightarrow cr'$ |
| $o, rn, \overline{tf}, \overline{br}, a, \overline{d}, \overline{r} \rightarrow cr'$ |
| $in, ig, rn, \overline{tf}(\overline{br}, \overline{r} \lor a, \overline{r} \lor \overline{d}) \rightarrow cr'$ |
| $in, ig(rn, \overline{br}(a \lor \overline{d}) \lor tf, \overline{br}, \overline{d}, r \lor tf, rn, a, \overline{d} \lor \overline{rn}, \overline{tf}, \overline{br}, a, \overline{d}) \rightarrow cr'$ |
| $in, \overline{ig}, rn(tf, br, r \lor \overline{tf}, 2(\overline{br}, \overline{d}, \overline{r}), a \lor \overline{br}, a, \overline{d}) \rightarrow cr'$ |
| $in(tf, \overline{a}, r \lor rn, \overline{tf}, \overline{br}, a \lor rn, \overline{tf}, br, \overline{a}, r) \rightarrow cr'$ |
| $cr, \overline{ig}(\overline{br}, \overline{a} \lor \overline{rn}, \overline{tf}, \overline{a} \lor \overline{a}, \overline{d}), r \rightarrow cr'$ |
| $cr(\overline{rn}, \overline{tf} \lor \overline{br}), r \rightarrow cr'$ |
| $cr, \overline{rn}(tf \lor \overline{a}, \overline{d} \lor \overline{br}, \overline{a} \lor \overline{ig}, \overline{a}), r \rightarrow cr'$ |
| $cr, \overline{ig}, \overline{tf}(\overline{br}, \overline{a} \lor \overline{a}, \overline{d}), r \rightarrow cr'$ |
| $cr, \overline{tf}, \overline{br}, \overline{a}, \overline{d}, r \rightarrow cr'$ |
| $cr, ig, tf(\overline{br} \lor \overline{a}, \overline{d}), r \rightarrow cr'$ |
| $cr, \overline{ig}(\overline{tf}, \overline{a}, \overline{d} \lor tf \lor rn, \overline{br} \lor rn, \overline{d}), r \rightarrow cr'$ |
| $ov, ig, rn(\overline{tf}, a, \overline{r} \lor tf, br, r \lor \overline{br}, a \lor a, \overline{d}) \rightarrow cr'$ |
| $ov, rn(\overline{br}, a, \overline{d} \lor \overline{tf}, a, \overline{d}, \overline{r} \lor \overline{ig}, tf, r \lor a, \overline{d}) \rightarrow cr'$ |
| $ov, rn, \overline{tf}, \overline{br}(ig, d \lor ig, \overline{d}, \overline{r} \lor \overline{ig}, a, \overline{d} \lor a, \overline{r}) \rightarrow cr'$ |
| $ov, ig, \overline{rn}, \overline{tf}, \overline{br}, a, \overline{d}, \overline{r} \rightarrow cr'$ |
| $ov, tf, br(ig \lor \overline{ig}, \overline{d}), r \rightarrow cr'$ |
| $ov, \overline{rn}, tf, br, r \rightarrow cr'$ |
| $ov, \overline{a}, r \rightarrow cr'$ |
| $1(o, cr), ig, rn, \overline{tf}, \overline{br}, a, \overline{d}, \overline{r} \rightarrow cr'$ |
| $1(o, ov), \overline{ig}, rn, \overline{tf}, \overline{br}, a, \overline{d}, \overline{r} \rightarrow cr'$ |
| Rules for *override* |
| $in, ig, rn, \overline{tf}, br, d, \overline{r} \rightarrow ov'$ |
| $cr, ig, rn(2(\overline{tf}, br, \overline{r}) \lor \overline{tf}, \overline{br}, d \lor br, \overline{a}, d \lor br, a) \rightarrow ov'$ |
| $cr, \overline{ig}, rn, \overline{tf}, br(d \lor \overline{r}) \rightarrow ov'$ |
| $cr, ig, \overline{rn}, \overline{tf}, br(d \lor \overline{r}) \rightarrow ov'$ |
| $ov, ig, tf(br \lor \overline{br}, a) \rightarrow ov'$ |
| $ov, \overline{rn}, tf(br \lor \overline{br}, a) \rightarrow ov'$ |
| $ov, \overline{ig}, tf(rn \lor \overline{rn}, \overline{br}, \overline{a}(d \lor \overline{d}, r)) \rightarrow ov'$ |
| $ov, \overline{ig}, rn, \overline{tf}, br, a \rightarrow ov'$ |
| $ov, rn, tf, \overline{br}, \overline{a}(d \lor \overline{d}, r) \rightarrow ov'$ |

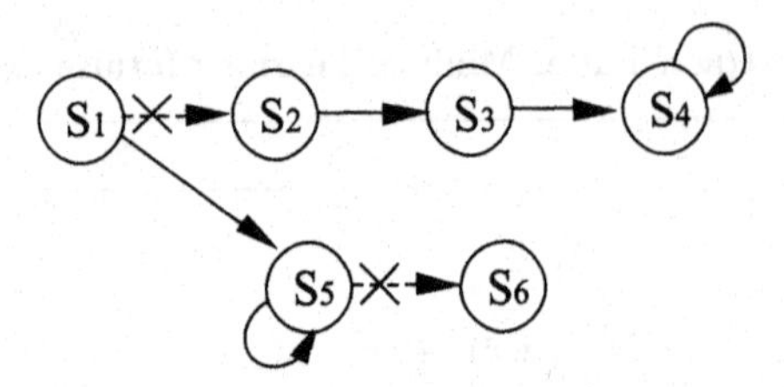


**Fig. 8.7.** Examples of state transitions in the cruise control network after minimal learning. Dotted arrows indicate incorrect transitions before minimal learning.

The trained cruise control network, extended with a minimum loss of information by *minimal learning*, executes the *desired behaviour* of the system when $S_1 = \{cruise, ignited, running, toofast, brake\}$ is given as initial state, by producing stable state $S_5 = \{override, ignited, running, toofast, brake\}$ as outcome, as discussed in Section 8.2.

## 8.3 Discussion

Comparing Figure 8.7 with Figure 7.3, one may verify that, in fact, an extension of the minimal learning procedure presented in Chapter 7 has been used in the above case study. In Figure 7.3, example $(i_1, o_1)$ is used to convert an (inconsistent) stable state into a non-stable one, leading to a new (consistent) state, while example $(i_2, o_2)$ converts such a new state into a stable one. In the above case study, safety properties are supposed to hold at any state, and therefore consistency must be checked at all states, as opposed to only at stable states, as in Chapter 7. Stable states are necessary, nevertheless, to provide information on the new valid stable state, as when $S_4$ was used together with $S_1$ to give the pair of training examples that led to $S_5$, complying with a revision policy. Note that $S_2$, $S_3$, $S_4$ and $S_6$ have become *unreachable states* from initial state $S_1$. However, there may be initial states, other than $S_1$, that lead to one or more of these inconsistent states, in which case *minimal learning* would need to be applied again. In practice, software engineering applications would contain a limited number of valid initial states.

Finally, note that requirements specifications evolve during the process of system development. For example, cost restrictions could have an impact on the system requirements. This would prompt changes either via new training examples or minimal learning. When new information becomes available, *C-IL²P* can be applied again, in a cycle of theory refinement, having the above set of rules as background knowledge. Although it is quite easy to check that certain extracted rules are wrong by inspection, finding out how to perform the necessary changes in a large specification is not straightforward. This is why inductive learning is useful here.

The combination of inductive and analytical learning, via the use of a hybrid machine learning technique such as *C-IL²P*, seems promising for evolving requirements specifications. Requirements evolution normally involve a limited number of training examples. This, however, may be compensated by the use of background knowledge, often available in requirements specifications. Nevertheless, such background knowledge may be incomplete and even incorrect, since we consider partial specifications. As a result, traditional techniques of inductive learning that perform training from examples only, do not seem adequate. This includes *Case Based Reasoning* [Mit97] and most models of *Artificial Neural Networks* [Hay99b]. We are left with (*i*) *Inductive Logic Programming* (*ILP*) techniques, (*ii*) *Hybrid Systems*; (*iii*) *Explanation Based Learning* (*EBL*) algorithms [MT93b], and their hybrids such as *Explanation-based Neural Networks* [Thr96] and combinations of *ILP* and *EBL* [MZ94]. *EBL* is not appropriate in the presence of incorrect domain knowledge [Mit97], as it relies heavily on the correctness of the background knowledge in order to generalise rules from fewer training examples. While the strength of *ILP* lies in the ability to induce relations due to the use of first-order logic, most *ILP* systems would suffer from *scalability* problems when applied to requirements engineering, as a result of the size of the *hypothesis space.* In the presence of incorrect background knowledge, such a problem may be further intensified. As a result, we believe that *Hybrid Systems* (including *Neuro-Fuzzy Systems* [CZ00] and *Evolutionary Systems* [Mit97]) are more appropriate for the task of requirements specifications evolution [dGZ99, TS94a, TBB+91].

Interestingly, the nature of the problem of requirements specification renders the use of theory refinement techniques, both appealing and challenging also from a machine learning perspective. Hence, while requirements specifications can benefit from the use of successful machine learning techniques, new and more efficient learning algorithms need to be developed to fulfil the needs of requirements engineering problems.

## 8.4 Appendix

**Table 8.4.** Complete set of integrity constraints derived from properties P1 to P5 of the Cruise Control Case Study

| $P_1$ |
|---|
| $ic_{1,1} : off' \wedge inactive' \rightarrow \perp$ |
| $ic_{1,2} : off' \wedge cruise' \rightarrow \perp$ |
| $ic_{1,3} : off' \wedge override' \rightarrow \perp$ |
| $ic_{1,4} : inactive' \wedge cruise' \rightarrow \perp$ |
| $ic_{1,5} : inactive' \wedge override' \rightarrow \perp$ |
| $ic_{1,6} : cruise' \wedge override' \rightarrow \perp$ |
| $ic_{1,7} : \neg off' \wedge \neg inactive' \wedge \neg cruise' \wedge \neg override' \rightarrow \perp$ |
| $P_2$ |
| $ic_{2,1} : off' \wedge ignited' \rightarrow \perp$ |
| $ic_{2,2} : inactive' \wedge \neg ignited' \rightarrow \perp$ |
| $ic_{2,3} : cruise' \wedge \neg ignited' \rightarrow \perp$ |
| $ic_{2,4} : override' \wedge \neg ignited' \rightarrow \perp$ |
| $P_3$ |
| $ic_{3,1} : inactive' \wedge \neg ignited' \rightarrow \perp$ |
| $ic_{3,2} : inactive' \wedge running' \wedge activate' \rightarrow \perp$ |
| $P_4$ |
| $ic_{4,1} : cruise' \wedge \neg ignited' \rightarrow \perp$ |
| $ic_{4,2} : cruise' \wedge \neg running' \rightarrow \perp$ |
| $ic_{4,3} : cruise' \wedge brake' \rightarrow \perp$ |
| $P_5$ |
| $ic_{5,1} : override' \wedge \neg ignited' \rightarrow \perp$ |
| $ic_{5,2} : override' \wedge \neg running' \rightarrow \perp$ |

Note: Simplification rules *MofN* and *MofN Subsumption* are applied on the *simplified* set of rules of Table 8.6 in order to derive the rules presented in Tables 8.7 and 8.8.

**Table 8.5.** Training examples $[i_i, o_i]$ $(1 \leq i \leq 10)$ for the cruise control neural network

| | *o* | *in* | *cr* | *ov* | *ig* | *rn* | *tf* | *br* | *a* | *d* | *r* |
|---|---|---|---|---|---|---|---|---|---|---|---|
| $i_1$ | −1 | 1 | −1 | −1 | −1 | −1 | −1 | −1 | −1 | −1 | −1 |
| $o_1$ | 1 | −1 | −1 | −1 | −1 | −1 | −1 | −1 | −1 | −1 | −1 |
| $i_2$ | −1 | 1 | −1 | −1 | 1 | 1 | −1 | −1 | 1 | −1 | −1 |
| $o_2$ | −1 | −1 | 1 | −1 | 1 | 1 | −1 | −1 | 1 | −1 | −1 |
| $i_3$ | −1 | −1 | 1 | −1 | −1 | −1 | −1 | −1 | −1 | −1 | −1 |
| $o_3$ | 1 | −1 | −1 | −1 | −1 | −1 | −1 | −1 | −1 | −1 | −1 |
| $i_4$ | −1 | −1 | 1 | −1 | 1 | −1 | −1 | −1 | −1 | −1 | −1 |
| $o_4$ | −1 | 1 | −1 | −1 | 1 | −1 | −1 | −1 | −1 | −1 | −1 |
| $i_5$ | −1 | −1 | 1 | −1 | 1 | −1 | 1 | −1 | −1 | −1 | −1 |
| $o_5$ | −1 | 1 | −1 | −1 | 1 | −1 | 1 | −1 | −1 | −1 | −1 |
| $i_6$ | −1 | −1 | 1 | −1 | 1 | 1 | −1 | −1 | −1 | 1 | −1 |
| $o_6$ | −1 | −1 | −1 | 1 | 1 | 1 | −1 | −1 | −1 | 1 | −1 |
| $i_7$ | −1 | −1 | −1 | 1 | −1 | −1 | −1 | −1 | −1 | −1 | −1 |
| $o_7$ | 1 | −1 | −1 | −1 | −1 | −1 | −1 | −1 | −1 | −1 | −1 |
| $i_8$ | −1 | −1 | −1 | 1 | 1 | −1 | −1 | −1 | −1 | −1 | −1 |
| $o_8$ | −1 | 1 | −1 | −1 | 1 | −1 | −1 | −1 | −1 | −1 | −1 |
| $i_9$ | −1 | −1 | −1 | 1 | 1 | 1 | −1 | −1 | 1 | −1 | −1 |
| $o_9$ | −1 | −1 | 1 | −1 | 1 | 1 | −1 | −1 | 1 | −1 | −1 |
| $i_{10}$ | −1 | −1 | −1 | 1 | 1 | 1 | −1 | −1 | −1 | −1 | 1 |
| $o_{10}$ | −1 | −1 | 1 | −1 | 1 | 1 | −1 | −1 | −1 | −1 | 1 |

**Table 8.6.** Cruise control system before minimal learning: number of input vectors queried, rules extracted and rules left after simplification (except MofN) for each output neuron

| | *Queried* | *Extracted* | *Simplified* |
|---|---|---|---|
| *o* | 2048 | 886 | 146 |
| *in* | 2048 | 1041 | 154 |
| *cr* | 2048 | 1018 | 225 |
| *ov* | 2048 | 1053 | 143 |
| *ig* | 2048 | 1029 | 10 |
| *rn* | 2048 | 1028 | 10 |
| *tf* | 2048 | 856 | 106 |
| *br* | 2048 | 1026 | 10 |
| *a* | 2048 | 941 | 58 |
| *d* | 2048 | 692 | 91 |
| *r* | 2048 | 1427 | 200 |

**Table 8.7.** Rules extracted before Minimal Learning (off and inactive)

| Rules for $off$ |
|---|
| $o, \overline{ig}(\overline{tf} \vee rn, 1(\overline{br}, \overline{a}, \overline{d}) \vee rn, \overline{r})) \rightarrow o'$ |
| $o, \overline{ig}, \overline{rn}(tf, d \vee \overline{br}, a, d) \rightarrow o'$ |
| $o, rn, \overline{tf}, \overline{br}, \overline{a}, d, \overline{r} \rightarrow o'$ |
| $in, \overline{ig}(\overline{rn} \vee \overline{tf} \vee \overline{r} \vee 2(\overline{br}, \overline{a}, \overline{d})) \rightarrow o'$ |
| $in, \overline{rn}, \overline{tf}, \overline{br}, 1(\overline{a}, \overline{d}, \overline{r}) \rightarrow o'$ |
| $in, \overline{rn}, \overline{tf}(\overline{a}, \overline{r} \vee \overline{d}, \overline{r}) \rightarrow o'$ |
| $in, \overline{tf}, \overline{br}, \overline{a}, \overline{d}, \overline{r} \rightarrow o'$ |
| $in, ig, \overline{rn}, \overline{br}(\overline{tf} \vee \overline{a}), d, \overline{r} \rightarrow o'$ |
| $in, \overline{ig}(\overline{br}, 1(\overline{a}, \overline{d}, \overline{r}) \vee 2(\overline{a}, \overline{d}, \overline{r})) \rightarrow o'$ |
| $cr, \overline{ig}, \overline{tf} \rightarrow o'$ |
| $cr, \overline{ig}, \overline{rn}(\overline{a} \vee \overline{d}) \rightarrow o'$ |
| $ov, \overline{ig}(\overline{rn}, \overline{br} \vee \overline{tf} \vee \overline{r}) \rightarrow o'$ |
| $ov, \overline{tf}, \overline{br}, \overline{a}, \overline{r}(rn, \overline{d} \vee \overline{rn}, d) \rightarrow o'$ |
| $ov, ig, \overline{rn}, \overline{tf}, \overline{br}, d, \overline{r} \rightarrow o'$ |
| $ov, \overline{ig}(\overline{br}, 1(\overline{a}, \overline{d}, \overline{r}) \vee 2(\overline{a}, \overline{d}, \overline{r})) \rightarrow o'$ |
| $1(o, in), \overline{ig}, \overline{br}, \overline{r} \rightarrow o'$ |
| $1(in, cr, ov), \overline{ig}, \overline{tf}(\overline{rn} \vee \overline{br}, \overline{a} \vee \overline{br}, \overline{d} \vee \overline{a}, \overline{d} \vee \overline{r}) \rightarrow o'$ |
| $1(in, cr, ov), \overline{ig}(\overline{rn}(\overline{br}, \overline{a} \vee \overline{br}, \overline{d} \vee \overline{r}) \vee \overline{br}(\overline{a}, \overline{r} \vee \overline{d}\overline{r}) \vee \overline{a}, \overline{d}, \overline{r}) \rightarrow o'$ |
| Rules for *inactive* |
| $o, ig(tf \vee \overline{rn} \vee \overline{tf}, 1(\overline{a}, \overline{d}) \vee \overline{tf}, br, \overline{d} \vee \overline{tf}, rn, \overline{br}, \overline{d}, r) \rightarrow in'$ |
| $o, \overline{ig}, \overline{rn}, tf(br, \overline{d} \vee br, \overline{d}, \overline{r} \vee \overline{br}, \overline{d}, r) \rightarrow in'$ |
| $in, ig, \overline{rn}(\overline{tf}, br, 1(a, \overline{d}, \overline{r}) \vee tf \vee \overline{a}, \overline{r} \vee \overline{tf}, \overline{br}, \overline{a}, d, \overline{r}) \rightarrow in'$ |
| $in, ig, tf(br, \overline{a}, 1(\overline{d}, \overline{r}) \vee \overline{br}, \overline{a}, \overline{d}, r \vee \overline{br}, \overline{a}, d, \overline{r}) \rightarrow in'$ |
| $in, \overline{ig}, \overline{rn}, tf(br, \overline{d}, \overline{r} \vee a, \overline{r}) \rightarrow in'$ |
| $in, ig, rn, tf, br, \overline{a}, d, \overline{r} \rightarrow in'$ |
| $cr, ig(tf \vee \overline{rn} \vee \overline{a}, \overline{d}, \overline{r}) \rightarrow in'$ |
| $cr(ig, \overline{tf}, br \vee \overline{ig}, \overline{rn}, tf), \overline{d}, \overline{r} \rightarrow in'$ |
| $ov, ig(\overline{rn} \vee tf, br \vee tf, \overline{a}, \overline{d}) \rightarrow in'$ |
| $ov, \overline{ig}, \overline{rn}, tf, a, \overline{d}, \overline{r} \rightarrow in'$ |
| $1(o, cr, ov)(ig, tf, \overline{a}, \overline{r}) \rightarrow in'$ |
| $1(o, cr, ov)(ig, \overline{rn}, \overline{d}) \rightarrow in'$ |
| $1(o, in, cr, ov)(ig, \overline{rn}, \overline{a}, \overline{d}) \rightarrow in'$ |

**Table 8.8.** Rules extracted before Minimal Learning (cruise and override)

| Rules for *cruise* |
|---|
| $o, ig, rn, 3(\overline{tf}, \overline{br}, \overline{d}, \overline{r}), a \rightarrow cr'$ |
| $o, rn, \overline{tf}, \overline{br}, a, \overline{d}, \overline{r} \rightarrow cr'$ |
| $o, \overline{ig}, \overline{rn}, \overline{tf}, br, \overline{a}, r \rightarrow cr'$ |
| $o, rn(tf, br, r \vee \overline{a}, r) \rightarrow cr'$ |
| $o, \overline{ig}, rn, \overline{tf}, \overline{br}, a, \overline{d}, \overline{r} \rightarrow cr'$ |
| $in, ig, rn, \overline{tf}(\overline{br}, \overline{r} \vee \overline{d}) \rightarrow cr'$ |
| $in, ig, rn, \overline{br}(a \vee \overline{d}) \rightarrow cr'$ |
| $in, ig, rn, a, \overline{d} \rightarrow cr'$ |
| $in, ig, \overline{rn}, \overline{tf}, \overline{br}, a, \overline{d} \rightarrow cr'$ |
| $in, ig, tf, 2(\overline{br}, \overline{a}, \overline{d}) \rightarrow cr'$ |
| $in, tf, \overline{a}, r \rightarrow cr'$ |
| $in, rn, br(\overline{ig}, tf \vee \overline{tf}, \overline{a}), r \rightarrow cr'$ |
| $in, \overline{ig}, rn, \overline{tf}(\overline{br}, a, \overline{r} \vee a, \overline{d}, \overline{r}) \rightarrow cr'$ |
| $in, \overline{ig}, rn, \overline{br}, a, \overline{d} \rightarrow cr'$ |
| $cr, ig, tf(\overline{br}, r \vee \overline{a}, \overline{d}, r) \rightarrow cr'$ |
| $cr, \overline{ig}(rn(\overline{br}, r \vee \overline{d}, r) \vee tf, r \vee \overline{tf}, \overline{a}, \overline{d}, r) \rightarrow cr'$ |
| $cr, \overline{rn}(tf \vee \overline{ig}, \overline{a} \vee 2(\overline{br}, \overline{a}, \overline{d})) \rightarrow cr'$ |
| $cr, \overline{a}, r, \overline{tf}(\overline{ig}, \overline{rn} \vee \overline{ig}, \overline{br} \vee \overline{rn}, \overline{d} \vee \overline{br}, \overline{d} \vee \overline{ig}, \overline{d}) \rightarrow cr'$ |
| $cr, \overline{a}, r(\overline{br}(\overline{ig} \vee \overline{rn} \vee \overline{d}) \vee \overline{ig}, \overline{d}) \rightarrow cr'$ |
| $ov, ig, rn(tf, br, r \vee \overline{br}, a \vee a, \overline{d} \vee \overline{tf}, \overline{br}, \overline{d} \vee \overline{tf}, a, \overline{r}) \rightarrow cr'$ |
| $ov, ig, \overline{rn}, \overline{tf}, \overline{br}, a, \overline{d}, \overline{r} \rightarrow cr'$ |
| $ov, rn, \overline{tf}(\overline{ig}, \overline{br}, \overline{d} \vee \overline{br}, \overline{r} \vee \overline{d}, \overline{r}), a \rightarrow cr'$ |
| $ov, \overline{ig}, tf, br, \overline{d}, r \rightarrow cr'$ |
| $ov, tf(br(ig \vee \overline{rn}) \vee \overline{ig}, rn), r \rightarrow cr'$ |
| **Rules for *override*** |
| $in, ig, rn, \overline{tf}, br, d, \overline{r} \rightarrow ov'$ |
| $cr, ig, \overline{rn}, \overline{tf}, br(d \vee \overline{r}) \rightarrow ov'$ |
| $cr, ig, rn, \overline{tf}(br \vee \overline{br}, d \vee \overline{r}) \rightarrow ov'$ |
| $cr, ig, rn, br(a \vee \overline{a}, d \vee \overline{r}) \rightarrow ov'$ |
| $cr, \overline{ig}, rn, \overline{tf}, br(d \vee \overline{r}) \rightarrow ov'$ |
| $ov, \overline{ig}, \overline{rn}, tf, \overline{br}(\overline{a}, d \vee \overline{a}, \overline{d}, r) \rightarrow ov'$ |
| $ov, ig, tf, \overline{br}, a \rightarrow ov'$ |
| $ov, \overline{rn}, tf, \overline{br}, a \rightarrow ov'$ |
| $ov, rn, tf, \overline{br}(\overline{a}, d \vee \overline{a}, \overline{d}, r) \rightarrow ov'$ |
| $ov(\overline{rn}, tf, br \vee \overline{ig}, rn, tf \vee ig, tf, br) \rightarrow ov'$ |

# 9. Neural-Symbolic Integration: The Road Ahead

In this book, we have seen that the integration of Connectionist Logic Systems (CLS) and Hybrid Systems by Translation (HST) may provide effective Neural-Symbolic Learning Systems. *C-IL$^2$P* is an example of such a system. In order to enable effective learning from examples and background knowledge, the main insight was to keep the network structure as simple as possible, and try to find the best symbolic representation for it. We have done so by presenting a Translation Algorithm from logic programs to single hidden layer neural networks. It was essential, though, to show the equivalence between the symbolic representation and the neural network, in order to ensure a sound translation of the background knowledge into a connectionist representation. Such a theorem also rendered *C-IL$^2$P* as a massively parallel model of symbolic computation, as in CLSs, with two corollaries showing that the neural network computes, respectively, the stable model semantics of the general logic program given as background knowledge, and the answer set semantics of the extended logic program given as background knowledge. An extension of the system to accommodate superiority relations between rules, finally provided the symbolic representation that best fits into single hidden layer networks; that of prioritised extended logic programming.

After empirically testing *C-IL$^2$P* in two well-known real-world problems of molecular biology, and in a safety-critical application domain on power systems fault diagnosis, we proceeded to investigate the problem of the converse translation; that is, the problem of extracting the symbolic representation encoded in trained neural networks. The motivation for investigating this problem was the intention to remedy one of the strongest criticisms of neural networks: their inability to provide the explanation for the answers computed. The converse translation would also allow the creation of a learning cycle, as occurs in most HSTs, composed of the insertion, refinement and extraction of (incomplete) domain theories in a neural environment. Efforts were concentrated on ensuring that the extraction was sound, as done for the insertion of knowledge. In doing so, the observation that neural networks might have a nonmonotonic behaviour and, thus, that the extraction of knowledge should be able to reflect this aspect, was fundamental for the definition of the extraction method and the subsequent proof of soundness. An empirical analysis indicated, then, that very high fidelity between the extracted set of rules and the network can be achieved even for very large networks, and that a poorer readability is the price that one has to pay in exchange for soundness. Nevertheless, the use of the full power of *M of N* rules as simplifications, taking advantage of the relation between *M of N* rules and the valid subsets of the partial ordering on the network's input space, seems to be promising to tackle the problem of readability, without the loss of soundness.

Finally, by comparing the theory inserted in the network with the theory extracted from the network in the power systems application, we suggested that the process of refining the network could be seen as a process of theory revision. This empirical evidence, together with the observation that $C\text{-}IL^2P$ networks and Truth Maintenance Systems are, in fact, equivalent, led to the idea of using inductive learning to solve inconsistencies in the answer sets of trained networks. We needed, though, an appropriate learning algorithm, which should not allow the loss of information represented in the network. We defined a training algorithm, which we called Minimal Learning, used to evolve neural networks. Such an algorithm switches the answer set of a network, precisely as prescribed by a given set of examples, with a minimal loss of information. Of course, it can also be used to refine the network, regardless of the presence of inconsistencies, whenever the presence or absence of a concept in the network's answer set is desirable. Either when a revision is necessary or when a refinement is desirable, the key problem is how to define the set of examples that would allow minimal learning to take place accordingly. In the case of revision, this will depend on the revision policy used. For the foundational and compromise policies, we have presented a Dependency Finding Algorithm, which automatically gives the set of examples for revision by minimal learning. Interestingly, due to the paraconsistency of $C\text{-}IL^2P$ networks, the same algorithm also allows one to reason about inconsistent answer sets, in order to derive some plausible conclusions from it when no further information is available to solve the inconsistency. The Foundational Policy for Theory Revision has been applied in a case study to evolve a requirements specification of an automobile cruise control system.

Both paradigms of computational intelligence, symbolic and connectionist, have virtues and deficiencies. Research into the integration of the two has important implications, in that one might be able to benefit from the advantages that each confers. Concomitantly, the limits of effective integration should also be pursued, in that it might become disadvantageous in comparison with purely symbolic or purely connectionist systems. We believe that this book contributes to the synergetic integration of connectionism and symbolism.

There are many avenues of research yet to be explored in the relatively new area of neural-symbolic integration. In what follows, we present some of them, which we believe are interesting pursuing in an attempt to find the limits of effective neural-symbolic integration. We concentrate on the subjects of knowledge extraction and on the following possible extensions of the *C-IL²P* framework: the addition of disjunctive information, the representation of modalities, and *C-ILP*: *Connectionist Inductive Logic Programming*, the extension to the first-order case. We also consider more general notions of acceptable logic programs for neural networks computation.

## 9.1 Knowledge Extraction

There are some interesting open questions relating to the *explanation capability* of neural networks, specifically to the trade-off between the *complexity* and *quality* of rule extraction methods. One way to reduce this trade-off would be to investigate more efficient heuristics to prune the network's search space. To this end, the partial ordering on the network's input vectors could be valuable. For example, the extension of the extraction system to perform a stochastic search, as opposed to a deterministic search, in the lattice of input vectors seems promising. Stochastic searches have outperformed deterministic ones in a variety of logic and AI tasks, starting with the work of Selman, Levesque and Mitchell on satisfiability [SLM92]. Consequently, we believe that a stochastic search of the frontier of activations in the lattice of input vectors could improve the experimental results obtained with the current (deterministic) implementation of the extraction algorithm. Another way to reduce the *quality* $\times$ *complexity* trade-off would be to investigate more efficient ways of performing the simplification of extracted rules. In particular, due to its compactness, $MofN$ rules are of great interest here, and again the partial ordering on the input vectors can be very handy. Take, for example, a regular network $N$ with three input neurons $\{A, B, C\}$ and one output neuron $X$. The partial ordering $\preceq_{\mathbf{I}}$ on the set of input vectors $\mathbf{I}$ of $N$ is shown in Figure 9.1. Let the maximum element $(1,1,1)$ represent $[A, B, C]$, in this order. We say that a subset $\mathbf{U}$ of $\mathbf{I}$ is *valid* iff for any input vectors $\mathbf{i}_i \in \mathbf{U}$, $\mathbf{i}_j \in \mathbf{I}$, if $\mathbf{i}_i \preceq_{\mathbf{I}} \mathbf{i}_j$ then $\mathbf{i}_j \in \mathbf{U}$. For example, in the ordering of Figure 9.1, $\mathbf{U_1} = \{4,5,6,7,8\}$ is a valid subset of $\mathbf{I}$, while $\mathbf{U_2} = \{5,7,8\}$ is not because $5 \preceq_{\mathbf{I}} 6$ and $6 \notin \mathbf{U_2}$. The interest in this definition is that any $MofN$ simplification corresponds to a valid subset of $\mathbf{I}$. In the case of $3-ary$ input vectors, there are 9 valid subsets, namely, $\{8\}$, $\{7,8\}$, $\{6,7,8\}$, $\{4,5,6,7,8\}$, $\{5,6,7,8\}$, $\{4,6,7,8\}$, $\{3,4,5,6,7,8\}$, $\{2,3,4,5,6,7,8\}$, and $\mathbf{I}$ itself. Given $(1,1,1) = [A, B, C]$, these subsets correspond, respectively, to the following sets of rules: $\{(A \wedge B \wedge C) \rightarrow X\}$, $\{(A \wedge B) \rightarrow X\}$, $\{(A \wedge 1(BC)) \rightarrow X\}$, $\{(A \vee (B \wedge C)) \rightarrow X\}$, $\{2(ABC) \rightarrow X\}$, $\{A \rightarrow X\}$, $\{1(AB) \rightarrow X\}$, $\{1(ABC) \rightarrow X\}$, and $\{\rightarrow X\}$. Hence, in this case, $2(ABC)$ is associated with $\{5,6,7,8\}$, $1(AB)$ is associated with $\{3,4,5,6,7,8\}$, and so on. As a

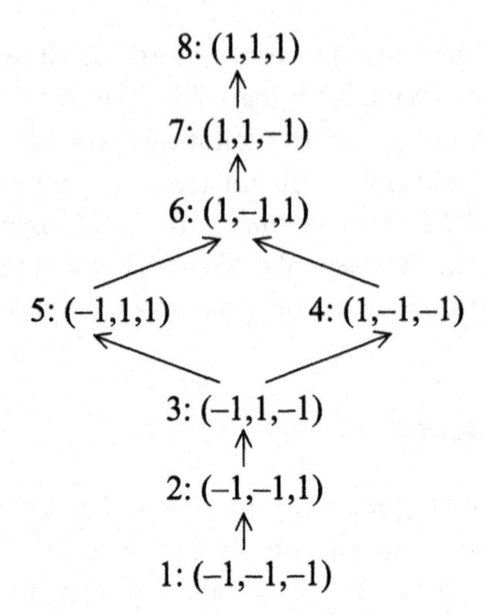


**Fig. 9.1.** The relation between *M of N* rules and $\preceq_{\mathbf{I}}$.

result, if we could formally express the relation between $MofN$ rules and valid subsets of **I**, we would be one step ahead in the simplification of rules, which could eventually lead to the extraction of fully simplified rules directly from the ordering on **I**.

In addition, by manipulating $MofN$ rules, the readability of the extracted set of rules can be enhanced further. Some examples include:

1. $1(ab) \vee 1(cd) = 1(abcd)$
2. $(a \wedge 1(cd)) \vee (b \wedge 1(cd)) = 1(ab) \wedge 1(cd)$
3. $2(\sim a \sim b \sim c \sim d) = \sim 3(abcd)$
4. $1(abc) \vee 2(abc) = 1(abc)$
5. $1(abc) \wedge 2(abc) = 2(abc)$

To this end, one could develop an algebra of $MofN$ rules. For instance, in what follows, let $S_n$ denote the set of literals $\{L_1, L_2, ..., L_n\}$, where $n \in \aleph$. Let $x \in \aleph$. We use $\binom{x}{S_n}$ to compactly represent $x(L_1, L_2, ..., L_n)$, and $\overline{S_n}$ to denote the set of negated propositional symbols in $S_n$, i.e. $\{\neg L_1, \neg L_2, ..., \neg L_n\}$.

It is clear that the usual distributivity laws involving negation hold, namely:

$$\neg\left(\binom{x}{S_n} \wedge \binom{y}{S_m}\right) = \neg\binom{x}{S_n} \vee \neg\binom{y}{S_m} \tag{9.1}$$

and

$$\neg\left(\binom{x}{S_n} \vee \binom{y}{S_m}\right) = \neg\binom{x}{S_n} \wedge \neg\binom{y}{S_m} \tag{9.2}$$

In addition, associativity, commutativity and distributivity of conjunction and disjunction extend to $MofN$ rules. Some of the properties found to hold for $MofN$ expressions so far include:

$$\binom{x}{S_n} \wedge \binom{y}{S_m} =$$

$$\left(a \wedge \binom{x-1}{S_n - a} \wedge \binom{y-1}{S_m - a}\right) \vee \left(\binom{x}{S_n - a} \wedge \binom{y}{S_m - a}\right)$$

And-Extract

$$\binom{x}{S_n} \vee \binom{y}{S_m} =$$

$$\left(a \wedge \left(\binom{x-1}{S_n - a} \vee \binom{y-1}{S_m - a}\right)\right) \vee \left(\binom{x}{S_n - a} \wedge \binom{y}{S_m - a}\right)$$

Or-Extract

$$\neg \binom{x}{S_n} = \binom{(n+1) - x}{\overline{S_n}}$$

Negation

$$\binom{0}{S_n} = True$$

True

$$\binom{n}{S_n} = S_n$$

Simple

$$\binom{n+1}{S_n} = False$$

False

$$\binom{x}{S_n} \vee \binom{x}{S_m} = \binom{x}{S_n \cup S_m}$$

Single

In addition to $MofN$ rules, further improvement of *readability* could be achieved by the extraction of metalevel priorities from the network. At the present stage, any metalevel priority relation is extracted into the object level as default negation. Either by post-processing the set of extracted rules, or somehow by directly extracting rules with a preference relation, a more readable set of rules could be obtained. Consider, for example, a non-regular network $N$ from which the following set of rules is extracted: $R = \{ab \rightarrow h_1; c \rightarrow h_2; h_1 \rightarrow \neg x; \sim h_1 h_2 \rightarrow x\}$. When hidden neurons $h_1$ and $h_2$ are eliminated, we obtain $R' = \{ab \rightarrow \neg x; \sim ac \rightarrow x; \sim bc \rightarrow x\}$. However, by associating $h_1$ with $r_1 : ab \rightarrow \neg x$ and $h_2$ with $r_2 : c \rightarrow x$, we find out that $R'$ is equivalent to $R'' = \{r_1 : ab \rightarrow \neg x; r_2 : c \rightarrow x\}$ together with the preference relation $r_1 \succ r_2$, indicating that $r_1$ has priority over $r_2$. Clearly, $R''$ is more readable than $R'$.

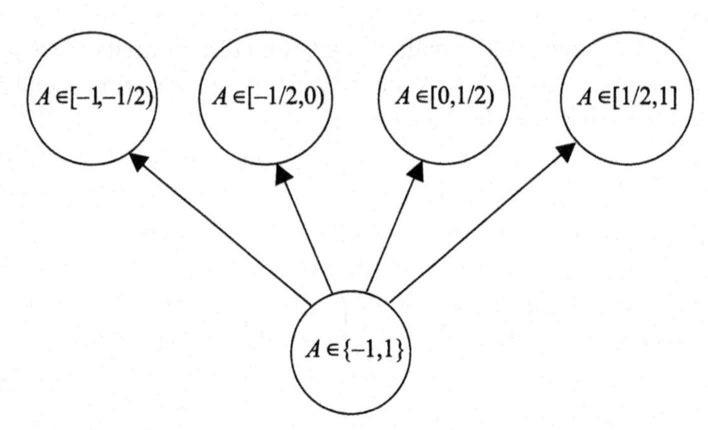


**Fig. 9.2.** Converting numerical attributes into discrete attributes prior to learning.

Finally, let us briefly consider the problem of rule extraction from networks with *continuous inputs*. When the Translation Algorithm of *C-IL$^2$P* is used, one can convert numerical attributes into discrete ones, using any desired degree of accuracy, as done in [Set97a], for example. Figure 9.2 illustrates this process for a concept $A \in \{-1, 1\}$, which is converted into four discrete attributes by dividing the input space uniformly. In this case, the extraction algorithm of *C-IL$^2$P* can be applied directly, without any modifications. The interesting case for future investigation arises when a network trained with continuous inputs is simply given, and we want to extract rules from it; i.e. instead of the whole system, we can only use the extraction module of *C-IL$^2$P*. In this case, it seems that the process used for the extraction of rules from hidden to output subnetworks should be applied also for input to hidden subnetworks. As before, the ordering on the input vectors of regular (sub)networks is valid for any activation values chosen. The problem, though, lies in the choice of "good" activation values. It is similar to the problem of defining a fuzzification scheme and its membership functions, as in [LZ00]. The proof of soundness in this case, however, seems to be a big challenge, and, in our point of view, soundness should be regarded as the minimum requirement of any rule extraction method.

## 9.2 Adding Disjunctive Information

There has been a large amount of work on Disjunctive Logic Programming and Abductive Logic Programming. As a starting point, we refer the reader to [Dix92] for a classification scheme of the semantics of disjunctive programs, and to [KKT94] for a study of the role of abduction in Logic Programming. In [SI94], Sakama and Inoue showed that abduction, i.e. the ability to perform reasoning with hypotheses[1], has very strong links with extended disjunctive logic programming.

Briefly, in a disjunctive logic program (resp. extended disjunctive logic program) there may be a disjunction of atoms (resp. literals) in the head of a clause. Let us take Gelfond and Lifschitz's approach to disjunctive logic programming [GL91], which they call *disjunctive databases.*

**Definition 9.2.1.** *An* Extended Disjunctive Database *is a set of rules of the form* $L_1 \mid ... \mid L_k \leftarrow L_{k+1}, ..., L_m, \sim L_{m+1}, ..., \sim L_n$*, where* $L_1, ..., L_n$ *are literals and* $k \leq m \leq n$*.*[2]

The following definition of answer sets for extended disjunctive databases generalises the definition of answer sets given in Chapter 3, Definition 3.4.4.

**Definition 9.2.2.** *An* Answer Set *of an* Extended Disjunctive Database $P$ *that does not contain* $\sim$ *is any minimal subset* $S$ *of Lit such that: for each rule* $L_1 \mid ... \mid L_k \leftarrow L_{k+1}, ..., L_m$ *from* $P$*, if* $L_{k+1}, ..., L_m \in S$ *then for some* $1 \leq i \leq k$*,* $L_i \in S$*. If* $S$ *contains a pair of complementary literals, then* $S = Lit$*.*

Let $P$ be an extended disjunctive database. For any set $S \subset Lit$, let $P^+$ be the extended disjunctive database obtained from $P$ by deleting: ($i$) each rule that has a formula $\sim L$ in its body with $L \in S$, and ($ii$) all formulas of the form $\sim L$ in the bodies of the remaining rules. As before, $P^+$ does not contain $\sim$, so that its answer sets are given by Definition 3.4.3 (Chapter 3). As before, if $S$ is one of them, then we say that $S$ is an answer set of $P$.

*Example 9.2.1.* [GL91] The disjunctive database $P = \{a \mid b \leftarrow; c \leftarrow a; c \leftarrow b\}$ has two answer sets, namely, $\{a, c\}$ and $\{b, c\}$. Now, the answer to a query may depend on which answer set is selected. For instance, the answer to the query $a$, relative to the first answer set, is *true*, while the answer to the same query relative to the second answer set is *unknown*. The answer to the query $c$ is unconditionally *true*.

Clearly, the addition of disjunctive information into neural networks implies the loss of an important property, namely, the uniqueness of the network's stable state. In a sense, a disjunctive database can be translated into a non-acceptable general logic program, such that each clause with $k > 1$

[1] "The logic of abduction examines all of the norms which guide us in formulating new hypotheses and deciding which of them to take seriously." Charles Pierce

[2] Gelfond and Lifschitz use $\mid$ rather than $\vee$ in the head of the disjunctive rule to differentiate between the meaning of disjunction in disjunctive databases and the meaning of disjunction in Classical Logic. For example, consider $P = \{p \mid \neg p \leftarrow; q \leftarrow p\}$. Unlike the law of the excluded middle in Classical Logic, the first rule of $P$ cannot be dropped without changing the meaning of $P$.

generates $k$ clauses in which only one of the disjuncts at a time is left in its head and all the others disjuncts are shifted to its body preceded by $\sim$. For example, the disjunctive database $D = \{a \mid b \leftarrow; c \leftarrow a; c \leftarrow b\}$ can be translated to the program $P = \{a \leftarrow\sim b; b \leftarrow\sim a; c \leftarrow a; c \leftarrow b\}$. The answer sets of $D$ are equal to the answer sets of $P$. However, since $P$ is not an acceptable program, a network $N$ that encodes $P$ may, depending on its initial input vector, (1) settle down in the stable state that represents the answer set $\{a, c\}$, (2) settle down in the stable state that represents the answer set $\{b, c\}$, or (3) loop. The behaviour of the network can be associated with its input vectors, if we consider that input vectors are *assumptions* about the truth values of the associated literals. In the case of $P$, for example, $N$ converges to $\{a, c\}$ when $a$ is assumed *true* and $b$ is assumed *false*, $N$ converges to $\{b, c\}$ when, conversely, $a$ is assumed *false* and $b$ is assumed *true*, and $N$ loops otherwise; that is, when $a$ and $b$ are both *true* or both *false*.[3]

Now, let us see what happens if we try to encode $a \vee b$ as $b \leftarrow \neg a$ and $a \leftarrow \neg b$, i.e. using "classical" negation, instead of default negation. Differently from the previous case, now the network will not loop. This can be easily seen by converting $\neg a$ into $a'$ and $\neg b$ into $b'$. Nevertheless, are we really adding disjunctive information in this case? Take, for example, $P = \{\neg a \leftarrow; b \leftarrow \neg a; a \leftarrow \neg b\}$. The network $N$, encoding $P$, will converge to a unique stable state $\{\neg a, b\}$, given any initial input vector, and such a stable state complies with what we expect from $\{a \vee b, \neg a\}$. However, take $P' = \{c \leftarrow a; c \leftarrow b; b \leftarrow \neg a; a \leftarrow \neg b\}$. The network $N'$, encoding $P'$, will not be able to derive $c$, as one would expect, from $\{a \vee b \leftarrow; c \leftarrow a; c \leftarrow b\}$. In other words, the translation of $a \vee b$ into $b \leftarrow \neg a$ and $a \leftarrow \neg b$ suffers from the lack of distributivity, which is a major drawback.

It seems that an alternative to representing Gelfond and Lifschitz's disjunctive databases in neural networks that do not loop is to encode disjuncts explicitly as neurons labelled $a_1 \vee ... \vee a_n$. Such neurons would need to perform a different task, maybe with some kind of external control. If, for example, an output neuron labelled $a \vee b$ fires, it should signal that input neuron $a$ has to be clamped as *true* until the network finds a stable state, and then that input neuron $b$ has to be clamped as *true*, also until the network gets stable. In other words, we would be assuming $a$ and using the network to try to find an answer set, and then repeating the process for the case of hypothesis $b$. Figure 9.3 illustrates the idea for the program $P = \{x \leftarrow; a \vee b \leftarrow x; y \leftarrow a; y \leftarrow b\}$. Also in Figure 9.3, the zoomed input neuron $a$ shows all the possible connec-

[3] Note that in the case of positive non-acceptable programs, like $P_1 = \{a \leftarrow b; b \leftarrow a\}$, the use of initial input vector $\mathbf{i} = (-1, ..., -1)$ is sufficient to guarantee the convergence of the equivalent network to a unique stable state. This is so because of the model intersection property of Horn clauses. In the case of general or extended programs, though, the solution to the loop problem is not simple, as $P_2 = \{b \leftarrow\sim a; a \leftarrow\sim b\}$, which has two minimal models, indicates.

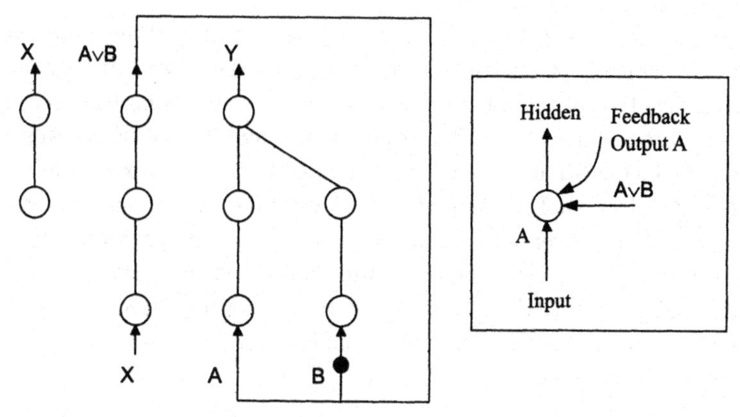


**Fig. 9.3.** An example of a possible implementation of disjunctive information in neural networks.

tions to any input neuron. Initially, the input value is taken as the activation of $a$. After the first iteration in the network, either the feedback from output $a$ or the one from $a \vee b$ should provide the activation of $a$. In this example, by assuming either $a$ or $b$, and from $x$ and $a \vee b \leftarrow x$, a unique answer set is derived, namely, $S = \{x, a \vee b, y\}$.

Now, when multiple answer sets $S_i$, $i \in \aleph$, are derived from the assumptions, e.g. when $P = \{x \leftarrow; a \vee b \leftarrow x; y \leftarrow a; z \leftarrow b\}$, $S_1 = \{x, a \vee b, y\}$ and $S_2 = \{x, a \vee b, z\}$, we can either (1) be sceptical about $S_i$, and take the intersection $\bigcap_i S_i$ as the answer set of $P$, or (2) use a credulous policy, and take $\bigcup_i S_i$ as the answer set of $P$, or even (3) try to reason about the answer sets, and, instead of taking, for example, $S_1 \cap S_2 = \{x, a \vee b\}$ or $S_1 \cup S_2 = \{x, a \vee b, y, z\}$, conclude that $x$ and $a \vee b$ are unconditionally *true,* while $y$ is *true,* subject to hypothesis $a$, and $z$ is *true,* subject to hypothesis $b$.

A drawback of the above approach for computing the answer sets of disjunctive databases is that all the combinations of the disjuncts in the program have to be considered. If, for example, $P = \{a \vee b \leftarrow; b \vee c \leftarrow; x \leftarrow c\}$, we would have to compute four answer sets by assuming, respectively, $\{a, b\}$, $\{a, c\}$, $\{b\}$ and $\{b, c\}$. When there are many disjuncts in a program, such a computation may clearly become intractable.

Another problem occurs when certain answer sets are inconsistent. In [GL91], Gelfond and Lifschitz solve this problem by simply taking the whole set of literals of the program (*Lit*) as its answer set. However, take, for example, $P = \{a \vee b \leftarrow; \neg b \leftarrow\}$. It is clear that the unique answer set of $P$ should be $\{a, \neg b\}$. From $P' = \{\neg b \leftarrow; b \leftarrow\sim a;\ a \leftarrow\sim b\}$, however, two answer sets can be obtained, namely, $S_1 = \{a, \neg b\}$ and $S_2 = \{b, \neg b\}$. In this case, we do not want to end up with $S_1 \cap S_2 = \{\neg b\}$, or with $S_1 \cup S_2 = \{a, b, \neg b\}$, or, worse, with $Lit = \{a, \neg a, b, \neg b\}$, as the answer set of $P$. We simply want to discard $S_2$, which is inconsistent, and keep $S_1$ as the unique answer set of $P$.

An alternative to the above problems would be to simply choose a disjunct, e.g. proceeding from left to right, and make it *true*. If the answer set obtained is inconsistent then we have made the wrong "execution" choice. We need to select another disjunct, until we have the information we want in its associated answer set. In this case, we stop. This procedure resembles the *imperative future approach* to disjunctions in executable temporal logic [Gab89] (see also [GHR94]), where $exec(a \vee b)$ should read "*make a true, unless it conflicts with the database, in which case make b true*".

## 9.3 Extension to the First-Order Case

The extension of *C-IL²P* to the first-order case is a complex and vast area for further research.

> "Despite the progress in *knowledge-based neurocomputing* (*KBN*), many open problems remain. *KBN* can not yet harness the full power of predicate logic representations frequently used in AI. Although good progress has been made to represent symbolic structures in KBN, the dynamic variable binding problem remains to be solved. Further questions to address include the development of new integration strategies and more diverse types of knowledge representations, e.g. procedural knowledge, methods to exchange knowledge, and reasoning capabilities." [CZ00]

In [McC88], John McCarthy argued that neural networks applications use unary predicates only, and that the concepts they compute are propositional functions of these predicates. Thus, he argues, neural networks cannot perform concept description, but only discrimination. Nevertheless, there has been a considerable amount of work on representing first-order theories in artificial neural networks [Ajj89, Ajj97, Bal86, GW98, Hol90, HK92, Pin91, Pin95, SA90, Sha99, BZB01] (see [HU94] for a collection of works on first-order logic representation in artificial neural networks). In line with *C-IL²P*, [HZB98] and [HKS99] try to keep the network's structure simple in order to benefit from efficient inductive learning algorithms, at the same time using more powerful languages. In [HKS99], for example, it is shown that, for any first-order acyclic logic program $\mathcal{P}$, there exists a single hidden layer neural network $\mathcal{N}$ that *approximates* the calculation of $T_{\mathcal{P}}$. The theorem does not state, however, how $\mathcal{N}$ could be constructed from $\mathcal{P}$, but it indicates that the use of simple neural networks is possible. To this end, in [BZdG98], we have used the *LINUS* system [LD94, LDG90] to translate first-order concepts into a propositional attribute-value language, and applied *C-IL²P*. The idea was to induce relational concepts with neural networks, using *LINUS* as the "front end" for *C-IL²P*.[4] Some first-order inductive learning tasks, taken from the literature on symbolic machine learning, were applied successfully, thus indicating that *C-IL²P* coupled with *LINUS* may induce relations correctly.

Of course, one may argue that the combination of *C-IL²P* and *LINUS* is not different from what we have done in the experiments of Chapters 4 and 6 using grounded programs, and thus that the neural learning process is essentially propositional. In addition, in comparison with Inductive Logic Programming (*ILP*), *LINUS* and *C-IL²P* may be viewed as rather roundabout, unless experimental results could be shown unequivocally favorable. The interesting question here, yet to be answered, is: are neural networks capable of generalising rules of the form $p(x, y) \leftarrow q(y, z)$ from instances such as $p(a, b) \leftarrow q(b, c), p(c, d) \leftarrow q(d, e)$, etc.? Due to the complexity of this research subject, we believe that the way forward, towards a truly *Connectionist-ILP* system, is to progressively study fragments of first-order logic and more expressive propositional logics, such as propositional modal logic.

## 9.4 Adding Modalities

We believe that *ensembles* of neural networks could be useful to enhance the expressive power of *C-IL²P*, at the same time maintaining its simplicity. Figure 9.4 shows an ensemble of three *C-IL²P* networks $(W_1, W_2, W_3)$ which might communicate in different ways. If, for example, we look at $W_1$, $W_2$ and $W_3$ as possible worlds, we might be able to incorporate modalities, such as necessity ($\Box$) and possibility ($\Diamond$), into the language of *C-IL²P*. For example, assume that $\mathcal{R}(W_1, W_2)$, where $\mathcal{R}$ is an accessibility relation. The rule "(1) If $W_1 : \Box A$ then $W_2 : A$" (read as "if $A$ is necessarily *true* in $W_1$ then $A$ is *true* in $W_2$") could be communicated from $W_1$ to $W_2$ by connecting $\Box A$ in the output layer of $W_1$ to $A$ in the output layer of $W_2$ such that, whenever $\Box A$ is activated in $W_1$, $A$ is activated in $W_2$. In addition, analogously to the feedback of *C-IL²P* networks, we could have feedback among *C-IL²P* networks. In this case, the rule "(2) If $(W_2 : A) \vee (W_3 : A)$ then $W_1 : \Diamond A$" (read as "if $A$ is *true* in $W_2$ or $W_3$ then $A$ is possibly *true* in $W_1$") could be implemented by connecting output neurons $A$ of $W_2$ and $W_3$ into output neuron $\Diamond A$ of $W_1$, through two hidden neurons $(h_1, h_2)$ in $W_1$ such that

[4] In fact, our interest in *LINUS* originated from the fact that it provides a translation from first-order, non-recursive, logic programs to the form of attribute-value propositions, and vice versa; the idea being that any efficient machine learning method could be used in the process of hypothesis generation, in between these translations. In [BZdG98], we have simply chosen to use neural networks as such a method.

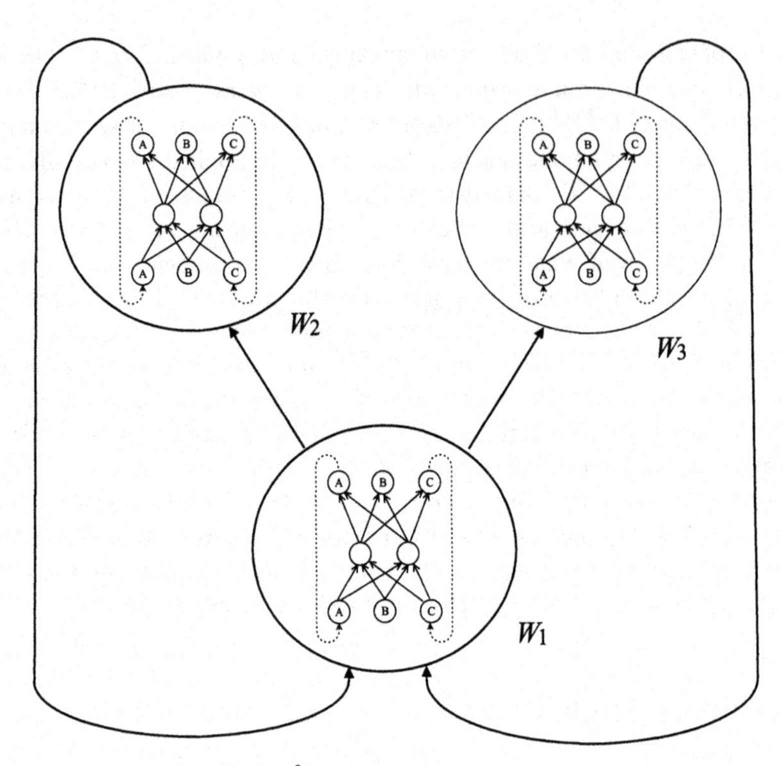


**Fig. 9.4.** An ensemble of *C-IL²P* networks.

$W_1 : h_1 \vee h_2 \rightarrow \Diamond A$. Similarly, "(3) If $(W_2 : A) \wedge (W_3 : A)$ then $W_1 : \Box A$" would be implemented by the feedback from neurons $A$ of $W_2$ and $W_3$ into output neuron $\Box A$ of $W_1$, through a single hidden neuron $h_3$ in $W_1$ such that $(W_2 : A) \wedge (W_3 : A) \rightarrow W_1 : h_3$ and $W_1 : h_3 \rightarrow \Box A$.

Rules (1), (2) and (3) above use, respectively, the following rules of natural deduction for propositional modal logic:

$$\frac{W_1.\mathcal{R}(W_1, W) : \Box A}{W : A} \qquad (\Box \text{ Elimination})$$

$$\frac{\exists W.\mathcal{R}(W_1, W) : A}{W_1 : \Diamond A} \qquad (\Diamond \text{ Introduction})$$

$$\frac{\forall W.\mathcal{R}(W_1, W) : A}{W_1 : \Box A} \qquad (\Box \text{ Introduction})$$

where $\Box$ Elimination states that if $\Box A$ is *true* in $W_1$ then $A$ is *true* in any world accessible from $W_1$, $\Diamond$ Introduction says that if $A$ is *true* in some world accessible from $W_1$ then $\Diamond A$ is *true* in $W_1$, and $\Box$ Introduction states that if $A$ is *true* in any world accessible from $W_1$ then $\Box A$ is *true* in $W_1$.

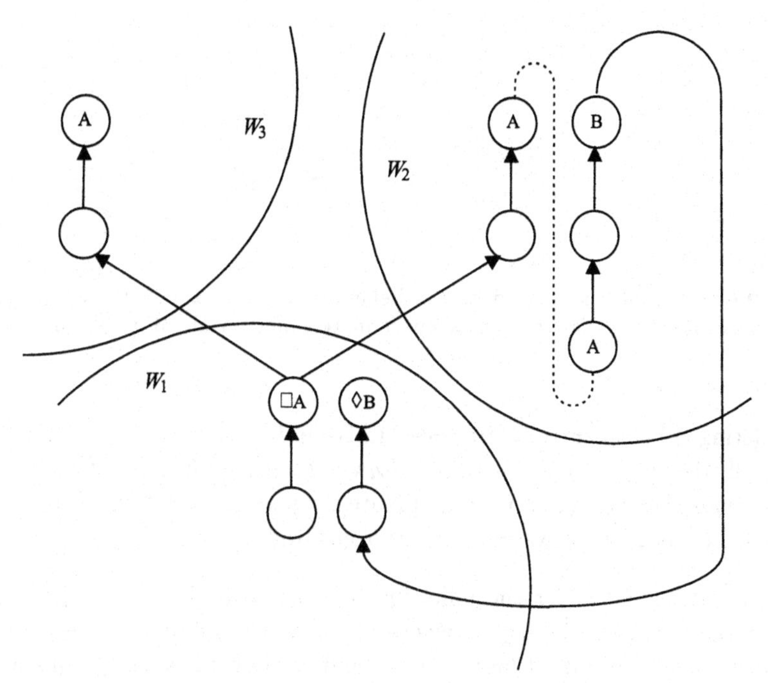


**Fig. 9.5.** Representing modalities in an ensemble of *C-IL²P* networks.

Let us give a simple example of how an ensemble of networks could be used to represent $\Box$ and $\Diamond$. Let $P = \{W_1 : \Box A,\ W_2 : A \rightarrow B\}$, $\mathcal{R}(W_1, W_2)$, $\mathcal{R}(W_1, W_3)$. For any atom $X$ in $B_P$, $\Box X$ and $\Diamond X$ are also treated as atoms, and thus are represented as neurons in each network of the ensemble. The semantics of $\Box$ and $\Diamond$, however, is modelled through connections between networks. For example, from $W_1 : \Box A$, we should have $W_2 : A$ and $W_3 : A$ (see Figure 9.5). From $W_2 : A$ and $W_2 : A \rightarrow B$, we derive $W_2 : B$. If $(W_2 : B) \vee (W_3 : B)$ then $W_1 : \Diamond B$. Similarly, if $(W_2 : B) \wedge (W_3 : B)$ then $W_1 : \Box B$ (not shown in Figure 9.5).

In this framework, learning within each possible world $(W_1, W_2,\ W_3)$ seems straightforward, while the links between *C-IL²P* networks would be defined according to the rules of natural deduction for propositional modal logic.

## 9.5 New Preference Relations

Nute's superiority relation, which we have studied in Chapter 3, is used to indicate which rules should *override* which other rules. In this case, it is reasonable to restrict the applicability of the preference relation to rules with

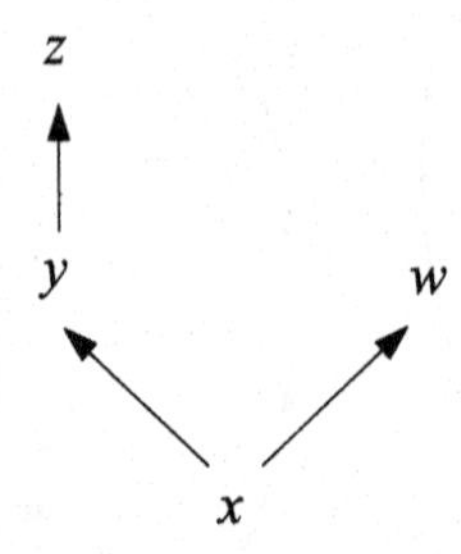


**Fig. 9.6.** An agent's relative degree of belief in rules $x$, $y$, $z$ and $w$, where $x \longrightarrow y$ indicates that $x \preceq y$, i.e. that the agent is more confident about $x$ than $y$.

conflicting conclusions only. However, in Belief Revision, preference relations normally describe an agent's *relative degree of belief* in the rules of a database. In this case, the above restriction of Nute's superiority relation might be too strong. Let us give an example, taken from [Rod97].

*Example 9.5.1.* Consider the following labelled belief base $B = \{y : pt \rightarrow \neg ft; z : bt \rightarrow ft; w : bt\}$, $y \preceq z$, where $pt$, $ft$ and $bt$ abbreviate, respectively, `Tweety is a penguin`, `Tweety flies` and `Tweety is a bird`, and $y \preceq z$ indicates that the agent is more confident about rule $y$ than about rule $z$. The agent also believes that Tweety is a bird ($bt$), although he does not know anything about the reliability of this information w.r.t. his other beliefs. Now, suppose that the agent is given the information $x : pt$, and that this information should be given the highest priority in $B$. In other words, $x \preceq z$, $x \preceq y$ and $x \preceq w$. The diagram of Figure 9.6 shows the agent's relative degree of belief in $x, y, z$ and $w$.

In order to accommodate $pt$ consistently, the agent must give up at least one of the three sentences he had accepted before. The retraction of each of these sentences corresponds to each of the maximal consistent subsets of $B$, namely, $\{x, y, z\}$, $\{x, y, w\}$ and $\{x, z, w\}$ [Bre90]. From these possibilities, the agent needs to rule out $\{x, z, w\}$, because it violates the preference for $y$ over $z$ by rejecting $y$ and keeping $z$ in $B$. The other two possibilities are plausible, because the agent does not know anything about the relative priority between $z$ and $w$. Thus, either $\{x, y, w\}$ or $\{x, y, z\}$ are maximal consistent subsets of $B$ which do not violate the preference relation $\preceq$ on $B$.

The above example shows that when $\preceq$ indicates the relative degree of belief in the rules of $B$, $x \preceq w$ should not mean that $pt$ has to *always* override $bt$. As a result, neural networks with *long-term memory* would be necessary to simulate many different scenarios and then choose the most appropriate ones.

## 9.6 A Proof Theoretical Approach

When we extract rules from a trained network, we obtain a database which can be used instead. By querying the database, we might obtain an explanation for the network's answers. However, if we are not concerned about the "big picture", that is, about finding out what is computed by the network, but only about providing an explanation for a particular answer, could we not use the network as our database?

The idea of *querying* a neural network was very useful in the extraction of symbolic knowledge. In addition, we have seen that querying $A$ in a network is similar to assuming $A$. Could we not use this to apply the *Deductive Theorem* in the network? In fact, this is what we have done in Chapter 5. When, for example, we wanted to check whether $a \rightarrow b$, we assumed $a$ and tried to derive $b$.

Let us consider a very restricted fragment of Gabbay's goal-directed algorithmic proof system [Gab98], which is based on Natural Deduction, where embedded implications are not allowed in the database. We also discard negation by default ($\sim$) and metalevel priorities from the language. Classical negation ($\neg$) is encoded in the network as $a \rightarrow \perp$. Now, let $DB = \{a \rightarrow b; b \rightarrow \perp\}$ and $X = \{a \rightarrow \perp\}$. We want to know whether $DB \vdash X$. $DB$ is encoded in the network as usual. To prove $X$, we assume $a$, i.e. we clamp $a = 1$ in the network's input layer, and try to prove $\perp$, i.e. we check if neuron $\perp$ is activated in the network's output layer when it reaches a stable state. In our example, assuming $a$, we obtain $b$, and from $b$ we derive $\perp$. So, $DB \vdash a \rightarrow \perp$. Now, let $Y = \{a\}$. Does $DB \vdash Y$? By assuming $\emptyset$, i.e. taking input vector $(-1, -1, ..., -1)$, we do not derive $a$, that is, $a$ does not belong to the stable state of the network. Thus, $DB \nvdash Y$.

Let us try another example. Let $DB = \{a \rightarrow b\}$ and $X = \{\neg b \rightarrow \neg a\}$. We want to know if $DB \vdash X$, where $X$ is translated into $(b \rightarrow \perp) \rightarrow (a \rightarrow \perp)$. Here, in order to assume $b \rightarrow \perp$, we need to add $b \rightarrow \perp$ in the network. Then, we can assume $a$, by setting $a = 1$ in the network's input layer, and see if $\perp$ follows. Indeed, from $a$ and $a \rightarrow b$, we obtain $b$, and, from $b$ and $b \rightarrow \perp$, we derive $\perp$. In other words, $a \rightarrow b \vdash \neg b \rightarrow \neg a$.

Note that we have used *forward chaining* in the above example, while Gabbay's proof system performs *backward computation.* His system's computation for the above example would be as follows:

| | |
|---|---|
| $a \rightarrow b \vdash (b \rightarrow \perp) \rightarrow (a \rightarrow \perp)$ | ? |
| $a \rightarrow b, b \rightarrow \perp \vdash a \rightarrow \perp$ | Rule for $\rightarrow$ |
| $a \rightarrow b, b \rightarrow \perp, a \vdash \perp$ | Rule for $\rightarrow$ |
| $a \rightarrow b, b \rightarrow \perp, a \vdash b$ | Rule for atoms using $b \rightarrow \perp$ |
| $a \rightarrow b, b \rightarrow \perp, a \vdash a$ | Rule for atoms using $a \rightarrow b$ |
| *sucess* | $a \in DB$ |

Note also that the converse $\neg b \rightarrow \neg a \vdash a \rightarrow b$ cannot be shown using neural networks, because we do not know yet how to represent $(b \rightarrow \perp) \rightarrow (a \rightarrow \perp)$. The question here is how to represent embedded implications in neural networks.

## 9.7 The "Forbidden Zone" $[A_{max}, A_{min}]$

In Chapter 3, we have seen that, in order to encode a logic program $\mathcal{P}$ into a neural network $\mathcal{N}$ correctly, we need to ensure that there is an interval $[A_{max}, A_{min}]$ such that the neurons of $\mathcal{N}$ do not present activation values in that interval. This is so because, from the proof of Theorem 3.1.1, we should have $A_{min} > \frac{k-1}{k+1}$, where $k$ is the number of literals in the body of a clause of $\mathcal{P}$. Since we assume that $A_{max} = -A_{min}$, whenever $k > 1$, the non-empty interval $[A_{max}, A_{min}]$ is necessary for $\mathcal{N}$ to correctly encode $\mathcal{P}$.

Thus, throughout this book, we have assumed that no hidden or output neuron of a trained network $\mathcal{N}$ presents activation value within $[A_{max}, A_{min}]$. In the experiments of Chapters 4 and 6, we have observed that, unless $A_{min} \simeq 1$, the above assumption can be verified. In other words, activation values within $[A_{max}, A_{min}]$ are unlikely to occur, specially when training data are discrete. However, when the learning process is not successful, some neurons might present activation values within such an interval. Thus, if one wants to make sure that activation values within $[A_{max}, A_{min}]$ never occur, a penalty function needs to be added to the learning algorithm. In [MZB98], such a penalty function was added to the learning process of *C-IL²P*. Despite the fact that it might prevent the use of *off the shelf* learning algorithms, the penalty function managed to avoid activation values within $[A_{max}, A_{min}]$ in different application domains.

Still in what regards learning, in Chapter 7 we have made the assumption that the set of training examples could be symbolically represented as an acceptable program. In other words, we have assumed that, after a network $\mathcal{N}$ is trained, it still encodes an acceptable program. Although the assumption seems reasonable for "well-behaved" application domains, a proof of "acceptability", given a set of examples, would be desirable.

## 9.8 Acceptable Programs and Neural Networks

In Chapter 3, we saw that the class of acceptable programs provides nice properties to the equivalent neural network. However, as pointed out in Chapter 2, neural networks that always converge to a unique stable state are not restricted to the class of acceptable programs. This violates the following conjecture.

*Conjecture 9.8.1.* [HK94] Networks that always converge to a unique stable state, given any initial input vector, are equivalent to, and describe, the class of acceptable programs.

We give a counter-example. For convenience, we repeat the programs $\mathcal{P}_1$, $\mathcal{P}_2$ and $\mathcal{P}_3$ of Example 2.5.2.

$$\mathcal{P}_1 = \{a \leftarrow c, \sim a\}$$
$$\mathcal{P}_2 = \{a \leftarrow \sim a, c\}$$
$$\mathcal{P}_3 = \{b \leftarrow a; \quad a \leftarrow \sim b; \quad b \leftarrow \sim a\}$$

Clearly, $\mathcal{P}_2$ is not an acceptable program, since it requires $|a| < |\sim a|$, while the definition of acceptable programs given in Chapter 2 states that $|a| = |\sim a|$ where $|x|$ is the *level* of literal $x$. Given $\mathcal{P}_2$ and its translation to a network $\mathcal{N}_2$, one can easily check that $\mathcal{N}_2$ settles down into a unique stable state, namely $\{\sim a, \sim c\}$, given any initial input vector. Table 9.1 illustrates this process.

**Table 9.1.** All possible input-output vectors of network $N2$

| $a$ | $c$ | $a$ | $c$ |
|---|---|---|---|
| −1 | −1 | −1 | −1 |
| −1 | 1 | 1 | −1 |
| 1 | −1 | −1 | −1 |
| 1 | 1 | −1 | −1 |

In view of the above we propose a more general definition of acceptability that is appropriate for neural networks, namely that a program $\mathcal{P}$ is *NN-Acceptable* w.r.t. $\mathcal{I}$ and $||$ if there is some ordering of the literals in the bodies of the clauses in $\mathcal{P}$ such that the resulting program is acceptable w.r.t. $\mathcal{I}$ and $||$. By this criterion, both programs $\mathcal{P}_1$ and $\mathcal{P}_2$ are *NN-Acceptable.* However, $\mathcal{P}_3$ is neither acceptable nor *NN-Acceptable*, since it requires $|a| < |\sim a| < |b|$ and $|b| < |\sim b| < |a|$. Given $\mathcal{P}_3$ and its translation to a network $\mathcal{N}_3$, one can easily check that $\mathcal{N}_3$ also settles down into a unique stable state, namely $\{\sim a, b\}$, for any initial input vector. Table 9.2 illustrates this process.

**Table 9.2.** All possible input-output vectors of network $N3$

| $a$ | $b$ | $a$ | $b$ |
|---|---|---|---|
| −1 | −1 | 1 | 1 |
| −1 | 1 | −1 | 1 |
| 1 | −1 | 1 | 1 |
| 1 | 1 | −1 | 1 |

Therefore, $\mathcal{P}_3$ would be a counterexample to Conjecture 9.8.1 even if acceptable were replaced by *NN-Acceptable.* The question of which class of programs can be described by single hidden layer networks, though, remains without an answer. We end this discussion by conjecturing the following generalisation of Conjecture 9.8.1.

*Conjecture 9.8.2.* Networks that always converge to a unique stable state, given any initial input vector, are equivalent to, and describe, the class of programs that admit a *simplification* that is *NN-Acceptable.* A program $P'$ is a *simplification* of a program $P$ iff $P'$ is obtained from $P$ by applying the simplification rules of *Complementary Literals* and *Subsumption* (Definitions 5.2.6 and 5.2.5) until no more applications are possible.

## 9.9 Epilogue

In this chapter we have briefly described only a few of the possible avenues of research into neural symbolic learning systems. Other interesting topics include: changing the definition of $T_{\mathcal{P}}$ in order to derive variants of *Default Logic* and Logic Programming semantics; incorporating *probabilities* into *C-IL²P*, thus allowing an explicitly quantitative approach [Pea00]; studying the learning and generalisation capabilities of *C-IL²P* networks w.r.t. the logical foundations of the system; exploring different learning algorithms such as *Cascade Adaptive Resonance Theory Mapping* [CGR91, Tan97], *Radial-Basis Function* and *Support Vector Machines* [Hay99a], and their impact on the task of rule extraction; as well as performing more applications on *Bioinformatics* and *Software Engineering.*

In conclusion, human cognition successfully integrates the connectionist and symbolic paradigms of Artificial Intelligence (AI). Yet the modelling of cognition develops these separately in neural computation and symbolic logic/AI areas. There is now a movement towards a fruitful midpoint between these extremes, in which the study of logic is combined with connectionism. It is essential that these be integrated in order to enable technologies for building more effective intelligent systems. We believe that the approach presented in this book, in which theoretical developments and practical applications are combined, together with the investigation of the above identified subjects for future work, will provide a more effective way of developing the research into neural-symbolic integration.

# References


[AB94] K. R. Apt and N. Bol. Logic programming and negation: a survey. *Journal of Logic Programming*, 19-20:9–71, 1994.

[ABM98] G. Antoniou, D. Billington, and M. J. Maher. Sceptical logic programming based default reasoning: defeasible logic rehabilitated. In Rob Miller and Murray Shanahan, editors, *COMMONSENSE 98, The Fourth Symposium on Logical Formalizations of Commonsense Reasoning*, pages 1–20, London, 1998.

[ADT95] R. Andrews, J. Diederich, and A. B. Tickle. A survey and critique of techniques for extracting rules from trained artificial neural networks. *Knowledge-based Systems*, 8(6):373–389, 1995.

[AFB+92] T. A. Alspaugh, S. R. Faulk, K. H. Britton, R. A. Parker, D. L. Parnas, and J. E. Shore. Software requirements for the A-7E aircraft. Technical Report 3876, Naval Research Laboratory, Washington, DC, 1992.

[AG95] R. Andrews and S. Geva. Inserting and extracting knowledge from constrained error backpropagation networks. In *Proceedings of the Sixth Australian Conference on Neural Networks ACNN95*, 1995.

[AGM85] C. A. Alchourron, P. Gardenfors, and D. Makinson. On the logic of theory change: partial meet contraction and revision functions. *The Journal of Symbolic Logic*, 50:510–530, 1985.

[Ajj89] V. Ajjanagadde. Reasoning with function symbols in a connectionist system. In *Proceedings of the eleventh Annual Conference of the Cognitive Science Society*, pages 388–395, 1989.

[Ajj97] V. Ajjanagadde. *Rule-Based Reasoning in Connectionist Networks*. PhD thesis, University of Minnesota, 1997.

[AMB00] G. Antoniou, M. J. Mahar, and D. Billington. Defeasible logic versus logic programming without negation as failure. *The Journal of Logic Programming*, 42:47–57, 2000.

[Ant97] G. Antoniou. *Nonmonotonic Reasoning*. MIT Press, Cambridge, MA, 1997.

[AP93] K. R. Apt and D. Pedreschi. Reasoning about termination of pure prolog programs. *Information and Computation*, 106:109–157, 1993.

[Bal86] D. H. Ballard. Parallel logical inference and energy minimization. In *Proceedings of the National Conference on Artificial Intelligence AAAI86*, pages 203–208, 1986.

[BCC+99] D. S. Blank, M. S. Cohen, M. Coltheart, J. Diederich, B. M. Garner, R. W. Gayler, C. L. Giles, L. Goldfarb, M. Hadeishi, B. Hazlehurst, M. J. Healy, J. Henderson, N. G. Jani, D. S. Levine, S. Lucas, T. Plate, G. Reeke, D. Roth, L. Shastri, J. Sougne, R. Sun, W. Tabor, B. B. Thompson, and S. Wermter. Connectionist symbol processing: Dead or alive? *Neural Computing Surveys*, 2:1–40, 1999.

[BE99] G. Brewka and T. Eiter. Preferred answer sets for extended logic programs. *Artificial Intelligence*, 109:297–356, 1999.

[BG91] C. Balkenius and P. Gardenfors. Nonmonotonic inference in neural networks. In *Principles of Knowledge Representation and Reasoning, Proceedings of the Second International Conference KR91*, pages 32–39, 1991.

[BL96] N. K. Bose and P. Liang. *Neural Networks Fundamentals with Graphs, Algorithms, and Applications.* McGraw-Hill, 1996.

[Bre90] G. Brewka. *Nonmonotonic Reasoning: Logical Foundations of Commonsense*, volume 12 of *Cambridge Tracts in Theoretical Computer Science.* Cambridge University Press, 1990.

[Bre91] G. Brewka. Cumulative default logic: in defense of nonmonotonic inference rules. *Artificial Intelligence*, 50, 1991.

[BS89] H. Blair and V. S. Subrahmanian. Paraconsistent logic programming. *Theoretical Computer Science*, 68:135–154, 1989.

[BV01] B. Boutsinas and M. N. Vrahatis. Artificial nonmonotonic neural networks. *Artificial Intelligence*, 132:1–38, 2001.

[BZB01] R. Basilio, G. Zaverucha, and V. C. Barbosa. Learning logic programs with neural networks. In *11th International Conference on Inductive Logic Programming ILP01*, pages 15–26. Springer-Verlag, LNCS 2157, 2001.

[BZdG98] R. Basilio, G. Zaverucha, and A. S. d'Avila Garcez. Inducing relational concepts with neural networks via the LINUS system. In *Proceedings of the Fifth International Conference on Neural Information Processing ICONIP98*, pages 1507–1510, 1998.

[CGR91] G. A. Carpenter, S. Grossberg, and J. H. Reynolds. Artmap: Supervised real-time learning and classification of nonstationary data by a self-organizing neural network. *Neural Networks*, 4:565–588, 1991.

[Che80] B. Chellas. *Modal Logic: An Introduction.* Cambridge University Press, 1980.

[Cla78] K. L. Clark. Negation as failure. In H. Gallaire and J. Minker, editors, *Logic and Databases*, pages 293–322, Plenum Press, New York, 1978.

[CR95] Y. Chauvin and D. Rumelhart, editors. *Backpropagation: Theory, Architectures and Applications.* Lawrence Erlbaum, 1995.

[Cra96] M. W. Craven. *Extracting Comprehensible Models from Trained Neural Networks.* PhD thesis, University of Wisconsin, Madison, 1996.

[CS94] M. W. Craven and J. W. Shavlik. Using sampling and queries to extract rules from trained neural networks. In *Proceedings of the International Conference on Machine Learning ICML94*, pages 37–45, 1994.

[Cyb89] G. Cybenco. Approximation by superposition of sigmoidal functions. In *Mathematics of Control, Signals and Systems 2*, pages 303–314. 1989.

[CZ97] A. Chagrov and M. Zakharyaschev. *Modal Logic*. Clarendom Press, Oxford Logic Guides, Vol. 35, 1997.

[CZ00] I. Cloete and J. M. Zurada, editors. *Knowledge-Based Neurocomputing*. The MIT Press, 2000.

[dC74] N. C. A. da Costa. On the theory of inconsistent formal systems. *Notre Dame Journal of Formal Logic*, 15:497–510, 1974.

[dGBG00] A. S. d'Avila Garcez, K. Broda, and D. M. Gabbay. Metalevel priorities and neural networks. In P. Frasconi, M. Gori, F. Kurfess, and A. Sperduti, editors, *Proceedings of ECAI2000, Workshop on the Foundations of Connectionist-Symbolic Integration*, Berlin, Germany, 2000.

[dGBG01] A. S. d'Avila Garcez, K. Broda, and D. M. Gabbay. Symbolic knowledge extraction from trained neural networks: A sound approach. *Artificial Intelligence*, 125:155–207, 2001.

[dGBGdS99] A. S. d'Avila Garcez, K. Broda, D. M. Gabbay, and A. F. de Souza. Knowledge extraction from trained neural networks: a position paper. In *Proceedings of 6th IEEE International Conference on Neural Information Processing ICONIP99*, pages 685–691, Perth, Australia, 1999.

[dGRNK01] A. S. d'Avila Garcez, A. Russo, B. Nuseibeh, and J. Kramer. An analysis-revision cycle to evolve requirements specifications. In *Proceedings of the 16th IEEE Automated Software Engineering Conference ASE01*, page to appear, 2001.

[dGZ99] A. S. d'Avila Garcez and G. Zaverucha. The connectionist inductive learning and logic programming system. *Applied Intelligence Journal, Special Issue on Neural Networks and Structured Knowledge*, 11(1):59–77, 1999.

[dGZC97] A. S. d'Avila Garcez, G. Zaverucha, and L. A. Carvalho. Logic programming and inductive learning in artificial neural networks. In C. Herrman, F. Reine, and A. Strohmaier, editors, *Knowledge Representation in Neural Networks*, pages 33–46. Logos-Verlag, Berlin, 1997. Also In Proceedings of the Workshop on Knowledge Representation and Neural Networks, XX German Conference on Artificial Intelligence KI96, Dresden, Germany, pages 9-18, September 1996.

[dGZdS97] A. S. d'Avila Garcez, G. Zaverucha, and V. N. L. da Silva. Applying the connectionist inductive learning and logic programming system to power systems' diagnosis. In *Proceedings of the IEEE International Joint Conference on Neural Networks ICNN97*, pages 121–126, Houston, Texas, USA, June 1997.

[Dix92] J. Dix. Classifying semantics of disjunctive logic pograms. In K. R. Apt, editor, *Proceedings of the Joint International Conference and Symposium on Logic Programming*, pages 589–603, Washington, 1992. MIT Press.

[DL01] S. Dzeroski and N. Lavrac, editors. *Relational Data Mining*. Springer-Verlag, September 2001.

[Doy79] J. Doyle. A truth maintenance system. *Artificial Intelligence*, 12:231–272, 1979.

[DP90] B. A. Davey and H. A. Priestley. *Introduction to Lattices and Order*. Cambridge University Press, 1990.

[DS96] B. DasGupta and G. Schinitger. Analog versus discrete neural networks. *Neural Computation*, 8:805–818, 1996.

[Elm90] J. L. Elman. Finding structure in time. *Cognitive Science*, 14(2):179–211, 1990.

[End72] H. B. Enderton. *A Mathematical Introduction to Logic*. Academic Press, 1972.

[FGH+94] A. Finkelstein, D. Gabbay, A. Hunter, J. Kramer, and B. Nuseibeh. Inconsistency handling in multi-perspective specifications. *IEEE Transactions on Software Engineering*, 20(8):569–578, August 1994.

[Fis87] D. H. Fisher. Knowledge acquisition via incremental conceptual clustering. *Machine Learning*, 2:139–172, 1987.

[Fit85] M. Fitting. A kripke-kleene semantics for general logic programs. *Journal of Logic Programming*, 2:295–312, 1985.

[Fit94] M. Fitting. Metric methods: Three examples and a theorem. *Journal of Logic Programming*, 21:113–127, 1994.

[FK00] P. A. Flach and A. C. Kakas. On the relation between abduction and inductive learning. In D. M. Gabbay and R. Kruse, editors, *Handbook of Defeasible Reasoning and Uncertainty Management Systems, Volume 4: Abductive Reasoning and Learning*, pages 1–33. 2000.

[Fu89] L. M. Fu. Integration of neural heuristics into knowledge-based inference. *Connection Science*, 1:325–340, 1989.

[Fu91] L. M. Fu. Rule learning by searching on adapted nets. In *Proceedings of the National Conference on Artificial Intelligence AAAI91*, pages 590–595, 1991.

[Fu94] L. M. Fu. *Neural Networks in Computer Intelligence*. McGraw Hill, 1994.

[Gab89] D. M. Gabbay. The declarative past and imperative future. In H. Barringer, editor, *Proceedings of the Colloquium on Temporal Logic and Specifications*, LNCS 398. Springer-Verlag, 1989.

[Gab98] D. M. Gabbay. *Elementary Logics: a Procedural Perspective*. Prentice Hall, London, 1998.

[Gab99] D. M. Gabbay. Compromise update and revision: a position paper. In R. Pareschi and B. Fronhoefer, editors, *Dynamic Worlds*, pages 111–148. Kluwer Academic Publishers, 1999.

[Gal88] S. I. Gallant. Connectionist expert systems. *Communications of the ACM*, 31(2):152–169, 1988.

[Gal93] S. I. Gallant. *Neural Networks Learning and Expert Systems*. MIT Press, 1993.

[Gar88] P. Gardenfors. *Knowledge in Flux: Modeling the Dynamics of Epistemic States*. Bradford Books, MIT Press, 1988.

[Gar92] P. Gardenfors, editor. *Belief Revision.* Tracts in Theoretical Computer Science, Cambridge University Press, 1992.

[Gar96] M. Garzon. *Models of Massive Parallelism: Analysis of Cellular Automata and Neural Networks.* Texts in Theoretical Computer Science, Springer, 1996.

[Gar00] P. Gardenfors. *Conceptual Spaces: The Geometry of Thought.* MIT Press, 2000.

[Ger93] J. L. Gersting. *Mathematical Structures for Computer Science.* Computer Science Press, 1993.

[GH91] D. M. Gabbay and A. Hunter. Making inconsistency respectable: a logical framework for inconsistency in reasoning. part 1: a position paper. In *Fundamentals of AI Research.* Springer-Verlag, 1991.

[GHR94] D. M. Gabbay, I. Hodkinson, and M. Reynolds. *Temporal logic: mathematical foundations and computational aspects.* Oxford University Press, 1994.

[GL88] M. Gelfond and V. Lifschitz. The stable model semantics for logic programming. In *Proceedings of the fifty Logic Programming Symposium*, 1988.

[GL91] M. Gelfond and V. Lifschitz. Classical negation in logic programs and disjunctive databases. *New Generation Computing*, 9:365–385, 1991.

[GL97] D. Gamberger and N. Lavrac. Conditions for occams razor applicability and noise elimination. In M. Someren and G. Widmer, editors, *Proceedings of the European Conference on Machine Learning*, pages 108–123, Prague, Czech Republic, 1997.

[GM91] L. Giordano and A. Martelli. Truth maintenance systems and belief revision. In J. P. Martins and M. Reinfrank, editors, *Truth Maintenance Systems*, Springer-Verlag, LNAI 515, pages 71–86, Berlin, 1991.

[GN99] C. Ghezzi and B. Nuseibeh, editors. *Special Issue of IEEE Transactions on Software Engineering on Managing Inconsistency in Software Development.* November 1999.

[GO93] C. L. Giles and C. W. Omlin. Extraction, insertion and refinement of production rules in recurrent neural networks. *Connection Science, Special Issue on Architectures for Integrating Symbolic and Neural Processes*, 5(3):307–328, 1993.

[Goo87] J. Goodwin. A theory and system for nonmonotonic reasoning. In *Linkoping Studies in Science and Technology*, number 165. University of Linkoping, 1987.

[GR94] P. Gardenfors and H. Rott. Belief revision. In D. M. Gabbay, C. Hogger, and J. Robinson, editors, *Handbook of Logic in Artificial Intelligence and Logic Programming*, volume 4, pages 35–132. Oxford University Press, 1994.

[GW98] R. W. Gayler and R. Wales. Connections, binding, unification and analogical promiscuity. In K. Holyoak, D. Gentner, and B. Kokinov, editors, *Advances in Analogy Research: Integration of Theory and Data from the Cognitive, Computational and Neural Sciences*, pages 181–190. Sofia, Bulgaria, 1998.

[Hay99a] S. Haykin. *Neural Networks: A Comprehensive Foundation.* Prentice Hall, 1999.

[Hay99b] S. Haykin. *Neural Networks: A Comprehensive Foundation.* Prentice Hall, 2nd edition edition, 1999.

[Hil95] M. Hilario. An overview of strategies for neurosymbolic integration. In *Proceedings of the Workshop on Connectionist-Symbolic Integration: from Unified to Hybrid Approaches, IJCAI 95*, 1995.

[HK92] S. Holldobler and F. Kurfess. CHCL: A connectionist inference system. In B. Fronhofer and G. Wrightson, editors, *Parallelization in Inference Systems*, pages 318–342. Springer, 1992.

[HK93] T. M. Heskes and B. Kappen. On-line learning processes in artificial neural networks. In J. G. Taylor, editor, *Mathematical Approaches to Neural Networks*, pages 199–233. Elsevier Science Publishers, 1993.

[HK94] S. Holldobler and Y. Kalinke. Toward a new massively parallel computational model for logic programming. In *Proceedings of the Workshop on Combining Symbolic and Connectionist Processing, ECAI 94*, 1994.

[HKL96] C. Heitmeyer, J. Kirby, and B. Labaw. Automated consistency checking of requirements specifications. *ACM Transactions on Software Engineering and Methodology*, (5(3)):231–261, June 1996.

[HKP91] J. Hertz, A. Krogh, and R. G. Palmer. *Introduction to the Theory of Neural Computation.* Studies in the Science of Complexity. Addison-Wesley, Santa Fe Institute, 1991.

[HKS99] S. Holldobler, Y. Kalinke, and H. P. Storr. Approximating the semantics of logic programs by recurrent neural networks. *Applied Intelligence Journal, Special Issue on Neural Networks and Structured Knowledge*, 11(1):45–58, 1999.

[HMS66] E. B. Hunt, J. Marin, and P. J. Stone. *Experiments in Induction.* Academinc Press, New York, 1966.

[HN94] H. Hirsh and M. Noordewier. Using background knowledge to improve inductive learning: a case study in molecular biology. *IEEE Expert*, 10:3–6, 1994.

[HN98] A. Hunter and B. Nuseibeh. Managing inconsistent specifications: Reasoning, analysis and action. *Transactions on Software Engineering and Methodology, ACM Press*, 1998.

[Hol90] S. Holldobler. A structured connectionist unification algorithm. In *Proceedings of the National Conference on Artificial Intelligence AAAI90*, pages 587–593, 1990.

[Hol93] S. Holldobler. Automated inferencing and connectionist models. Postdoctoral Thesis, Intellektik, Informatik, TH Darmstadt, 1993.

[Hop82] J. J. Hopfield. Neural networks and physical systems with emergent collective computational abilities. In *Proceedings of the National Academy of Sciences of the U.S.A.*, volume 79, pages 2554–2558, 1982.

[HSW89] K. Hornik, M. Stinchcombe, and H. White. Multilayer feedforward networks are universal approximators. *Neural Networks*, 2:359–366, 1989.

[HU94] V. Honavar and L. Uhr, editors. *Artificial Intelligence and Neural Networks: Steps Toward Principled Integration.* Academic Press, 1994.

[HZB98] N. Hallack, G. Zaverucha, and V. Barbosa. Towards a hybrid model of first-order theory refinement. In *Workshop on Hybrid Neural Symbolic Integration, Neural Information Processing Systems NIPS98*, Breckenridge, Colorado, USA, 1998.

[Jor86] M. I. Jordan. Attractor dynamics and parallelisms in a connectionist sequential machine. In *Proceedings of the Eighth Annual Conference of the Cognitive Science Society*, pages 531–546, 1986.

[Kas98] N. Kasabov. The ECOS framework and the ECO training method for evolving connectionist systems. *Journal of Advanced Computational Intelligence*, 2(6):195–202, 1998.

[Kas99] N. Kasabov. Evolving connectionist systems for on-line knowledge-based learning: principles and applications. Technical Report TR-99-02, Department of Information Science, University of Otago, Dunedin, New Zealand, 1999.

[Kat89] B. F. Katz. EBL and SBL: A neural network synthesis. In *Proceedings of the Eleventh Annual Conference of the Cognitive Science Society*, pages 683–689, Ann Arbor, 1989.

[Kir87] J. Kirby. Example NRL/SCR software requirements for an automobile cruise control and monitoring system. Technical Report TR-87-07, Wang Institute of Graduate studies, 1987.

[KKT94] A. C. Kakas, R. A. Kowalski, and F. Toni. The role of abduction in logic programming. In D. M. Gabbay, C.J. Hogger, and J. A. Robinson, editors, *Handbook of Logic in Artificial Intelligence and Logic Programming*, volume 5, pages 235–324. Oxford Science Publications, 1994.

[KLF01] S. Kramer, N. Lavrac, and P. Flach. Propositionalization approaches to relational data mining. In S. Dzeroski and N. Lavrac, editors, *Relational Data Mining*, pages 262–291. Springer-Verlag, September 2001.

[KMD94] A. C. Kakas, P. Mancarella, and P. M. Dung. The acceptability semantics of logic programs. In *Proceedings of the eleventh International Conference on Logic Programming ICLP*, pages 504–519. MIT Press, Cambridge, MA, 1994.

[Kol94] J. F. Kolen. *Exploring the Computational Capabilities of Recurrent Neural Networks.* PhD thesis, The Ohio State University, 1994.

[KP92] N. K. Kasabov and S. H. Petkov. Neural networks and logic programming: A hybrid model and its applicability to building expert systems. In *Proceedings of the European Conference on Artificial Intelligence ECAI92*, pages 287–288, 1992.

[Kri96] R. Krishnan. A systematic method for decompositional rule extraction from neural networks. In R. Andrews and J. Diederich, editors, *Proceedings of the NIPS96 Rule Extraction from Trained Artificial Neural Networks Workshop*, 1996.

[Kur97] F. J. Kurfess. Neural networks and structured knowledge. In C. Herrmann, F. Reine, and A. Strohmaier, editors, *Knowledge Representation in Neural Networks*, pages 5–22. Logos-Verlag, Berlin, 1997.

[Kur00] F. Kurfess, editor. *Applied Intelligence Journal, Special Issue on Neural Networks and Structured Knowledge: Rule Extraction and Applications*, volume 12. Kluwer, January 2000.

[LD92] N. Lavrac and S. Dzeroski. Background knowledge and declarative bias in inductive concept learning. In K. Jantke, editor, *Proceedings of the International Workshop on Analogical and Inductive Inference*, pages 51–71, Springer, Berlin, 1992.

[LD94] N. Lavrac and S. Dzeroski. *Inductive Logic Programming: Techniques and Applications.* Ellis Horwood Series in Artificial Intelligence, 1994.

[LDG90] N. Lavrac, S. Dzeroski, and M. Grobelnik. Experiments in learning nonrecursive definitions of relations with LINUS. Technical report, Josef Stefan Institute, Yugoslavia, 1990.

[LGT96] S. Lawrence, C. Lee Giles, and A. Chung Tsoi. Lessons in neural networks training: overfitting may be harder than expected. In *Proceedings of the National Conference in Artificial Intelligence AAAI96*, pages 540–545, 1996.

[Llo87] J. W. Lloyd. *Foundations of Logic Programming.* Springer-Verlag, 1987.

[Luk90] V. Lukaszewicz. *Nonmonotonic Reasoning: Formalization of Commonsense Reasoning.* Ellis Horwood, 1990.

[LZ00] A. Lozowski and J. M. Zurada. Extraction of linguistic rules from data via neural networks and fuzzy approximation. In I. Cloete and J. M. Zurada, editors, *Knowledge-Based Neurocomputing*, pages 403–417. The MIT Press, 2000.

[Mai97] F. Maire. A partial order for the M of N rule extraction algorithm. *IEEE Transactions on Neural Networks*, 8(6):1542–1544, 1997.

[Mai98] F. Maire. Rule-extraction by backpropagation of polyhedra. Technical report, Machine Learning Research Center, School of Computing Science, Queensland University of Technology, Brisbane, Australia, 1998.

[Mak93] D. Makinson. Five faces of minimality. *Studia Logica*, 52:339–379, 1993.

[McC80] J. McCarthy. Circunscription: A form of nonmonotonic reasoning. *Artificial Intelligence*, 13:27–39, 1980.

[McC88] J. McCarthy. Epistemological challenges for connectionism. In *Behavior and Brain Sciences*, volume 11, page 44. 1988.

[Med94] L. Medsker. Neural networks connections to expert systems. In *Proceedings of the World Congress on Neural Networks*, 1994.

[Men96] T. Menzies. Applications of abduction: Knowledge level modeling. *International Journal of Human Computer Studies*, 1996.

[Mic87] R. S. Michalski. Learning strategies and automated knowledge acquisition. In *Computational Models of Learning*, Symbolic Computation. Springer-Verlag, 1987.

[Min91] M. Minsky. Logical versus analogical, symbolic versus connectionist, neat versus scruffy. *AI Magazine*, 12(2), 1991.

[Mit97] T. M. Mitchell. *Machine Learning.* McGraw-Hill, 1997.

[MMHL86] R. S. Michalski, I. Mozetic, J. Hong, and N. Lavrac. The multi-purpose incremental learning system AQ15 and its testing application to three medical domains. In *Proceedings of the National Conference in Artificial Intelligence AAAI86*, volume 2, pages 1041–1045, 1986.

[MMS92] C. McMillan, M. C. Mozer, and P. Smolensky. Rule induction through integrated symbolic and subsymbolic processing. In J. Moody, S. Hanson, and R. Lippmann, editors, *Advances in Neural Information Processing Systems NIPS92*, volume 4. Morgan Kaufmann, San Mateo, CA, 1992.

[Moo85] R. Moore. Semantical considerations on nonmonotonic logic. *Artificial Intelligence*, 25(1):75–94, 1985.

[MR91] R. J. Machado and A. F. Rocha. The combinatorial neural network: a connectionist model for knowledge based systems. In B. Bouchon-Meunier, R. R. Yager, and L. A. Zadeh, editors, *Uncertainty in Knowledge Bases: Proceedings of the third International Conference on Information Processing and Management of Uncertainty in Knowledge-based Systems IPMU91*, pages 578–587, Springer, Berlin, 1991.

[MR94] S. Muggleton and L. Raedt. Inductive logic programming: theory and methods. *Journal of Logic Programming*, 19:629–679, 1994.

[MT93a] W. Marek and M. Truszczynski. *Nonmonotonic Logic: Context Dependent Reasoning*. Springer-Verlag, 1993.

[MT93b] T. M. Mitchell and S. B. Thrun. Explanation-based learning: A comparison of symbolic and neural network approaches. In *Tenth International Conference on Machine Learning*, Amherst, MA, 1993.

[MW01] T. L. McCluskey and M. M. West. The automated refinement of a requiremens domain theory. *Journal of Automated Software Engineering*, 8(2), 2001.

[MZ94] R. J. Mooney and J. M. Zelle. Integrating ILP and EBL. In *SIGART Bulletin*, volume 5, pages 12–21. 1994.

[MZB98] R. Menezes, G. Zaverucha, and V. Barbosa. A penalty-function approach to rule extraction from knowledge-based neural networks. In *Workshop on Hybrid Neural Symbolic Integration, Neural Information Processing Systems NIPS98*, Breckenridge, Colorado, USA, 1998.

[NR92] H. Narazaki and A. L. Ralescu. A connectionist approach for rule-based inference using an improved relaxation method. *IEEE Transactions on Neural Networks*, 3(5):741–751, 1992.

[NTS91] M. O. Noordewier, G. G. Towell, and J. W. Shavlik. Training knowledge-based neural networks to recognize genes in DNA sequences. In *Advances in Neural Information Processing Systems NIPS91*, volume 3, pages 530–536, 1991.

[Nut87] D. Nute. Defeasible reasoning. In *Proceedings of the Hawaii International Conference on Systems Science*, pages 470–477. IEEE Press, 1987.

[Nut94] D. Nute. Defeasible logic. In D. M. Gabbay, C.J. Hogger, and J. A. Robinson, editors, *Handbook of Logic in Artificial Intelligence and Logic Programming*, volume 3, pages 353–396. Oxford Science Publications, 1994.

[OM94] D. Ourston and R. J. Mooney. Theory refinement combining analytical and empirical methods. *Artificial Intelligence*, 66:273–310, 1994.

[OM00] R. C. O'Reilly and Y. Munakata. *Computational Explorations in Cognitive Neuroscience: Understanding the Mind by Simulating the Brain.* MIT Press, 2000.

[O'N89] M. C. O'Neill. Escherichia coli promoters: consensus as it relates to spacing class, specificity, repeat substructure, and three dimensional organization. *Journal of Biological Chemistry*, 264:5522–5530, 1989.

[Opi95] D. W. Opitz. *An Anytime Approach to Connectionist Theory Refinement: Refining the Topologies of Knowledge-Based Neural Networks.* PhD thesis, University of Wisconsin, Madison, 1995.

[OS93] D. W. Opitz and J. W. Shavlik. Heuristically expanding knowledge-based neural networks. In *Proceedings of the International Joint Conference on Artificial Intelligence IJCAI93*, pages 1360–1365, 1993.

[Pea00] J. Pearl. *Causality: Models, Reasoning and Inference.* Cambridge University Press, 2000.

[Per96] D. Perlis. Sources of, and exploiting, inconsistency: preliminary report. In *Symposium of Logical Formalization of Commonsense Reasoning*, pages 169–175. Stanford University, 1996.

[PHD94] E. Pop, R. Hayward, and J. Diederich. RULENEG: Extracting rules from a trained ANN by stepwise negation. Technical report, Neural Research Center, School of Computing Science, Queensland University of Technology, 1994.

[Pin91] G. Pinkas. Energy minimization and the satisfiability of propositional calculus. *Neural Computation*, 3(2), 1991.

[Pin95] G. Pinkas. Reasoning, nonmonotonicity and learning in connectionist networks that capture propositional knowledge. *Artificial Intelligence*, 77:203–247, 1995.

[PK92] M. Pazzani and D. Kibler. The utility of knowledge in inductive learning. *Machine Learning*, 9:57–94, 1992.

[PM95] D. L. Parnas and J. Madey. Functional documentation for computer systems. Technical Report CRL 309, McMaster University, 1995.

[PMG98] D. Poole, A. Mackworth, and R. Goebel. *Computational Intelligence: A Logical Approach.* Oxford University Press, 1998.

[Poo87] D. Poole. A logical framework for default reasoning. *Artificial Intelligence*, 36:27–47, 1987.

[Poo91] D. Poole. Compiling a default reasoning system into prolog. *New Generation Computing*, 9:3–38, 1991.

[Prz88] T. C. Przymusinski. On the declarative semantics of logic programs with negation. In J. Minker, editor, *Foundations of Deductive Databases and Logic Programming*, pages 193–216, Morgan-Kaufmann, Los Altos, 1988.

[PS97] H. Prakken and G. Sartor. Argument-based extended logic programming with defeasible priorities. *Journal of Applied Non-Classical Logic*, 7:25–75, 1997.

[PY73] F. P. Preparata and R. T. Yeh. *Introduction to Discrete Structures.* Addison-Wesley, 1973.

[Qui86] J. R. Quinlan. Induction of decision trees. *Machine Learning*, 1:81–106, 1986.

[Rei80] R. Reiter. A logic for default reasoning. *Artificial Intelligence*, 13:81–132, 1980.

[RHW86] D. E. Rumelhart, G. E. Hinton, and R. J. Williams. Learning internal representations by error propagation. In D. E. Rumelhart and J. L. McClelland, editors, *Parallel Distributed Processing: Explorations in the Microstructure of Cognition*, volume 1. MIT Press, 1986.

[RMNK00] A. Russo, R. Miller, B. Nuseibeh, and J. Kramer. An abductive approach for handling inconsistencies in SCR specifications. In *Proceedings of the 3rd ICSE Workshop on Intelligent Software Engineering*, Limerick, 2000.

[RMNK01] A. Russo, R. Miller, B. Nuseibeh, and J. Kramer. An abductive approach for analysing event-based requirements specifications. Technical Report TR2001/7, Department of Computing, Imperial College, 2001.

[Rod97] O. T. Rodrigues. *A Methodology for Iterated Information Change.* PhD thesis, Department of Computing, Imperial College, 1997.

[Ros62] F. Rosenblatt. *Principles of Neurodynamics: Perceptrons and the Theory of Brain Mechanisms.* Spartan, New York, 1962.

[Rus96] S. Russell. Machine learning. In M. A. Boden, editor, *Handbook of Perception and Cognition*, Artificial Intelligence, chapter 4. Academic Press, 1996.

[SA90] L. Shastri and V. Ajjanagadde. From simple associations to semantic reasoning: A connectionist representation of rules, variables and dynamic binding. Technical report, University of Pennsylvania, 1990.

[Set97a] R. Setiono. Extracting rules from neural networks by pruning and hidden-unit splitting. *Neural Computation*, 9:205–225, 1997.

[Set97b] R. Setiono. A penalty-function for pruning feedforward neural networks. *Neural Computation*, 9:185–204, 1997.

[SH97] R. Setiono and L. Huan. NEUROLINEAR: from neural networks to oblique decision rules. *Neurocomputing*, 17:1–24, 1997.

[Sha88] L. Shastri. A connectionist approach to knowledge representation and limited inference. *Cognitive Science*, 12(13):331–392, 1988.

[Sha95] M. Shaw. Comparing architectural design styles. *IEEE Software*, 1995.

[Sha96] J. W. Shavlik. An overview of research at Wisconsin on knowledge-based neural networks. In *Proceedings of the International Conference on Neural Networks ICNN96*, pages 65–69, Washington, DC, 1996.

[Sha99] L. Shastri. Advances in SHRUTI: a neurally motivated model of relational knowledge representation and rapid inference using temporal synchrony. *Applied Intelligence Journal, Special Issue on Neural Networks and Structured Knowledge*, 11:79–108, 1999.

[SI94] C. Sakama and K. Inoue. On the equivalence between disjunctive and abductive logic programs. In *Proceedings of the International Conference on Logic Programming ICLP94*, pages 489–503, 1994.

[SLM92] B. Selman, H. Levesque, and D. Mitchell. A new method for solving hard satisfiability problems. In *Proceedings of the Tenth National Conference on Artificial Intelligence AAAI92*, 1992.

[SMT91] J. W. Shavlik, R. J. Mooney, and G. G. Towell. Symbolic and neural net learning algorithms: an empirical comparison. *Machine Learning*, 6:111–143, 1991.

[SN88] K. Saito and R. Nakano. Medical diagnostic expert system based on PDP model. In *Proceedings of the IEEE International Conference on Neural Networks*, pages 255–262, San Diegp, USA, 1988.

[SS95] H. T. Siegelmann and E. D. Sontag. On the computational power of neural nets. *Journal of Computer and System Sciences*, 1995.

[Sto90] G. D. Stormo. Consensus patterns in DNA. In *Methods in Enzymology*, volume 183, pages 211–221. Academic Press, Orlando, 1990.

[Sun95] R. Sun. Robust reasoning: integrating rule-based and similarity-based reasoning. *Artificial Intelligence*, 75(2):241–296, 1995.

[SZ01] G. Spanoudakis and A. Zisman. Inconsistency management in software engineering: Survey and open research issues. *Journal of Systems and Software*, 56(11), 2001.

[SZS95] V. N. Silva, G. Zaverucha, and G. Souza. An integration of neural networks and nonmonotonic reasoning for power systems' diagnosis. In *Proceedings of IEEE International Conference on Neural Networks ICNN95*, Perth, Australia, 1995.

[Tan97] A. H. Tan. Cascade artmap: Integrating neural computation and symbolic knowledge processing. *IEEE Transactions on Neural Networks*, 8(2):237–250, 1997.

[Tay93] J. G. Taylor. *The Promise of Neural Networks.* Springer-Verlag, London, 1993.

[TBB+91] S. B. Thrun, J. Bala, E. Bloedorn, I. Bratko, B. Cestnik, J. Cheng, K. De Jong, S. Dzeroski, S. E. Fahlman, D. Fisher, R. Haumann, K. Kaufman, S. Keller, I. Kononenko, J. Kreuziger, R. S. Michalski, T. Mitchell, P. Pachowicz, Y. Reich, H. Vafaie, K. Van de Welde, W. Wenzel, J. Wnek, and J. Zhang. The MONK's problems: A performance comparison of different learning algorithms. Technical Report CMU-CS-91-197, Carnegie Mellon University, 1991.

[Thr94] S. B. Thrun. Extracting provably correct rules from artificial neural networks. Technical report, Institut fur Informatik, Universitat Bonn, 1994.

[Thr96] S. Thrun. *Explanation-Based Neural Network Learning: A Lifelong Learning Approach.* Kluwer Academic Publishes, Boston, MA, 1996.

[TLI91] K. Thompson, P. Langley, and W. Iba. Using background knowledge in concept formation. In *Proceedings of the eighth International Machine Learning Workshop*, pages 554–558, 1991.

[Tow92] G. G. Towell. *Symbolic Knowledge and Neural Networks: Insertion, Refinement and Extraction.* PhD thesis, University of Wisconsin, Madison, 1992.

[TS93] G. G. Towell and J. W. Shavlik. The extraction of refined rules from knowledge-based neural networks. *Machine Learning*, 13(1):71–101, 1993.

[TS94a] G. G. Towell and J. W. Shavlik. Knowledge-based artificial neural networks. *Artificial Intelligence*, 70(1):119–165, 1994.

[TS94b] G. G. Towell and J. W. Shavlik. Using symbolic learning to improve knowledge-based neural networks. In *Proceedings of the National Conference on Artificial Intelligence AAAI94*, 1994.

[TSN90] G. G. Towell, J. W. Shavlik, and M. O. Noordewier. Refinement of approximately correct domain theories by knowledge-based neural networks. In *Proceedings of the National Conference on Artificial Intelligence AAAI90*, pages 861–866, Boston, USA, 1990.

[Val84] L. G. Valiant. A theory of the learnable. *Communications of the ACM*, 27(11), 1984.

[vEK76] M. H. van Emden and R. A. Kowalski. The semantics of predicate logic as a programming language. *Journal of the ACM*, 23(4):733–742, 1976.

[vGRS91] A. van Gelder, K. Ross, and J. Schlipf. The well-founded semantics for general logic programs. *Journal of the ACM*, 38(3):620–650, 1991.

[Vin94] R. Vingralek. Connectionist approach to finding stable models and other structures in nonmonotonic reasoning. Technical report, Computer Science Department, University of Kentucky, 1994.

[vL00] A. van Lamsweerde. Requirements engineering in the year 00: A research perspective. In *Invited Paper for ICSE'2000 - 22nd International Conference on Software Engineering*, Limerick, 2000. ACM Press.

[vLDL98] A. van Lamsweerde, R. Darimont, and E. Letier. Managing conflicts in goal-driven requirement engineering. *IEEE Transaction on Software Engineering*, 1998.

[vLW98] A. van Lamsweerde and L. Willemet. Inferring declarative requirements specifications from operational scenarios. *IEEE Transactions on Software Engineering, Special Issue on Scenario Management*, 1998.

[Wer74] P. J. Werbos. *Beyond Regretion: New Tools for Prediction and Analysis in the Behavioral Sciences.* PhD thesis, Harvard University, Cambridge, MA, 1974.

[Wer90] P. J. Werbos. Backpropagation through time: what does it mean and how to do it. In *Proceedings of the IEEE*, volume 78, pages 1550–1560, 1990.

[WHR+87] J. D. Watson, N. H. Hopkins, J. W. Roberts, J. A. Steitz, and A. M. Weiner. *Molecular Biology of the Gene*, volume 1. Benjamin Cummings, Menlo Park, 1987.

[Wil97] J. R. Williamson. A constructive, incremental-learning network for mixture, modeling and classification. *Neural Computation*, 9(7):1517–1543, 1997.

# Index